CIRCULATION IN THE COASTAL OCEAN

ENVIRONMENTAL FLUID MECHANICS

G. T. CSANADY

Woods Hole Oceanographic Institution

Circulation in the Coastal Ocean

D. REIDEL PUBLISHING COMPANY

DORDRECHT : HOLLAND / BOSTON : U.S.A.

LONDON : ENGLAND

Library of Congress Cataloging in Publication Data

Csanady, G. T.
Circulation in the coastal ocean.

(Environmental fluid mechanics)
Bibliography: p.
Includes index.
1. Ocean circulation. I. Title. II. Series.
GC228.5.C74 1982 551.47 82-11287
ISBN 90-277-1400-2

Published by D. Reidel Publishing Company,
P.O. Box 17, 3300 AA Dordrecht, Holland

Sold and distributed in the U.S.A. and Canada
by Kluwer Boston Inc.,
190 Old Derby Street, Hingham, MA 02043, U.S.A.

In all other countries, sold and distributed
by Kluwer Academic Publishers Group,
P.O. Box 322, 3300 AH Dordrecht, Holland

D. Reidel Publishing Company is a member of the Kluwer Group

Printed in The Netherlands

Table of Contents

Foreword

For some time there has existed an extensive theoretical literature relating to tides on continental shelves and also to the behavior of estuaries. Much less attention was traditionally paid to the dynamics of longer term, larger scale motions (those which are usually described as 'circulation') over continental shelves or in enclosed shallow seas such as the North American Great Lakes. This is no longer the case: spurred on by other disciplines, notably biological oceanography, and by public concern with the environment, the physical science of the coastal ocean has made giant strides during the last two decades or so.

Today, it is probably fair to say that coastal ocean physics has come of age as a deductive quantitative science. A well developed body of theoretical models exist, based on the equations of fluid motion, which have been related to observed currents, sea level variations, water properties, etc. Quantitative parameters required in using the models to predict e.g. the effects of wind or of freshwater influx on coastal currents can be estimated within reasonable bounds of error. While much remains to be learned, and many exciting discoveries presumably await us in the future, the time seems appropriate to summarize those aspects of coastal ocean dynamics relevant to 'circulation' or long-term motion.

At the Woods Hole Oceanographic Institution I have for several years offered a graduate course on 'coastal dynamics', which was more or less such an account of the fluid dynamics of circulation or non-tidal motion in enclosed and open shallow seas. The present text grew out of the lecture notes for that course, although the topic evolved considerably in the course of writing. It owes much to two review articles completed recently (Csanady 1981a, c). The end product is more nearly a monograph than a text, being more narrowly focused on long period motions than a reasonable coastal dynamics course should be.

In writing a synthesis of a subject that represented for some years one's own central research interest it is apparently unavoidable to end up with a considerable bias toward using one's own results and illustrations. Legitimate reasons for this are intimate familiarity and easy availability, but an unfortunate side effect is to endow those results and illustrations with an importance which is quite undeserved. The reader should make appropriate allowances for this. I have tried to quote key original papers wherever appropriate, but probably missed some. The literature of coastal oceanography is now so large that to quote all important contributions would have interfered with the central goal of this monograph, the synthesis of knowledge.

The philosophy of the approach adopted in this monograph to understanding circulation in the coastal ocean is briefly as follows. The observed total behavior of any shallow sea is far too complex for the human mind to comprehend in all detail, let alone to relate it to simple dynamical notions. However, it is possible to distil from reality gross flow characteristics which occur repeatedly under similar external stimuli (e.g. in response to storms), and which may legitimately be regarded as distinct phenomena in the same sense as a tornado or a drought. Once recognized, such phenomena are taken to be 'understood', if they are quantitatively reproducible in a drastically simplified analytical model ocean, governed by Newtonian dynamics. The analytical models often have straight shores, or constant depth or contain homogeneous fluid, and have generally the minimum number of physical features necessary to reproduce the phenomenon in question. Essential is quantitative correspondence between model and observation within limits of error reasonably ascribed to the simplified nature of the model.

It is perhaps necessary to justify the emphasis on analytical rather than numerical models. To some extent the preference is clearly personal bias. It is also generally true, however, that an analytical result establishes more conclusively the intricate interaction between complex physical factors than a series of numerical simulations. Furthermore, there is often a difficult choice made in numerical modelling between resolution and area coverage. High resolution implies for practical reasons small area coverage and sometimes requires the acceptance of open boundaries, where arbitrary conditions must be imposed. Models of low resolution, on the other hand, may well miss some important sub-grid scale phenomena which are not readily fed into the model through the equations and the boundary conditions. Numerical models come into their own in synthesizing a variety of phenomena, *after* those phenomena have been identified and understood on the basis of simple analytical models. Quite realistic simulations of complex coastal ocean behavior then become possible. On the other hand, when a numerical model shows behavior not previously understood in simpler terms, one first has to eliminate the possibility that the behavior is a consequence of some unrealistic assumption made in modelling, and then to understand the (otherwise puzzling) behavior in simpler (analytical) terms. While not everybody will accept this assessment of the role of numerical modelling, at least it should explain why not more use was made in this monograph of numerically calculated results.

(*Satellite photo of Southern California Bight.*)

Infrared image of coastal ocean off southern California coast. Cold, southward-flowing coastal current separates from the coast at Pt. Conception and breaksdown into large eddies. Infrared 'visibility' is caused by sharp thermal contrasts. (Courtesy R. L. Bernstein, Scripps Institution of Oceanography.)

CHAPTER 1

Fundamental Equations and Their Simplification

1.0. INTRODUCTION

Dynamical processes in shallow seas and over continental shelves differ markedly from those in the deep ocean for several reasons. The horizontal scales of motion are much smaller and the presence of coasts is a strong constraining influence in most locations. The depths involved are only of the order of 100 m, so that surface effects such as wind stress or surface cooling or heating extend to a large fraction of the water column, sometimes to all of it, whereas in the deep ocean the same influences reach what amounts to only a thin skin at the surface. At the same time, basins or continental shelves with characteristic horizontal dimensions of the order of 100 km behave in an 'oceanic' manner in the sense that motions in them are strongly affected by the Earth's rotation. Seas of this size and larger, with depth ranges of up to a few hundred meters will be taken to constitute the coastal ocean. This definition includes enclosed shallow seas such as the North American Great Lakes, open seas such as the broad and flat continental shelves of 'Atlantic' type, or the narrow and steep shelves of 'Pacific' type, as well as semi-enclosed bodies of water such as the Gulf of Maine or the North Sea.

The dominant observable motions in shallow seas are rotary currents, associated with tides over continental shelves, with inertial oscillations in stratified, enclosed seas. Such motions in what one might call a pure form are characterized by the rotation of the current vector through 360° in a period not very different from the Earth's rotation rate, and illustrate the dynamical importance of rotation. Water particle motions during a full tidal or inertial cycle are along a closed ellipse of a typical longer axis length of a few kilometers, there being to net displacement in an idealized pure tidal or inertial oscillation. In reality, there is of course always some residual motion, which adds up cycle after cycle and produces fluid particle displacements over the longer term much larger than the diameter of the tidal or inertial ellipse. The problem of 'circulation' is to describe and understand the pattern of these longer term water particle displacements. The distribution of important water properties, such as temperature, salinity, the concentration of heavy metals or nutrients, and the transport of these properties or of life-forms incapable of locomotion depends critically on the pattern of circulation, but not very much on the oscillatory water motions, at least not in a direct way (indirect effects include, for example, turbulence and mixing produced by tidal currents).

The understanding and prediction of circulation requires the development of appropriate theoretical models based on the equations of fluid motion. Although a Lagrangian approach to these equations would be more appropriate when focussing on long-term particle displacements, this approach is not well enough developed at present, so that the conventional Eulerian equations will be used.

In this first chapter the fundamental equations will be reviewed and the idealizations will be introduced which lead to the simplification of these equations to the form found appropriate for coastal ocean dynamics. The same material is covered in greater detail in standard texts on oceanography (e.g., Krauss, 1966) or on geophysical fluid dynamics (e.g., Pedlosky, 1979). Although condensed, the development below is intended to be such that a reader with a reasonable background in fluid mechanics (e.g., an engineering graduate) should be able to follow it.

1.1. THE EQUATION OF CONTINUITY

In classical fluid mechanics the Eulerian 'equation of continuity' is usually written down for a simple substance in the following form:

$$\frac{\partial \rho}{\partial t} + \frac{\partial}{\partial x_i}(\rho u_i) = 0, \tag{1.1}$$

where ρ is density, t is time, and u_i are fluid velocity components along the three co-ordinate axes x_i (i = 1, 2, 3, and the summation convention applies*). The equation expresses the physical idea that local density changes are solely due to the convergence or divergence of the mass flux, ρu_i. A rearrangement of Equation (1.1) yields

$$\frac{1}{\rho}\frac{d\rho}{dt} + \frac{\partial u_i}{\partial x_i} = 0, \tag{1.1a}$$

where

$$\frac{d\rho}{dt} = \frac{\partial \rho}{\partial t} + u_i \frac{\partial \rho}{\partial x_i}$$

is the total derivative of density, or the rate of change of density following a fluid particle.

In applying this equation to the coastal ocean one notes first that sea water is not a simple substance, and that density changes may occur through diffusion of salt. To account for this effect a source term should be introduced on the right of (1), expressing convergence of mass through salt diffusion. In bodies of fresh water this complication is absent and the principal agency of density change is heating or cooling of the fluid, the effects of which are accounted for by Equation (1.1) if u_i are understood to include thermal expansions and contractions.

However, for realistic values of either heat or salt flux divergence, temperature and salinity changes take place very slowly. For example, even on a mid-summer day the surface layers of the sea do not heat up much faster than by about 1 °C in 3 hr, or $dT/dt = 10^{-4}$ °C s^{-1}. Given a typical value of $\rho^{-1}\ \partial\rho/\partial T$ of 10^{-4} °C^{-1} (this quantity

---

* The suffix notation and associated conventions are used here until simplified forms of the equations of motion are derived. A simple description of what this involves is given e.g. by Jeffreys and Jeffreys (1956). Batchelor (1967) uses this notation throughout.

varies with temperature, see standard thermodynamic texts), the rate of density change corresponding to intense solar heating is 10^{-8} s^{-1}, or 1 part in one thousand in about a day. Density changes due to salt flux are similarly sluggish: the molecular diffusivity of salt is very low (order 10^{-9} m^2 s^{-1}) and turbulence is effectively suppressed by gravity at the nearly horizontal separation surfaces one usually finds between more or less saline layers of water. Thus the term $\rho^{-1}\ d\rho/dt$ in Equation (1.1a) is of the order of 10^{-8} s^{-1} or less, and so is the diffusive mass flux divergence term that should be added on the right-hand side of Equation (1.1) to account for salinity changes.

By contrast, horizontal velocities in shallow seas are of order 0.1–1.0 m s^{-1}, maximum distances of interest of order 100 km. Thus velocity gradients of a magnitude 10^{-6} s^{-1} and greater commonly occur. Compared to these the $\rho^{-1}\ d\rho/dt$ term is negligible even under rather extreme conditions. The velocity derivatives must then balance between themselves to a high degree of approximation:

$$\frac{\partial u_i}{\partial x_i} = 0. \tag{1.2}$$

This is the standard form of the continuity equation as it applies to incompressible fluids. Physically it expresses the fact that the net total volume rate of flow into any control volume fixed in space is always zero, inflows exactly cancelling outflows. As the above quantitative estimates have shown, this simple form of the continuity equation applies with adequate accuracy in the coastal ocean.

1.2. MOMENTUM BALANCE

In the Eulerian approach to fluid mechanics, Newton's law of motion is expressed for a fluid particle by the following equation:

$$\frac{d(\rho u_i)}{dt} = -\frac{\partial p}{\partial x_i} + \frac{\partial \tau_{ij}}{\partial x_j} + \rho b_i. \tag{1.3}$$

The left-hand side is rate of change of momentum (per unit volume) following a fluid particle while the right-hand side is a sum of surface and body forces. The surface forces consist of hydrostatic pressure gradients and of the divergence of the other components of the stress tensor, τ_{ij}. Body forces ρb_i may be of a variety of origins, all characterized by the fact that they are proportional to particle mass, hence the appearance of density. Of the interior stresses the shear stress in horizontal planes is particularly important in shallow seas, because it transmits the driving force of the wind to the surface layers and the braking action of the sea floor to the bottom layers.

The first index in τ_{ij} will be taken to define the direction of the shearing force (parallel to x_i), the second index the plane of its action (perpendicular to x_j). The x_3 axis is conveniently laid along the local vertical; τ_{13} and τ_{23} are then the important shear stress components in horizontal planes. The stresses τ_{12} and τ_{21} may transmit forces near lateral boundaries, but their importance is usually limited.

The total derivative on the left of Equation (1.3) may also be written as

$$\frac{\mathrm{d}(\rho u_i)}{\mathrm{d}t} = \frac{\partial(\rho u_i)}{\partial t} + \frac{\partial}{\partial x_j}(\rho u_i u_j) \tag{1.4}$$

having used (1.2). The second term on the right is the divergence of the momentum flux brought about by the flow itself. This term is nonlinear in the velocity components and is a source of considerable mathematical complexity. However, having the form of a divergence it represents the redistribution of momentum in space, the effects of which are often only important in limited regions of the flow field.

1.2.1. Reynolds Stresses

The flow in shallow seas is usually turbulent, at least in surface and bottom 'mixed' layers. Motions to be discussed below have time scales greatly in excess of the typical period of turbulent velocity fluctuations. Under these circumstances it is convenient to regard the equations of continuity and motion, Equations (1.2) and (1.3), as applying to a suitably defined mean flow. All turbulent quantities which enter the equations are supposed resolved into 'mean' and 'fluctuating' components:

$$u_i = \bar{u}_i + u_i',$$

$$p = \bar{p} + p'.$$

The mean of u_i', p', etc. is zero by hypothesis. Taking means on the equations of continuity and motion, all terms containing fluctuations vanish, except the mean product terms $\rho\, \overline{u_i' u_j'}$ in Equation (1.3). The divergence of these plays a role in the momentum balance which is very similar to the role of viscous shear stresses. Hence $-\rho\, \overline{u_i' u_j'}$ is usually taken over to the right of the momentum equation and combined with the mean viscous shear stress $\bar{\tau}_{ij}$:

$$\tau_{ij} = \bar{\tau}_{ij} - \rho\, \overline{u_i' u_j'}\,.$$

The mean viscous shear stress is negligible except under some very unusual circumstances. In what follows, interior stresses will be understood to include the Reynolds stresses, $-\rho\, \overline{u_i' u_j'}$, and Equation (1.3) will apply to the mean flow. Overbars will not be shown henceforth and all flow quantities, u_i, p, etc. will be understood to refer to the mean flow.

1.2.2. The Body Forces

The velocity components u_i will be understood measured relative to a rotating earth, so that the body forces include the centrifugal and Coriolis forces associated with the Earth's rotation. As usual, the centrifugal force will be absorbed in the gravity force, and the total body force written as

$$\begin{aligned} b_i &= 2\epsilon_{ijk}\, u_j\, \Omega_k + g_i, \\ g_i &= -g\delta_{i3}, \end{aligned} \tag{1.5}$$

where Ω_k are components of the Earth's angular velocity in a local coordinate system, and g is the local acceleration of gravity. Of the other body forces, the attraction of Moon and Sun are important in the deep ocean, but not in shallow seas. The local coordinate system to be used in dealing with the coastal ocean is illustrated in Figure 1.1.

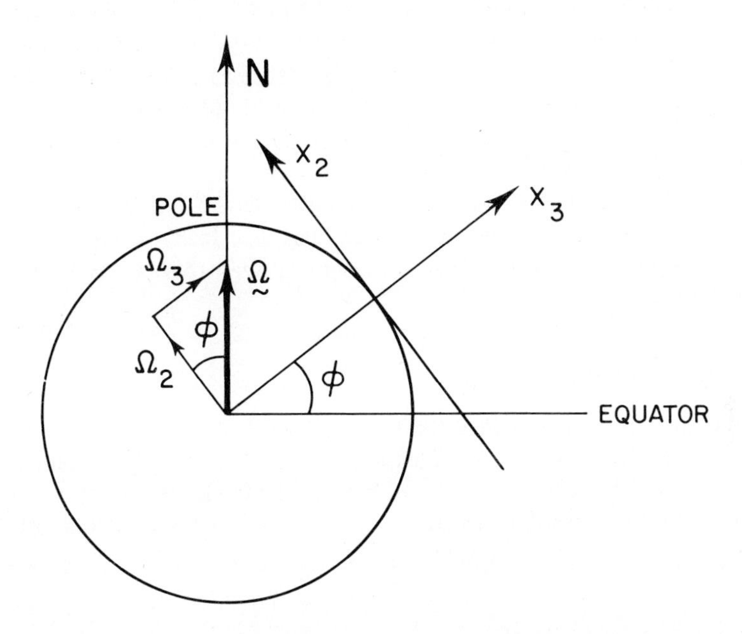

Fig. 1.1. Local horizontal (x_1, x_2) and vertical (x_3) coordinate system at latitude ϕ, and components of the Earth's angular velocity vector.

The magnitude of the Earth's angular velocity vector $(\Omega_i \Omega_i)^{1/2} = \Omega$ is 2π radians in 24 hr, or

$$\Omega = 0.7292 \times 10^{-4}\ \mathrm{s}^{-1}.$$

On substituting (1.4) and (1.5) into Equation (1.3) the following form of the equations of motion is obtained:

$$\frac{\partial(\rho u_i)}{\partial t} + \frac{\partial}{\partial x_j}(\rho u_i u_j) = -\frac{\partial p}{\partial x_i} - g\rho\delta_{i3} + 2\rho\epsilon_{ijk}u_j\Omega_k + \frac{\partial \tau_{ij}}{\partial x_j}. \tag{1.6}$$

In their general form these equations are more or less intractable. In order to arrive at suitably simple models of flow phenomena in the coastal ocean a number of further idealizations have to be made.

1.3. HYDROSTATIC APPROXIMATION

The equations of motion (1.6) apply separately in all three coordinate directions, but for a large class of motions the momentum balance is degenerate in the vertical, being dominated by the force of gravity. The full vertical momentum balance is:

$$\frac{du_3}{dt} = -\frac{1}{\rho}\frac{dp}{dx_3} + 2(u_1\Omega_2 - u_2\Omega_1) - g + \frac{1}{\rho}\frac{\partial \tau_{3j}}{\partial x_j}. \tag{1.7}$$

The gravitational acceleration g is close to 10 m s^{-2}. Typical horizontal velocities in the coastal ocean are 0.1–1.0 m s^{-1}, so that the Coriolis accelerations are of order 10^{-5}–10^{-4} m s^{-2}. Maximum stress gradients are of the order of 0.1 Pa* distributed over a 10 m deep water column, i.e. $\rho^{-1}\ \partial\tau/\partial z = 10^{-5}$ m s^{-2}. Vertical accelerations du_3/dt are comparable to g in surface wave orbital motions. However, in the flow phenomena to be discussed in this text vertical velocities are at most of order 10^{-2} m s^{-1}, time scales at least 10^3 s, i.e. the vertical accelerations are of order 10^{-5} m s^{-2} or less. It follows then that in such quasi-horizontal motions of long time scale the large acceleration of gravity must be balanced by the vertical pressure gradient $\partial p/\partial x_3$ to a high degree of approximation:

$$\frac{\partial p}{\partial x_3} = -\rho g. \tag{1.8}$$

This balance is the same as in a fluid at rest and its application to moving fluids is referred to as the hydrostatic approximation. The hydrostatic equilibrium position of the free surface will be taken to be $x_3 = 0$. In a moving fluid, the free surface is displaced slightly upward or downward, to a position $\zeta(x_1, x_2)$, see Figure 1.2. On integrating (1.8) from some level x_3 to the free surface one finds

$$p = p_a + \int_{x_3}^{\zeta} \rho g\ dx_3, \tag{1.9}$$

where p_a is atmospheric pressure.

The horizontal momentum equations, Equations (1.6) with $i = 1, 2$, contain pressure gradients which may be calculated on the basis of the hydrostatic approximation

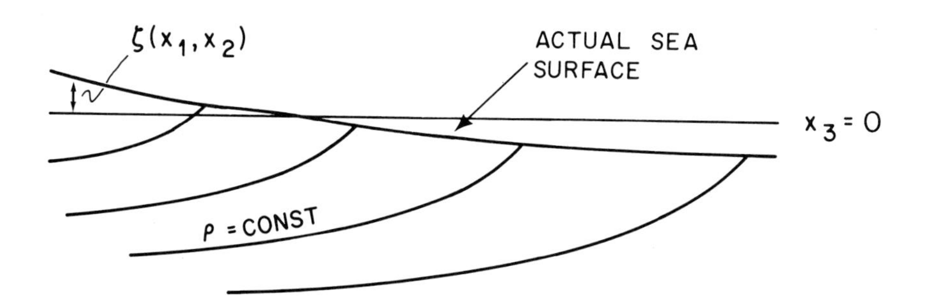

Fig. 1.2. Sea surface elevation $\zeta(x_1, x_2)$ and constant density surfaces below.

---

* 1 Pa (Pascal) = 1 Nm^{-2} = 10 dynes cm^{-2}.

Equation (1.9) as follows:

$$\frac{\partial p}{\partial x_i} = \frac{\partial p_a}{\partial x_i} + \rho_s g \frac{\partial \zeta}{\partial x_i} + \int_{x_3}^{\zeta} g \frac{\partial \rho}{\partial x_i} \, dx_3, \quad (i = 1, 2) \tag{1.10}$$

where ρ_s is surface density.

Equation (1.10) shows that pressure gradients within the water column arise on account of atmospheric pressure gradients, gradients of the sea surface elevation ζ, and interior horizontal density gradients. Atmospheric pressure gradients constitute external forcing of motions in a shallow sea. Compared to wind stress, they are relatively unimportant, so that the $\partial p_a/\partial x_i$ term will be dropped in subsequent discussion. The second term on the right of (1.10) contains acceleration of gravity times sea surface slope, as if water particles were rolling down on an incline sloping as the sea surface. The third term expresses the effects of buoyancy arising from differences in fluid density along horizontal (geopotential) surfaces. Given typical conditions in shallow seas, this last term is conveniently reformulated for greater simplicity.

1.3.1. The Boussinesq Approximation

Density variations in a given shallow sea are of the order of one part per thousand, so that a constant reference density ρ_0 may be substituted for ρ_s without serious error in (1.10). The horizontal scales of density variation are, on the other hand, such that the integral involving $\partial \rho/\partial x_i$ is generally *not* negligible. It is convenient to write

$$\rho = \rho_0(1 + \epsilon), \tag{1.11}$$

where ϵ is of order 10^{-3}. Quantities of order ϵ need only be retained in the buoyancy term, a step known as the 'Boussinesq approximation'.

Dropping the atmospheric pressure term, Equation (1.10) may then be expressed as

$$\frac{1}{\rho_0} \frac{\partial p}{\partial x_i} = g \frac{\partial}{\partial x_i} [\zeta - \zeta_d(x_3)], \quad (i = 1, 2) \tag{1.12}$$

with

$$\zeta_d(x_3) = -\int_{x_3}^{0} \epsilon \, dx_3.$$

The quantity ζ_d has been defined so as to express pressure in terms of the hydraulic 'head' of a fluid column. More accurately, ζ_d is a differential 'head' between a fluid column of reference density ρ_0, and one of the actual density $\rho(x_3)$. In oceanography $\zeta_d(x_3)$ is referred to as the steric height or dynamic height of the sea surface relative to level x_3. Although not explicitly shown, $\zeta_d(x_3)$ is also a function of horizontal position (x_1, x_2), as is the sea surface elevation ζ.

The reference density ρ_0 is conveniently chosen to be the density of the densest water present, in which case ϵ is negative everywhere and ζ_d is either zero or positive. Should the pressure gradient $\partial p/\partial x_i$ vanish at all x_1, x_2 at some deep enough level $x_3 = -h$,

Equation (1.12) yields

$$\zeta = \zeta_d(-h) \tag{1.13}$$

on dropping an integration constant, i.e. with an appropriate choice of zero for sea level. The topography of the sea surface can in this case be determined from the density distribution by a calculation standard in oceanography (see e.g., Sverdrup *et al.*, 1942). In shallow seas it is rarely reasonable to postulate that $\partial p/\partial x_i$ vanishes at all horizontal locations below some great enough depth, except sometimes in the central, deep portions of closed basins. The simple dynamic height calculation (according to (1.13)) can only yield sea surface topography where the total depth is greater than the reference depth h. This leaves out the entire coastal region, which is usually the dynamically most active portion of a shallow sea, as the analysis of later chapters will show.

In the general case, $\zeta_d(-h)$, with some appropriate choice of reference level h, not necessarily the same at all horizontal locations, may be regarded as a density-related contribution to the sea surface elevation field. On top of this, one should generally suppose that there is a distribution $\zeta - \zeta_d(-h)$ which does not necessarily have anything to do with the internal distribution of mass in a shallow sea.

Typical velocities in shallow seas are of order 0.1 m s^{-1}, and they often develop in the course of a storm lasting for a day or so, i.e. 10^5 s. This corresponds to an average acceleration of 10^{-6} m s^{-2}. A sea surface slope of 10^{-7} (1 cm in 100 km) contributes the same acceleration to the horizontal momentum balances and is thus of significant magnitude.

A density anomaly ϵ of 1‰, constant over a surface layer of 30 m depth, yields a contribution of 3 cm to the surface dynamic height, relative to a depth greater than 30 m. Over typical horizontal distances of 100 km, which characterize the dimensions of shallow seas, a density-related sea level gradient of 10^{-7}, 1 cm in 100 km, results if the light surface layer is 30 m deep in one location, 20 m deep 100 km away. Density changes of this order of magnitude and more are almost always present in the coastal ocean so that the physical processes which control the density field affect horizontal momentum balances in an important way.

1.4. THE DENSITY FIELD

The density field may be determined from observed fluid property distributions and the known equation of state of seawater or fresh water. In the deep ocean it is necessary to take into account variations of density due to changes of pressure, temperature and salinity. In shallow seas where the maximum pressure is only some 10 atmospheres the density is effectively a function only of temperature (T) and salinity (S):

$$\rho = \rho\,(T, S). \tag{1.14}$$

Temperature variations in the coastal ocean are primarily due to heating or cooling of the surface, quantified by the heat flux, q_s, or heat transferred per unit area of the surface and unit time (kcal s^{-1} m^{-2}, or W m^{-2} in mechanical units). The temperature field $T(x_i, t)$ is described by a conservation law of a structure similar to (1.1), but with a flux-divergence term on the right:

$$\frac{\partial T}{\partial t} + \frac{\partial}{\partial x_i}(u_i T) = -\frac{\partial}{\partial x_i}\left(\frac{q_i}{\rho c_p}\right), \tag{1.15}$$

where c_p is specific heat and q_i are heat flux components (including radiative fluxes q_r and 'Reynolds' fluxes $\rho c_p\, \overline{u_i' T'}$). The surface heat flux q_s enters as a boundary condition:

$$q_s = q_3, \quad (x_3 = 0). \tag{1.15a}$$

At times of intense solar heating, the surface heat flux can be as high as 0.1 kcal s^{-1} m^{-2}. On a calm day and in relatively turbid water, most of this heat is absorbed by a layer only 1 m deep, giving $\partial T/\partial t$ of about 10^{-4} °C s^{-1}, or a 1 °C temperature increase in 3 hr, as mentioned in Section 1.1, if one supposes that none of the heat advects or diffuses away. This rate of heating or cooling is quite extreme, however, and in most places, most of the time the heat flux divergence on the right of (1.15) is several orders of magnitude less. For flow phenomena with time scales of the order of a day or two it is then reasonable to neglect the heat flux divergence:

$$\frac{dT}{dt} = \frac{\partial T}{\partial t} + \frac{\partial}{\partial x_i}(u_i T) = 0. \tag{1.16}$$

The physical meaning of this expression is that fluid particles conserve their temperature so that the temperature field is controlled entirely by advection.

Changes in salinity $S(x_i, t)$ within the coastal ocean are brought about mainly by freshwater inflow from rivers, although local rainfall and ice melt may also play a role. These changes are also subject to a conservation law:

$$\frac{\partial S}{\partial t} + \frac{\partial}{\partial x_i}(u_i S) = -\frac{\partial}{\partial x_i}\left(\frac{s_i}{\rho}\right), \tag{1.17}$$

where the salinity S is a mass fraction, and s_i are salt flux components in kg m^{-2} s^{-1} which again include Reynolds fluxes $\rho\, \overline{u_i' S'}$. In many cases the salt flux divergence is negligible compared to local changes and advection so that one may write in analogy with (1.16):

$$\frac{dS}{dt} = \frac{\partial S}{\partial t} + \frac{\partial}{\partial x_i}(u_i S) = 0 \tag{1.18}$$

characterizing the case when the salinity distribution is controlled entirely by advection, because fluid particles conserve salinity.

When both temperature and salinity are conserved by particles, the density also remains constant for a moving particle, and the density field is controlled by advection. For flow phenomena with a time scale of a day or two this is often a reasonable assumption. Over the long term, of course, large density variations are brought about by seasonal heating or cooling and by seasonal variations of freshwater runoff. Vigorous turbulent mixing associated with wind-driven or tidal shear flow is often the cause of more rapid temperature or salinity changes, sometimes as high as 10^{-5} °C s^{-1} for temperature or 10^{-8} s^{-1} (1 part in one thousand in about a day) for salinity. A storm often produces a vertically well mixed water column, where previously there has been pronounced stratification, with density differences of the usual order ($10^{-3}\ \rho_0$) between top and bottom layers. One must be careful to exclude such mixing events from a comparison with theoretical models based on particle density conservation.

1.5. QUASI-HORIZONTAL MOTIONS

The flow phenomena to be discussed in detail in this monograph are all characterized by a small ratio of vertical to horizontal velocities. Physically, this is partly because the aspect ratio, depth to width or length of a shallow sea, is usually small, of order 10^{-3}. Motions of basin-wide scale, extending over the entire depth must have vertical: horizontal velocity ratios of the same order, through simple geometrical constraint. The frequent presence of density stratification further impedes vertical movements so that the typical vertical velocity is much smaller than the typical horizontal velocity, in motions of moderate to large spatial scale.

Given $|u_3| \ll |u_1|$ or $|u_2|$, the components of the Coriolis acceleration in the horizontal momentum equations simplify as follows:

$$\begin{aligned} &\text{1st equation} \quad 2(u_2\Omega_3 - u_3\Omega_2) \cong 2u_2\Omega_3, \\ &\text{2nd equation} \quad 2(u_3\Omega_1 - u_1\Omega_3) \cong -2u_1\Omega_3. \end{aligned} \tag{1.19}$$

The geometry of the coordinate system shown above in Figure 1.1 gives

$$\Omega_3 = \Omega \sin\phi, \tag{1.20}$$

where ϕ is latitude. It is convenient to introduce the 'Coriolis parameter':

$$f = 2\Omega \sin\phi = 1.458 \times 10^{-4} \sin\phi \qquad (\text{s}^{-1}). \tag{1.21}$$

In terms of this parameter the Coriolis acceleration along x_1 is fu_2, along x_2, $-fu_1$. At mid-latitudes (near 41 °N) the value of f is close to $f = 10^{-4}\ \text{s}^{-1}$. In the southern hemisphere ϕ and therefore f may be taken to be negative. For brevity, however, unless otherwise stated, specific remarks concerning the direction of the Coriolis force, or other earth rotation effects, will refer to the northern hemisphere.

1.6. SURFACE AND BOTTOM STRESS

Of the Reynolds stresses the diagonal components $i = j$ are unimportant corrections to hydrostatic pressure. The component $\tau_{12} = \tau_{12} = -\rho\,\overline{u_1'u_2'}$ represents horizontal momentum flux, the divergence of which could be important in the first two momentum equations. In practice, however, the stress τ_{12} itself is rarely as high as 1 Pa and its horizontal scale of variation is rarely less than 10 km. The contribution of this stress to the kinematic stress divergence $\rho^{-1}\ \partial\tau_{ij}/\partial x_j$ is then of other 10^{-7} m s^{-2} or rather smaller than the typical vertical momentum flux divergence, $\rho^{-1}\ \partial\tau_{i3}/\partial x_3$ (cf., below). The difference is not large enough to rule out the importance of τ_{12} under all circumstances. However, no instance has so far come to light in which the braking action of horizontal eddy momentum transfer was a demonstrably significant dynamical factor in large scale circulation problems in the coastal ocean.

Vertical flux of horizontal momentum, $\rho\, \overline{u_1'u_3'}$ and $\rho\, \overline{u_2'u_3'}$ (i.e., Reynolds stress in horizontal planes, τ_{13} and τ_{23}) is very important in shallow seas. The force of the wind is the outstanding local driving force of the flow and this is communicated to layers below the surface through the stresses τ_{13} and τ_{23}. If the wind stress has components ρF_i ($i = 1, 2, F_i$ being the kinematic stress) continuity of stress at the air-sea interface requires

$$\tau_{i3} = \rho F_i, \quad (x_3 = 0). \tag{1.22}$$

The wind stress is exerted along the direction of the surface wind and its magnitude is usually specified by a quadratic drag law (see e.g., Roll, 1965):

$$F_i = \frac{\rho_a}{\rho_w} C_{10}\, W_i W, \tag{1.23}$$

where W is wind speed (magnitude), W_i wind velocity component along x_i, ρ_a/ρ_w is air-water density ratio and C_{10} is a drag coefficient referred to the conventional anemometer height of 10 m.

The value of C_{10} scatters somewhat, but recent data suggest (Amorocho and DeVries, 1980) the following:

$$\begin{aligned} C_{10} &= 1.6 \times 10^{-3}, \quad (W \leqslant 7\ \mathrm{m\,s^{-1}}) \\ C_{10} &= 2.5 \times 10^{-3}, \quad (W \geqslant 10\ \mathrm{m\,s^{-1}}) \end{aligned} \tag{1.24}$$

with a smooth transition for wind speeds between 7 and 10 m s^{-1}. The reason for the variation in C_{10} is that the air-side of the water surface becomes hydrodynamically rougher at higher wind speeds, owing to the formation of whitecaps. The Nikuradse sand-grain roughness corresponding to the above values of C_{10} is about 1.4 cm at low wind speeds, 10 cm at high wind speeds. Capillary ripples appear to be responsible for the former, white-caps for the latter. At the 'typical' wind speed of 7 m s^{-1} the drag law (1.23) gives a typical kinematic surface stress F of 10^{-4} m^2 s^{-2} (or $\tau = 0.1$ Pa).

Another reason why τ_{13} and τ_{23} are important is that these stresses also transmit upward the braking action of the sea floor. The typical bottom slope of shallow seas is 10^{-2}–10^{-3}, so that the sea floor is very nearly horizontal (over distances of kilometers and up: local irregularities will be regarded as 'roughness'). If the components of the kinematic bottom stress are B_i, one may write in analogy with (1.22):

$$\tau_{i3} = \rho B_i, \quad (x_3 = -H). \tag{1.25}$$

Bottom stress may also be expressed by a drag law similar to (1.23):

$$B_i = c_d\, u_i\, (u_1{}^2 + u_2{}^2)^{1/2}, \tag{1.26}$$

where c_d is a bottom drag coefficient and u_i are horizontal velocity components at a convenient reference height above the bottom, which could be chosen to be 1 m, for example. Experimental evidence on the bottom drag coefficient under field conditions is scant and in many calculations one is forced to rely on the order of magnitude estimate of $c_d = 2 \times 10^{-3}$.

This estimate nevertheless shows that the typical bottom stress magnitude corresponding to a near-bottom velocity of 0.2 m s^{-1} is about the same as the typical wind stress in moderate winds (0.1 Pa), while it is very much greater when the current speed approaches 1 m s^{-1}. Wind stress in hurricanes is about equivalent to the bottom stress induced by currents this strong. The stress divergence $\partial\tau_{i3}/\partial x_3$ due to bottom braking is correspondingly as important as the stress divergence due to wind stress.

1.7. INTERIOR STRESSES

Shear stress applied at the free surface or at the bottom is not distributed uniformly over the entire available water column in depths of order 100 m. Two physical mechanisms combine to prevent such an even distribution: a rotating fluid mechanism and stratification.

In a (homogeneous) rotating viscous fluid, stress applied at a boundary surface not parallel to the axis of rotation causes a shear layer of limited depth to form, within which the stress reduces to zero (Ekman, 1905). The depth of a viscous 'Ekman' layer is of order $\sqrt{2\nu/f}$, with ν = kinematic viscosity. Outside Ekman layers the flow remains essentially frictionless. When the Ekman layer is turbulent, as is invariably the case in shallow seas, Reynolds stresses remain similarly confined to a layer of finite depth, so that the action of turbulence in this respect is similar to the action of viscosity. However, the depth of a turbulent Ekman boundary layer is independent of fluid properties and is of the order of $0.1u_*f^{-1}$, where u_* is friction velocity, $u_* = (\tau/\rho)^{1/2}$, with τ the magnitude of the boundary stress. For the typical value of τ = 0.1 Pa, u_* = 1 cm s^{-1}, the Ekman boundary layer depth is of order 10 m at mid-latitudes. However, in a hurricane, which exerts stresses of 3 Pa and more, top and bottom Ekman layers may fully occupy a water column of about 100 m depth.

Stratification due either to surface heating or to freshwater influx into a saline sea further reduces the depth of surface and bottom shear layers, although not by very much, certainly not by an order of magnitude. The surface and bottom shear layers remain well mixed through the action of turbulence and the entire density variation in the water column is then compressed into the region in between the shear layers. Above or below mixed layers sharp density interfaces sometimes form. Although Reynolds stresses can be sustained by wave-like motions in the stratified interior, their intensity is typically two orders of magnitude less than surface or bottom stresses. Exceptions occur in sharp density interfaces, across which the velocity may change abruptly enough to induce significant stress.

In order to render the momentum equations tractable, it is necessary to parameterize the interior Reynolds stresses τ_{13} and τ_{23} in terms of the velocities. If a number of caveats are observed, this may be fairly effectively carried out using a gradient transport relationship

$$\tau_{i3} = \rho K \frac{\partial u_i}{\partial x_3}, \qquad (1.27)$$

where K is an eddy momentum diffusivity or (kinematic) viscosity.

In using this relationship the following must be kept in mind. The eddy viscosity K is not a fluid property, nor is it reasonable to use a constant value of K for the entire water column, or even for the entire surface or bottom shear layer. Within 'wall layers' of order 1 m thickness K varies rapidly with depth, above the bottom, below the free surface, and possibly also adjacent to sharp interfaces. The current velocity vector also varies rapidly across such wall layers, but the stress generally does not. When the velocity distribution within wall layers is of interest, it may be calculated using the well known logarithmic law. In the following, no special attention will be paid to wall layers: they only occupy a small fraction of the total water column, and our empirical knowledge of wall layers in the sea is very limited.

Outside wall layers, but within a homogeneous turbulent shear flow region, such as the top and bottom (mixed) shear layers, the eddy viscosity may be taken to be constant and proportional to the velocity and length scales of the flow (see e.g., Monin and Yaglom, 1971). In the regions influenced by surface and bottom stress these are in turn proportional to friction velocity u_* and mixed layer depth h (*not* the total water column depth). Thus one has

$$K = \frac{u_* h}{Re}, \tag{1.28}$$

where Re is an eddy Reynolds number. Data from field observations have suggested $Re = 12$ to 20.

In the continuously stratified interior of the water column (excluding sharp density interfaces) there is almost no observational basis for specifying K. An order of magnitude estimate is $K = 1\ \text{cm}^2\ \text{s}^{-1}$, or two orders of magnitude less than typical values following from (1.28) for the well-mixed shear layers.

At sharp density interfaces is appears best to use a drag law similar to (1.23) and (1.26). If the interface stress components are ρI_i ($i = 1, 2$), one may write

$$I_i = c_s\, \Delta u_i\, (\Delta u_1{}^2 + \Delta u_2{}^2)^{1/2}, \tag{1.29}$$

where c_s is a sharp-interface drag coefficient and Δu_i ($i = 1, 2$) are velocity differences across the interface, from outside a 'wall' layer on one side to outside a wall layer on the other side. A typical value of c_s deduced from laboratory observations is 0.5×10^{-3} (Csanady, 1978c), but there is no basis for judging whether such a value is realistic under field conditions.

By excluding wall layers from consideration, it is possible to use Equation (1.27) with K = constant from one boundary surface to another. Stress boundary conditions are to be applied then at these boundary surfaces, according to (1.23), (1.26), and (1.29). The calculation of interior stresses and the interior velocity distribution becomes in this way a procedure of considerable complexity, noting the quadratic drag laws and the scaling of eddy viscosity by boundary stresses.

While it is often possible to simplify the parametric relationships of boundary and interior stresses to velocities without serious error, it should be noted here that neglecting the presence of wall layers altogether is not a realistic step in approximate calculations of the interior velocity distribution. If, for example, the viscous fluid boundary condition of no slip at a solid boundary is imposed, together with a constant eddy viscosity hypothesis, the calculated velocity distribution shows little similarity to observed profiles, the difference being of the kind that exists between the quadratic (laminar flow) Poiseuille distribution and characteristic turbulent velocity profiles in a circular pipe (see e.g., Prandtl, 1965).

1.8. LINEARIZATION OF THE EQUATIONS

When the gradient transport relationship (1.27) is introduced into the horizontal momentum equations (1.6), these acquire the form of the Navier–Stokes equations, which in their nonlinear form are notoriously intractable. They may be simplified either by neglecting interior stresses, supposing them to be distributed in a specially simple way, or by neglecting the momentum advection (nonlinear) terms in the equations, a step known as 'linearization'.

It is usually supposed that momentum advection in geophysical problems may be neglected when the ratio of the nonlinear terms to the Coriolis force (measured by the Rossby number) is small:

$$u_1 \frac{\partial u_1}{\partial x_1} : fu_2 \cong \frac{u_s}{fL} \ll 1, \tag{1.30}$$

where u_s and L are appropriate velocity and horizontal length scales. This comparison focuses on the cross-stream momentum balance of the flow. The streamwise momentum balance, however, contains no Coriolis force and the key comparison is between nonlinear terms and the driving force of the wind or the braking force of bottom drag. Thus, if flow is mostly along the x_2 axis, the nonlinear terms are negligible in the streamwise momentum balance if:

$$u_1 \frac{\partial u_2}{\partial x_1} : \frac{u_*^2}{h} \cong \frac{u_s^2}{u_*^2} \cdot \frac{h}{L} \ll 1, \tag{1.31}$$

where h is the depth of the layer over which the boundary stress ρu_*^2 is distributed.

In the dynamics of coastal currents neither of the two criteria (1.30) or (1.31) is, in general, satisfied. Rossby numbers are commonly of order one, and so is the momentum flux ratio of (1.31). Momentum advection is thus at least potentially important. However, on examining the question, just what the precise effects of momentum advection are, one usually finds these to be relatively straightforward modifications of a pattern governed by the balance of Coriolis force, local acceleration and wind and bottom stress, i.e. by the linear terms in the equations of motion. The reasons for this are partly that the momentum advection terms have the form of a divergence, and that they vanish on approaching the boundaries. Thus their role is to transfer momentum from one region of the basin to another, without modifying the total momentum input.

To take a concrete example, wind stress often generates an intense, relatively narrow longshore current, which according to the balance of the linear terms in the equation of motion should remain attached to the coast. Momentum advection can under certain circumstances displace this current seaward, without otherwise significantly modifying its important characteristics, such as its horizontal scale or peak velocity. The effect is clearly important and its understanding desirable. However, it is also true that considerable insight may be gained into the generation of such currents by a theoretical model in which momentum advection is neglected. Such a linearized theory predicts accurately the modes and spatial scales of motion, along with its buildup or decay time, or the wavenumbers and frequencies of any oscillatory components of the motion. This is the principal use of linearized theory: to provide theoretical models of sufficient simplicity for the important physical relationships to become transparent in otherwise very involved problems.

The limitations of linearized theory are most serious in a stratified fluid, on account of large vertical and horizontal particle displacements which tend to be caused by wind or other external forcing in a relatively narrow coastal band. The effect of these excursions is not confined to the momentum transport terms in the equations of motion, but it includes first order changes in the pressure distribution through the distortion of the density field. As pointed out earlier, fluid particles tend to conserve their temperature and salinity for periods of the order of a day, so that rapid vertical particle displacements give rise to similarly rapid changes in the dynamic height ζ_d (Equation (1.12)). Given that the displacements occur in a narrow coastal band, it is clear that considerable horizontal pressure gradients result from such events. However, in linearized theory vertical particle displacements are postulated to be small. Linearized theory is still useful in understanding at least the initial phase of such motions, and it also gives fairly accurate predictions for time and space scales. The formal validity of the theory is limited to a short initial period, nevertheless.

1.9. SHALLOW WATER EQUATIONS

The linearized theory of a stratified water column is discussed further in Chapter 3 on the basis of appropriate further idealizations. Here and in Chapter 2 the case of a constant density fluid will be pursued further, setting $\zeta_d = 0$ in the calculation of the pressure gradients (Equation (1.12)). The linearized theory with this further simplification has some direct applicability to vertically well mixed continental shelves and other shallow seas. Furthermore, the behavior of a stratified sea may often be viewed as consisting of a homogeneous fluid response, and of a density-related response, which is simply additive. On top of that, stratified flow dynamics can under certain circumstances be reduced to a set of equations identical in form with those governing the motion of a homogeneous fluid.

At this point it is convenient to abandon suffix notation and use (x, y, z) coordinates with the z-axis vertically up, velocity components (u, v, w). The linearized horizontal momentum equations for a homogeneous fluid are then, using a gradient transport relationship (Equation (1.27)) to describe interior stresses:

$$\begin{aligned} \frac{\partial u}{\partial t} - fv &= -g\frac{\partial \zeta}{\partial x} + \frac{\partial}{\partial z}\left(K\frac{\partial u}{\partial z}\right), \\ \frac{\partial v}{\partial t} + fu &= -g\frac{\partial \zeta}{\partial y} + \frac{\partial}{\partial z}\left(K\frac{\partial v}{\partial z}\right). \end{aligned} \tag{1.32}$$

These momentum balances are supplemented by the continuity equation and the kinematic relationships defining the free surface and the bottom:

$$\frac{\partial u}{\partial x} + \frac{\partial v}{\partial y} = -\frac{\partial w}{\partial z},$$

$$w = \frac{\partial \zeta}{\partial t}, \quad (z = 0), \tag{1.32a}$$

$$-w = u\frac{dH}{dx} + v\frac{dH}{dy}, \quad (z = -H)$$

where $H(x, y)$ is the total depth of the water column at (x, y).

Boundary conditions at the free surface and at the bottom are

$$F_x = K\frac{\partial u}{\partial z}, \quad F_y = K\frac{\partial v}{\partial z}, \quad (z = 0)$$

$$B_x = K\frac{\partial u}{\partial z}, \quad B_y = K\frac{\partial v}{\partial z}, \quad (z = -H). \tag{1.33}$$

The kinematic wind-stress components (F_x, F_y) are external inputs, while the bottom stresses follow from a drag law:

$$B_x = c_d u(u^2 + v^2)^{1/2}, \quad B_y = c_d v(u^2 + v^2)^{1/2}, \quad (z = -H). \tag{1.33a}$$

The velocities at $z = -H$ are physically velocities above a wall layer, but in calculations they are usually taken to be velocities extrapolated from the constant eddy viscosity interior to the bottom. The drag coefficient must then be appropriate for such an extrapolated or 'slip' velocity.

The above set of equations adequately represents the dynamics of fluid motion in a well mixed, shallow sea. They are referred to for brevity as the 'shallow water equations'.

1.9.1. The 'Local' and the 'Global' Problem

When the above linearized equations apply, the problem of predicting the response of a shallow sea to some forcing effect is conveniently separated into a 'local' and a 'global' problem (Welander, 1957).

Once a suitable approximation to a distribution $K(z)$ has been adopted, Equations (1.32) and (1.33) may be solved for $u(z)$, $v(z)$, with the gradients of sea level, ζ, regarded as external inputs. The velocity distributions thus follow from a 'local' calculation (at given x, y), for given pressure gradients.

The remaining problem is to calculate the basin-wide or 'global' distribution of pressure (sea level), which is conveniently approached through the depth-integrated or 'transport' equations. Depth integration of the velocity components defines horizontal transports:

$$U = \int_{-H}^{0} u \, dz, \quad V = \int_{-H}^{0} v \, dz. \tag{1.34}$$

Depth integration of (1.32) and (1.33) yields

$$\frac{\partial U}{\partial t} - fV = -gH\frac{\partial \zeta}{\partial x} + F_x - B_x,$$

$$\frac{\partial V}{\partial t} + fU = -gH\frac{\partial \zeta}{\partial y} + F_y - B_y, \quad (1.35)$$

$$\frac{\partial U}{\partial x} + \frac{\partial V}{\partial y} = -\frac{\partial \zeta}{\partial t}.$$

In Equations (1.34) and (1.35) the small surface elevation changes $\zeta(x, y)$ have been neglected in comparison with $H(x, y)$. The surface stress (F_x, F_y) distribution is an external input. The bottom stress may be calculated using bottom velocities determined from a solution of the local problem. These depend on locally imposed external parameters such as surface stress and depth, as well as on local gradients of ζ, as already pointed out. Whatever the details, a closed set of three equations results for the three variables U, V, ζ, which, in principle, allows the calculations of $\zeta(x, y)$. This will be referred to below as the solution of the global problem. Equations (1.35) are called the transport equations.

In practice, the solution of the global problem is generally difficult, but made easier by neglecting bottom stresses altogether or representing them as a simple function of the transports (U, V). Such approximations turn out to be sufficiently accurate in a number of interesting and important problems.

Boundary conditions at coasts or at open boundaries are required in solving the transport equations. Along coasts, the transport normal to the boundary vanishes, a condition which will be referred to below as the 'coastal constraint'. At open boundaries the normal transport and the surface elevation are continuous. Open boundaries between a 'shallow' and a 'deep' sea pose some difficulties because a simultaneous solution of the equations of motion for the two regions separately is not always a practical possibility.

The solution of global and local problems will be discussed in this monograph on a number of occasions, for a variety of specific situations, beginning with some very simple examples later in this chapter.

1.10. POTENTIAL VORTICITY EQUATION

Turbulence tends to equalize not only the fluid properties temperature and salinity, and hence density, but also horizontal momentum. If the notion of a 'well mixed' fluid column is thus extended to momentum, an alternative simplification scheme for the equations of motion becomes accessible, which does not neglect the advection of either momentum or density. Let a stratified water column consist of a series of well mixed layers, and consider one such layer of depth h, with arbitrary stresses exerted at its top and bottom. The velocity is supposed independent of depth:

$$\frac{\partial u}{\partial z} = \frac{\partial v}{\partial z} = 0 \quad (1.36)$$

while the stress is taken to be evenly distributed:

$$\frac{1}{\rho}\frac{\partial \tau_x}{\partial z} = \frac{F_x - B_x}{h}, \qquad \frac{1}{\rho}\frac{\partial \tau_y}{\partial z} = \frac{F_y - B_y}{h}. \tag{1.37}$$

Note that this idealization of the interior stress and velocity distribution is inconsistent with Equation (1.27), representing an alternative (more or less opposite) parameterization scheme.

The depth of the layer h varies with time and with horizontal location. The continuity equation gives

$$\frac{\partial u}{\partial x} + \frac{\partial v}{\partial y} = -\frac{\partial w}{\partial z} = -\frac{1}{h}\frac{\mathrm{d}h}{\mathrm{d}t} \tag{1.38}$$

the second equality following from an integral application of this equation to an entire layer of depth h.

The equations of motion, with horizontal momentum advection and dynamic height terms retained are:

$$\begin{aligned} \frac{\partial u}{\partial t} + u\frac{\partial u}{\partial x} + v\frac{\partial u}{\partial y} - fv &= -g\frac{\partial(\zeta-\zeta_d)}{\partial x} + \frac{F_x - B_x}{h}, \\ \frac{\partial v}{\partial t} + u\frac{\partial v}{\partial x} + v\frac{\partial v}{\partial y} + fu &= -g\frac{\partial(\zeta-\zeta_d)}{\partial y} + \frac{F_y\ B_y}{h}. \end{aligned} \tag{1.39}$$

The pressure terms may be eliminated by differentiating the first equation with respect to y, the second with respect to x and substracting. Also making use of (1.38), the result may be put into the following form:

$$\frac{\mathrm{d}}{\mathrm{d}t}\left(\frac{\omega+f}{h}\right) = \frac{1}{h}\left[\frac{\partial}{\partial x}\left(\frac{F_y - B_y}{h}\right) - \frac{\partial}{\partial y}\left(\frac{F_x - B_x}{h}\right)\right], \tag{1.40}$$

where ω is the vertical component of vorticity:

$$\omega = \frac{\partial v}{\partial x} - \frac{\partial u}{\partial y}. \tag{1.41}$$

The quantity $(\omega+f)/h$ is known as 'potential vorticity'. Potential vorticity is seen to be generated by the curl of the net friction force affecting the layer. Where the latter quantity vanishes, potential vorticity is conserved, following a given fluid column. In this case, as the depth of a fluid column increases, $\omega + f$ must also increase. In seas of moderate horizontal extent, where $f \cong$ constant, this requires an increase of ω, i.e., the generation of positive or 'cyclonic' vorticity. One often expresses this result by the statement that stretching of fluid columns generates cyclonic vorticity, their squashing anticyclonic vorticity.

The potential vorticity equation (1.40) was derived from the equations of continuity and motion and it can be used as a substitute for one off them. Once the vorticity distribution is determined, the velocity and pressure fields can in principle be found. This approach is a useful supplement to the shallow water equations in the dynamics of shallow seas because it often allows the assessment of errors incurred through linearization.

1.11. SOME ELEMENTARY CONCEPTUAL MODELS

In the evolution of our understanding of atmospheric and oceanic motions on a rotating earth a few basic models have played an important role. They are also useful in aiding insight into the actual, complex behavior of the coastal ocean, which often exhibits approximate or partial resemblance to the elementary models. Three such elementary models are considered below; they are all based on the shallow water equations and serve to illustrate solutions of both the global and the local problem.

1.11.1. Geostrophic Balance

Consider the case when wind stress *and* bottom stress are negligible, the water is homogeneous and the motion is steady, so that all time derivatives in the transport equations (1.35) vanish. This case is clearly somewhat unrealistic, but one may think of a current established by some prior impulse, which is decaying very slowly. Without loss of generality the y-axis may be aligned with the direction of the transport vector, so that $U = 0$. The transport equations then reduce to

$$\begin{aligned} -fV &= -gH\frac{\partial\zeta}{\partial x}, \\ 0 &= -gH\frac{\partial\zeta}{\partial y}, \\ \frac{\partial V}{\partial y} &= 0. \end{aligned} \tag{1.42}$$

The first and third of these equations are consistent only if $\partial H/\partial y = 0$.* The physical problem is illustrated in Figure 1.3. Transport is along contours of constant depth (isobaths) and the sea level varies in the cross-isobath direction only. The balance of forces is non-trivial in the cross-flow direction only, where the Coriolis force is balanced by the pressure gradient. This is known as 'geostrophic balance'. It is also common to speak of geostrophic 'flow', but this can be confusing, because the first of Equations (1.42) may apply when the second does not.

With zero surface and bottom stress, the solution of the corresponding local problem is trivial

$$v = \frac{V}{H} = \frac{g}{f}\frac{\partial\zeta}{\partial x} \tag{1.43}$$

at all z and y, depending only on x. With $g \cong 10$ m s^{-2}, $f \sim 10^{-4}$ s^{-1} it is seen that a

---

* Actually $\partial(H/f)/\partial y = 0$; in shallow seas the variation of the Coriolis parameter with latitude is unimportant, and f will always be regarded as constant.

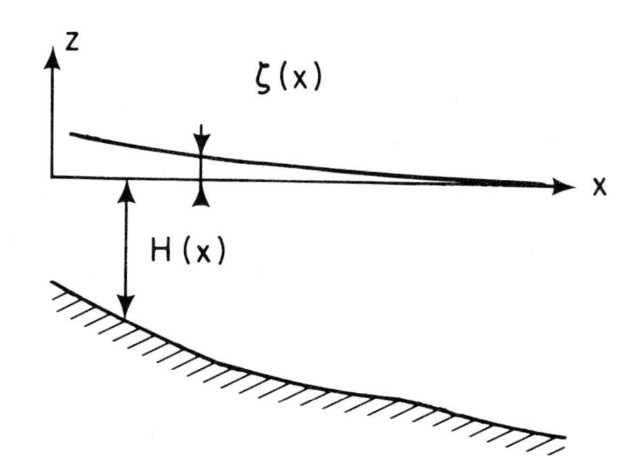

Fig. 1.3. Surface elevation and depth distributions in elementary coastal current model.

1 m s^{-1} velocity is in geostrophic balance with a surface level gradient of 10^{-5}, or 1 cm in 1 km.

This type of flow satisfies the coastal constraint at any x and is thus consistent with a coast located parallel to the y-axis. It may be regarded as an elementary coastal flow model.

1.11.2. Inertial Oscillations

Consider now motion far from any boundaries. Because it is the constraining action of coasts which may be thought responsible for the raising of sea level, it is reasonable to postulate $\zeta = 0$ everywhere in this model. Suppose also that wind and bottom stress vanish, the motion being a leftover of prior action. The remaining terms in the transport equation are

$$\frac{\partial U}{\partial t} - fV = 0,$$

$$\frac{\partial V}{\partial t} + fU = 0, \tag{1.44}$$

$$\frac{\partial U}{\partial x} + \frac{\partial V}{\partial y} = 0.$$

These have the solution

$$U = U_0 \cos ft, \qquad V = -U_0 \sin ft, \tag{1.45}$$

where U_0 = constant and the coordinate axes were chosen so that $V = 0$ at $t = 0$. The interior velocities are again simply

$$u = \frac{U}{H} = u_0 \cos ft,$$
$$v = \frac{V}{H} = -u_0 \sin ft. \tag{1.46}$$

The motion is periodic with period $T = 2\pi/f$, which is the half-pendulum day, or typically 17 h at mid-latitudes. The entire water mass oscillates in phase, particle paths being circles of radius u_0/f. With $u_0 = 1$ m s^{-1}, $f = 10^{-4}$ s^{-1} this radius is 10 km.

This type of motion, known as inertial oscillation is not consistent with the presence of a coast anywhere and is therefore quite unrealistic as a coastal ocean model. However, the force balance expressed by (1.44) prevails approximately in some more realistic coastal flow models. This is a balance between local acceleration and Coriolis force per unit mass.

1.11.3. Ekman Drift

Consider again motion far from any boundaries, supposing $\zeta = 0$ everywhere. Let the motion be forced by a wind-stress ρF applied along the y-axis, and suppose that any transients have decayed, leaving only steady flow. The water column will be supposed sufficiently deep for bottom stress to be negligible (this hypothesis will be examined *à posteriori*). The global equations (1.35) now reduce to

$$-fV = 0,$$
$$fU = F, \tag{1.47}$$
$$\frac{\partial U}{\partial x} + \frac{\partial V}{\partial y} = 0.$$

These are satisfied by

$$U = \frac{F}{f} = \text{constant},$$
$$V = 0. \tag{1.47a}$$

Transport is to the right of the wind (in the northern hemisphere, where f is positive) and is of a magnitude F/f. For 'typical' values of $F = 10^{-4}$ m^2 s^{-2}, $f = 10^{-4}$ s^{-1}, $U =$ 1 m^2 s^{-1}, or equal to the transport of a 0.1 m s^{-1} current over a 10 m depth range. This is known as 'Ekman transport', and the motion carrying this transport will also be described as 'Ekman drift'.

The solution of the local problem for this case is non-trivial. Let internal stresses be parameterized by the gradient transport relationships (1.27), with K = constant below a free surface wall layer of negligible thickness. Horizontal gradients vanish by hypothesis and the linearized momentum equations (1.32) apply in the following simple form:

$$-fv = K\frac{d^2u}{dz^2},\qquad fu = K\frac{d^2v}{dz^2},\tag{1.48}$$

with boundary conditions

$$\left.\begin{aligned}K\frac{du}{dz} &= 0,\\ K\frac{dv}{dz} &= u_*^2,\end{aligned}\right\}\ (z = 0)$$

$$\frac{du}{dz} = \frac{dv}{dz} = 0,\quad (z \to -\infty).\tag{1.49}$$

The vertical coordinate is conveniently scaled by the 'Ekman depth'

$$D = \sqrt{\frac{2K}{f}}.\tag{1.50}$$

In terms of this scale the solution of (1.48) and (1.49) is easily found to be:

$$\frac{u}{u_*} = \frac{u_*}{fD}e^{z/D}\left(\cos\frac{z}{D} - \sin\frac{z}{D}\right),\qquad \frac{v}{u_*} = \frac{u_*}{fD}e^{z/D}\left(\cos\frac{z}{D} + \sin\frac{z}{D}\right).\tag{1.51}$$

At $-z \gg D$ both the velocity and the stress vanish, so that it is in fact reasonable to suppose zero bottom stress for sufficiently deep water.

As already pointed out in Section 1.7, for a *turbulent* Ekman layer the depth scale D is empirically found to be approximately:

$$D \cong 0.1\frac{u_*}{f}.\tag{1.52}$$

In virtue of Equation (1.50) this is equivalent to

$$K = \frac{u_*^2}{200f}\tag{1.53}$$

or also

$$K = \frac{u_*D}{20}.\tag{1.54}$$

The last result is of the form of Equation (1.28), showing that the eddy Reynolds number of the homogeneous Ekman layer is about 20.

Equation (1.52) shows further that the nondimensional factor u_*/fD appearing in the solution (1.51) is approximately 10. Thus the velocity components at $z = 0$ (i.e., just below the free surface wall layer) are $u = v = 10u_*$. The velocity vector at this level points 45° to the right of the wind.

The properties of the oceanic free surface wall layer have not been very well explored so far, and they are not further discussed in this text. However, for the sake of illustrating here the entire velocity distribution at the sea surface, a logarithmic velocity distribution in the free surface wall layer will be assumed:

$$\frac{v_s - v}{u_*} = \frac{1}{k} \ln \frac{|z|}{r} + 8.5, \tag{1.55}$$

where k is Kàrman's constant ($\cong 0.4$) and r is equivalent 'sand-grain' roughness of the surface. The latter appears to be surprisingly large (of order 1 m in moderate winds), so that the logarithmic term in (1.55) applied to $-z = 1$ m does not contribute very much. The effective jump in velocity across the free surface wall layer is then close to $8.5u_*$, from the bottom of the wall layer to $z = -r$, which is the 'surface' in a wall layer model.

Figure 1.4 illustrates, in the form of a hodograph, the velocity distribution within an

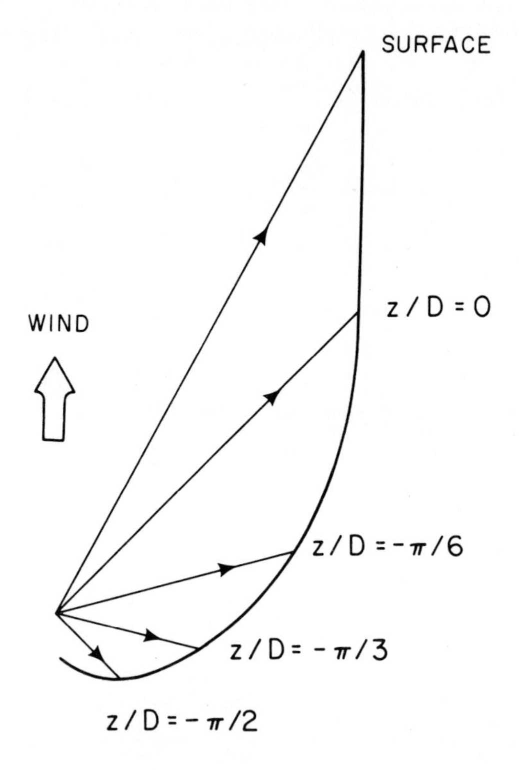

Fig. 1.4. Hodograph of horizontal velocity vector in sea surface Ekman layer; $z/D = 0$ is level just below surface wall layer, other levels as marked, with D = Ekman depth.

idealized turbulent Ekman layer at the free surface, including the wall layer model. The surface velocity is seen to be directed some 25° to the right of the wind, and to have a magnitude of about $20u_*$. As may be seen from Equation (1.23), this is about 3% of the wind speed, a rule of thumb frequently quoted in oceanographic texts.

For the 'typical' value of $u_* = 0.01$ m s^{-1} (corresponding to a wind speed $W = 7$ m s^{-1}) and with $f = 10^{-4}$ s^{-1} the Ekman depth D is 10 m. As may be seen from Equation (1.51) velocities become negligible at about $z = -3D$, or 30 m depth; The water column is frequently homogeneous enough to such a depth for the above theory to be realistic. However, it must be remembered that coasts have been supposed to be absent. A velocity distribution such as shown in Figure 1.4 is incompatible with a coast placed at any orientation, because the flow has a normal component to any imaginable coast at almost all levels within the Ekman layer. However, a thin lateral boundary layer is all that is required in principle to rectify this conflict.

The great Norwegian oceanographer, Valfried Ekman, has made many outstanding contributions to theoretical oceanography and it is appropriate that one speaks of an 'Ekman layer', 'Ekman depth' and 'Ekman transport' or 'Ekman drift'. These concepts have been defined above. Unfortunately, however, one also reads about 'Ekman models', 'Ekman flux', and 'Ekman dynamics', with the meaning of the expression not being always obvious. In the following, such loose usage of Ekman's name will be avoided.

CHAPTER 2

Inertial Response to Wind

2.0. INTRODUCTION

The stress of the wind exerted over the sea surface is the prime driving force of circulation in the coastal ocean. Yet, as already pointed out in the previous chapter, wind action only affects directly through internal friction a surface layer usually much thinner than the total water column. To the rest of the water mass the effects of the wind are communicated through pressure forces, arising in the first instance from the presence of coasts which 'dam up' the water. The response to these forces is controlled essentially by the inertia of the water mass, with the Earth's rotation playing an important role. In order to understand such effects it is necessary to study the pressure field which arises in response to forcing by wind. One would like to answer questions such as what the pattern of the pressure field is arising from given forcing by wind, by what precise physical mechanism it is established, how long the establishment takes and what quantitative parameters affect it.

In the present chapter this problem is approached through an analytical study of shallow sea models which are deliberately over-idealized for maximum simplicity. Linearized shallow water equations will be used, the water will be supposed of constant density, the depth constant and the geometry of the basin simple, contained by circular or straight shores. Except for a few brief comments, bottom friction will be neglected. The principal aim is to gain physical insight into the mechanism of pressure field generation by wind in a rotating shallow sea.

In a fluid of constant density the hydrostatic approximation connects the pressure field to the distribution of sea level, $\zeta(x, y)$. One practically well known manifestation of the wind-induced pressure field is known as 'wind setup' (Windstau, in older oceanographic texts) or piling up of water against a leeward coast. On the windward shore the level is depressed. In Lake Erie storms cause level fluctuations of the order of 1 m, a large enough change to affect the output of the Niagara power plant. Most serious are sea level fluctuations along flat coasts where they cause considerable damage through flooding. The investigations of this chapter are conveniently started by discussing the elementary model of wind setup.

2.1. WIND SETUP CLOSE TO COASTS

Consider a closed basin of constant depth H and suppose that a uniform wind stress $F_y = u_*^2$ acts at the surface, along the y direction. Coasts prevent normal transport, so that in a first approximation one may try to determine sea level changes at points close to coasts on the basis of the hypothesis that both components of the transport vanish everywhere:

$$U = V = 0. \tag{2.1}$$

A second hypothesis, reasonable in view of the first, is that bottom stresses remain negligible:

$$B_x = B_y = 0. \tag{2.1a}$$

For this case the transport equations (1.35) reduce to

$$\begin{aligned} 0 &= -gH\frac{\partial \zeta}{\partial x}, \\ 0 &= -gH\frac{\partial \zeta}{\partial y} + u_*^2, \\ 0 &= -\frac{\partial \zeta}{\partial t}. \end{aligned} \tag{2.2}$$

These equations have the solution

$$\zeta = \frac{u_*^2}{gH}y, \tag{2.3}$$

where the origin of the y axis is supposed placed in a manner consistent with conservation of total water mass. The solution is known as wind setup. It is characterized by a simple force balance between wind stress and horizontal pressure gradient associated with the sea level slope, independently of time, i.e., in steady state. In moderate winds, $u_* = 1$ cm s^{-1}, ($\tau_y = 0.1$ Pa) $g = 10$ m s^{-2}, $H = 30$ m, the surface slope $\partial\zeta/\partial y$ is 3×10^{-7}, so that in a basic of 100 km length a level difference of 3 cm appears between the upwind and the downwind ends. The large level changes in Lake Erie referred to above are associated with much stronger winds, exerting a stress of several pascals. Equation (2.3) with $\tau_y = 3$ Pa and for a lake 30 m deep, 300 km long, gives a level difference of 3 m.

2.1.1. Interior Velocity Distribution

The above is a classical, elementary solution of the global problem. Some of its limitations are exposed on considering the local problem, i.e., on calculating the interior velocity distribution associated with wind setup. Adopting the interior stress parameterization discussed in Section 1.7, for a homogeneous fluid and steady state the local problem is governed by the following forms of Equations (1.32):

$$\begin{aligned} -fv &= -g\frac{\partial \zeta}{\partial x} + \frac{\partial}{\partial z}\left(K\frac{\partial u}{\partial z}\right), \\ fu &= -g\frac{\partial \zeta}{\partial y} + \frac{\partial}{\partial z}\left(K\frac{\partial v}{\partial z}\right). \end{aligned} \tag{2.4}$$

Boundary conditions are a stress of ρu_*^2 along the y-axis at the surface and a drag law at the bottom:

$$\frac{\partial u}{\partial z} = 0, \quad K\frac{\partial v}{\partial z} = \rho u_*^2, \quad (z = -0)$$

$$\left.\begin{aligned} K\frac{\partial u}{\partial z} &= c_d\, u(u^2 + v^2)^{1/2}, \\ K\frac{\partial v}{\partial z} &= c_d\, v(u^2 + v^2)^{1/2}, \end{aligned}\right\} \quad (z = -H) \tag{2.4a}$$

where surface and bottom wall-layers have been supposed to be vanishingly thin. In order for these conditions to be consistent with the negligible bottom friction assumption made in solving the global problem, $(u^2 + v^2)^{1/2}$ must be small as $z \to -H$.

Substituting the global solution (2.3) into (2.4) one finds

$$\begin{aligned} -fv &= \frac{\partial}{\partial z}\left(K\frac{\partial u}{\partial z}\right), \\ f\left(u + \frac{u_*^2}{fH}\right) &= \frac{\partial}{\partial z}\left(K\frac{\partial v}{\partial z}\right). \end{aligned} \tag{2.4b}$$

These are the same as Equations (1.48), if only u is replaced by $u + (u_*^2/fH)$. For sufficiently great water depth H the solution is then

$$\begin{aligned} u &= -\frac{u_*^2}{fH} + u_E, \\ v &= v_E, \end{aligned} \tag{2.5}$$

where (u_E, v_E) is the Ekman layer solution given by Equations (1.51). The solution applies for $H >> D$. At $|z| >> D$ the velocity is constant:

$$\left.\begin{aligned} u &= -\frac{u_*^2}{fH}, \\ v &= 0, \end{aligned}\right\} \quad |z| >> D, \quad D = (2Kf^{-1})^{1/2}. \tag{2.6}$$

The flow at depth is in geostrophic balance with the pressure gradient $\partial\zeta/\partial y$. Closer to the surface the velocity is a sum of the geostrophic velocity (2.6) and the Ekman layer velocity. This velocity distribution is illustrated in Figure (2.1).

The hypothesis of negligible bottom stress is consistent with the bottom boundary condition expressed by (2.4a) only if velocities at depth are suitably small. If bottom stress is written as ρu_{**}^2, its ratio to the surface stress is of order (from (2.4a) and (2.6):

$$\frac{u_{**}^2}{u_*^2} \sim c_d\left(\frac{u_*}{fH}\right)^2. \tag{2.7}$$

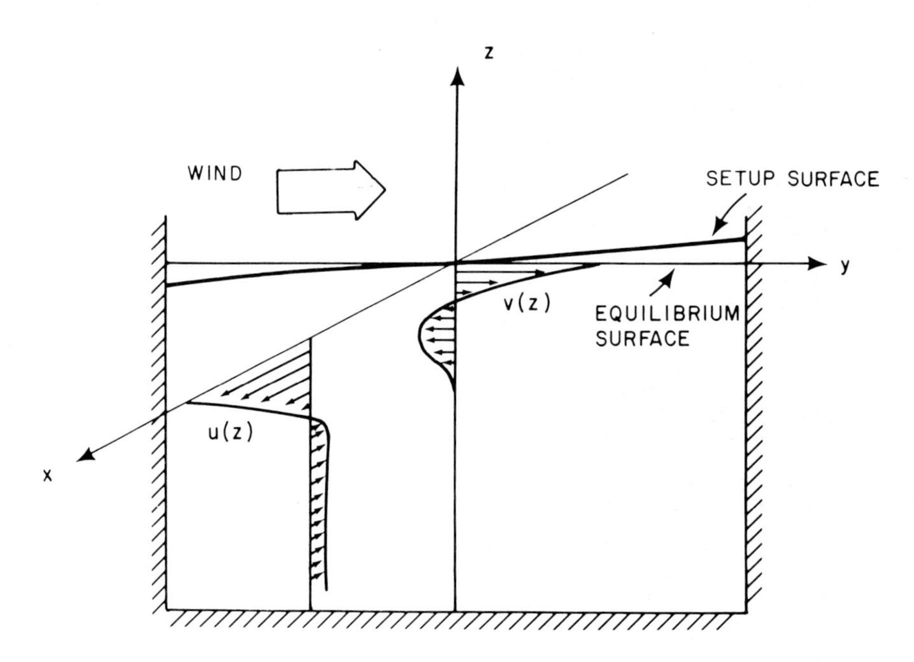

Fig. 2.1. Wind setup in constant depth basin, and associated interior velocities, an Ekman layer at the surface and geostrophic flow below.

This is of order 10^{-2} if u_*/fH is about 3, which is the case for the typical values of $u_* = 1$ cm s^{-1}, $f = 10^{-4}$ s^{-1}, $H = 30$ m. However, in very strong winds or very shallow water the hypothesis of negligible bottom stress fails. A more complete solution may be calculated and shows that the setup increases in such cases somewhat beyond the value given by Equation (2.1).

The global solution (2.1) expresses balance between the pressure gradient force, integrated over the water column, and the wind stress. The exact same balance applies in a laboratory flume or tank, where the Earth's rotation exerts no significant effect. Within the water column the details are seen to be more complicated on account of Earth rotation effects: the pressure gradient force is constant, but the wind stress is distributed only over a layer of depth comparable to the Ekman depth D. The Coriolis force of cross-wind flow in effect redistributes the wind-stress force over the entire water column by balancing the stress gradient near the surface, and the pressure gradient at depth. Corresponding to the depth-integrated force balance, the cross-wind Ekman transport near the surface is exactly balanced by opposite cross-wind transport over the entire water column due to the geostrophic velocity u_*^2/fH.

The calculated velocity distribution is again, as in the case of simple Ekman drift, not consistent with the presence of a coast of any orientation. However, a coast perpendicular to the wind conflicts only with velocities within the Ekman layer. Moreover, total transport is zero in any direction, so that all that is needed at a coast is a boundary layer within which vertical motions may somehow 'close' the streamlines. Further consideration of the coastal boundary layer problem is deferred to a later chapter, dealing with more realistic basins. In this chapter similar difficulties are ignored, because a discussion of lateral boundary layers in a constant depth basin would serve no useful purpose.

2.2. SETUP IN A BASIN OF ARBITRARY SIZE

Both the Ekman transport (Equation (1.47a)) and the wind setup global solutions were arrived at in an arbitrary manner, by postulating respectively vanishing pressure gradients and vanishing transports. These postulates were justified on an intuitive basis by supposing coasts to be 'far' or 'close'. The question arises, just how far or how close a coast has to be for these limiting hypotheses to apply.

For a constant depth basin containing homogeneous fluid, the transport equations will be written as

$$\frac{\partial U}{\partial t} - fV = -c^2 \frac{\partial \zeta}{\partial x} + F_x - B_x,$$

$$\frac{\partial V}{\partial t} + fU = -c^2 \frac{\partial \zeta}{\partial y} + F_y - B_y, \tag{2.8}$$

$$\frac{\partial U}{\partial x} + \frac{\partial V}{\partial y} = -\frac{\partial \zeta}{\partial t},$$

where $c^2 = gH$ is a constant. On differentiating the first equation with respect to y, the second with respect to x, subtracting the first from the second and substituting the continuity equation one arrives at

$$\frac{\partial}{\partial t}\left(\frac{\partial V}{\partial x} - \frac{\partial U}{\partial y}\right) = f\frac{\partial \zeta}{\partial t} + \frac{\partial}{\partial x}(F_y - B_y) - \frac{\partial}{\partial y}(F_x - B_x). \tag{2.9}$$

This is a depth integrated, linearized analog of the vorticity Equation (1.40). The left-hand side is rate of change of depth-integrated vorticity, the first term on the right is vorticity generation through vortex stretching by a rising surface, the remaining terms the curl of surface minus bottom stress.

For maximum simplicity, the spatial variability of wind- and bottom stress will in this chapter be ignored. In Equation (2.9) the curl of these stresses is correspondingly dropped. Because the time-integral of Equation (2.9) is required in the following argument, one must suppose here that the time-integral of the stress-curl terms remains suitably small. Because the neglected terms are not zero, only small, this is a restriction on the time period for which the argument made below can be taken to be valid. Given enough time, bottom friction certainly affects the pattern of flow set up initially by inertial forces.

Supposing that the motion starts from rest ($\zeta = 0$ at $t = 0$ at all x, y), the time integral of (2.9) with the stress-curl terms dropped is:

$$\frac{\partial V}{\partial x} - \frac{\partial U}{\partial y} = f\zeta. \tag{2.10}$$

This is a linearized expression of the conservation of potential vorticity. Next, differentiate the first of Equations (2.8) with respect to x, the second with respect to y and add, substituting also the continuity equation and (2.10):

$$\frac{\partial^2 \zeta}{\partial t^2} - c^2 \nabla_1^2 \zeta + f^2 \zeta = 0, \qquad \left(\nabla_1^2 = \frac{\partial^2}{\partial x^2} + \frac{\partial^2}{\partial y^2}\right). \tag{2.11}$$

A stress divergence term of the form

$$-\frac{\partial}{\partial x}(F_x - B_x) - \frac{\partial}{\partial y}(F_y - B_y)$$

has here been dropped on the right-hand side of this Equation. Having neglected the time integral of the stress-curl terms, it would be inconsistent to retain the stress divergence.

The wind setup solution, Equation (2.3) was derived also for spatially uniform wind and negligible bottom stress, postulating, in addition, steady state. The analogous problem in the more general case is to find a time independent solution of Equation (2.11), which must satisfy:

$$\nabla_1^2 \zeta - \frac{f^2}{c^2} \zeta = 0. \tag{2.12}$$

The boundary condition is vanishing normal transport at the coasts. This may be cast in terms of the surface elevation ζ by expressing cross-shore and longshore transports as functions of ζ. Let a coordinate system (n, s) be aligned with the local normal and tangent to the coast (Figure 2.2), with positive s measured counterclockwise around the basin perimeter, positive n outwards. Writing Equations (2.8) in terms of these coordinates for

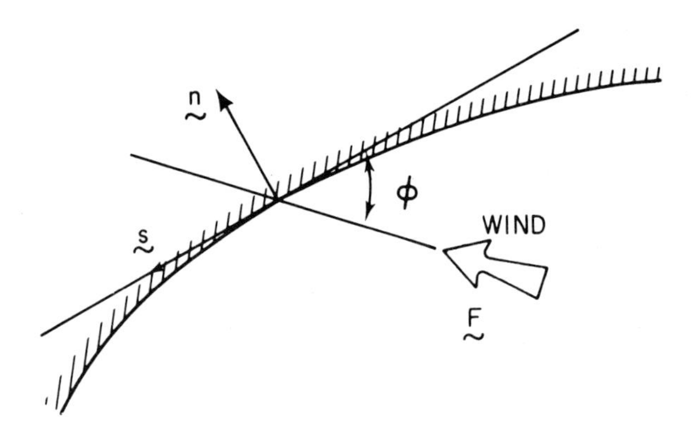

Fig. 2.2. Boundary conditions at a piece of coastline of arbitrary orientation.

transport components (V_n, V_s) it is easily shown that

$$\frac{\partial^2 V_n}{\partial t^2} + f^2 V_n = -c^2 \left(\frac{\partial^2 \zeta}{\partial n \partial t} + f \frac{\partial \zeta}{\partial s} \right) + f F_s. \tag{2.13}$$

At the coast $V_n = 0$. In steady state $\partial \zeta / \partial t = 0$ and Equation (2.13) reduces to

$$V_n = -\frac{c^2}{f} \frac{\partial \zeta}{\partial s} + \frac{F_s}{f} = 0, \tag{2.14}$$

or

$$\frac{\partial \zeta}{\partial s} = \frac{F_s}{c^2}.$$

The physical content of Equation (2.14) is balance between longshore wind stress and longshore pressure gradient at the coast, where the Coriolis force associated with the cross-shore transport vanishes. For uniform wind stress, the longshore component F_s varies simply with the orientation of the coastline:

$$F_s = F \cos \phi, \tag{2.15}$$

where ϕ is the angle included by the wind stress vector and the local tangent (Figure 2.2).

Thus the longshore surface level gradient is

$$\frac{\partial \zeta}{\partial s} = \frac{F \cos \phi}{c^2}. \tag{2.15a}$$

The solution of (2.12) and (2.15a) now only depends on basin geometry.

2.2.1. Setup in a Rectangular Basin

Consider next a rectangular basin for simplicity, side lengths 2a and 2b, the sides being aligned with the coordinate axes. The static setup solution (2.3) does *not* satisfy (2.12). However, searching for a solution of the form $\zeta = y$ func(x) one soon discovers that putting func(x) = $e^{\pm x/R}$ satisfies (2.12) if $R = c/f$, which is known as the 'radius of deformation'. The boundary conditions (2.15a) along the sides parallel with and perpendicular to the wind become:

$$\frac{\partial \zeta}{\partial y} = \frac{F}{c^2}, \quad (x = \pm a)$$
$$\frac{\partial \zeta}{\partial x} = 0, \quad (y = \pm b). \tag{2.16}$$

The first of these conditions may be met by a combination of two functions of the form $ye^{x/R}$ and $ye^{-x/R}$:

$$\zeta^* = \frac{Fy \cosh (x/R)}{c^2 \cosh (a/R)}. \tag{2.17}$$

However, ζ^* is not yet the required solution because it does not satisfy the second condition (2.16). According to this condition, the surface elevation at the end walls should be constant with x, whereas (2.17) shows a variation as $\cosh(x/R)$. One plausible fix is to add another solution which cancels the variation of the surface elevation along the end-walls. Any reasonable distribution of ζ along the end walls could be expanded into a trigonometric series, and it is therefore relevant to note that Equation (2.12) is also satisfied by

$$\zeta^{**} = e^{ly} \cos kx \tag{2.18}$$

provided that

$$-k^2 + l^2 - \frac{1}{R^2} = 0. \tag{2.19}$$

Choosing $k = k_n = (2n - 1)(\pi/2a)$, $n = 1, 2, 3, \ldots$ a trigonometric series is found in terms of which $\cosh x/R$ can be expanded in the interval $-a \leqslant x \leqslant a$ (this method is due to Taylor, 1920). The calculations are straightforward and yield finally a solution that does satisfy all boundary conditions (Csanady and Scott, 1974):

$$\zeta = \frac{F}{c^2}\left\{ y\,\frac{\cosh(x/R)}{\cosh(a/R)} + \frac{4b}{R^2}\sum_{n=1}^{\infty}(-1)^{(n-1)}\frac{1}{(2n-1)l_n^2}\,\frac{\sinh(l_n y)}{\sinh(l_n b)}\cos k_n x \right\}, \tag{2.20}$$

where l_n is calculated from (2.19), with $k = k_n = (2n - 1)(\pi/2a)$. All terms of the cosine series are seen to be multiplied by a factor of order a^2/R^2, so that they vanish as $a/R \to 0$. In the limit of the very narrow basin (in a direction normal to the wind) the simple static setup solution is thus recovered:

$$\zeta = \frac{Fy}{c^2}, \quad \left(\frac{a}{R} \to 0\right). \tag{2.21}$$

Substituting the definition of R, the criterion for the approximate validity of the simple static setup solution may be written as:

$$\frac{a}{R} = \frac{fa}{c} = \frac{fa}{\sqrt{gH}} \to 0. \tag{2.22}$$

This ratio may be regarded as a nondimensional basic size, or a nondimensional rotation rate, or else a measure of basin depth. Small basin size, or slow planetary rotation rate (closeness to the equator) or large basin depth all result in the approximate validity of the simple static setup solution.

In the opposite extreme of a large basin (*both* a, $b \gg R$) the exponentials in the denominators, $\cosh(a/R)$ and $\sinh(l_n b)$, become very large and reduce ζ to negligible values, except within distances of order R from the boundaries (where the numerators are also high in Equation (2.20)). This general result is suggested already by the form of Equation (2.12). Nondimensionalizing horizontal distances with the smaller side a (2.12) becomes

$$\frac{R^2}{a^2}\left(\frac{\partial^2 \zeta}{\partial (x/a)^2} + \frac{\partial^2 \zeta}{\partial (y/a)^2}\right) - \zeta = 0. \tag{2.23}$$

The second derivatives are seen to be multiplied by the factor $(R/a)^2$. If this factor is *small*, Equation (2.23) is approximately satisfied by the trivial solution $\zeta = 0$, except in boundary layers where the horizontal scale of variation is of order R instead of a (Carrier, 1953, has discussed similar problems of boundary layer character in some generality).

The third of Equations (2.8) shows that the volume transports in a *steady-state* problem are non-divergent and may be expressed in terms of a stream function

$$U = -\frac{\partial \psi}{\partial y}, \qquad V = \frac{\partial \psi}{\partial x}. \tag{2.24}$$

This automatically satisfies the third of Equation (2.8). The first two equations then yield

$$\psi = \frac{c^2}{f}\zeta - \frac{Fy}{f}, \tag{2.25}$$

where ζ is to be substituted from (2.20). The boundary condition at the shore is ψ = constant, conveniently $\psi = 0$. In the small basin limit the substitution of (2.25) into (2.20) yields $\psi \cong 0$, hence zero transport everywhere. In a large basin, outside boundary layers of scale width R one finds

$$\psi \cong -\frac{Fy}{f}, \quad \left(\frac{a}{R}, \frac{b}{R} >> 1\right) \tag{2.26}$$

from which

$$U = \frac{F}{f}, \qquad V = 0 \tag{2.27}$$

which is the Ekman transport corresponding to a wind stress along positive y. Figure 2.3 illustrates the transport streamline pattern in a basin of size $a/R = 15$, $b/a = 5$.

Under the idealizations made (most notably, negligible cumulative effects of bottom stress) these results answer the questions, how narrow a basin has to be for the simple static setup solution to be a valid approximation, and what happens when the basin is substantially wider. However, for a homogeneous fluid in a shallow sea the answer to the second question turns out to be academic. Given a basin as shallow as $H = 30$ m, $R = c/f$ at midlatitudes is of order 170 km. At least one dimension of shallow seas is usually of this order or less, so that the simple static setup solution is always a reasonable approximation. Even in the rather extreme case of $a/R = 1$, there is no boundary layer at the coasts, only a basinwide transport pattern involving some volume transport to the right of the wind near the basin center. The solution derived here will be found more useful later in describing the behavior of a stratified fluid.

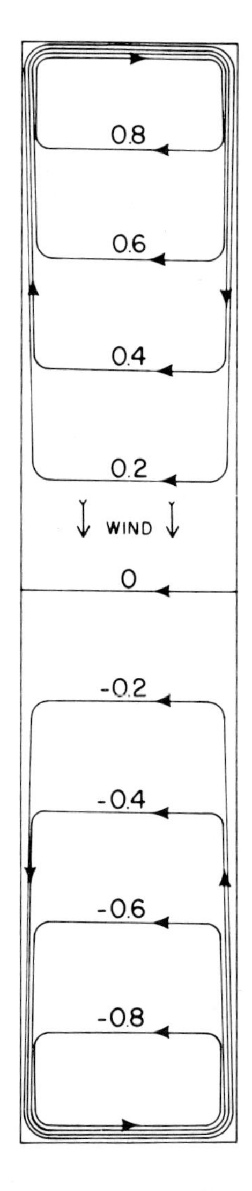

Fig. 2.3. Pattern of steady state wind-driven transport in a basin large compared to the radius of deformation (a/R = 15, b/a = 5). Ekman transport is released or accepted by coastal boundary layers. A setup opposing the wind is present along the long coasts, but this decays seaward within a distance of order R. From Csanady and Scott (1974).

2.3. SEICHES IN NARROW BASINS

The time-independent solutions discussed above, elucidating the properties of wind setup, do not say anything about how a given sea level distribution is generated. In the context of the circulation problem it is important to know how soon the pressure gradient will begin opposing the wind stress, after the latter is applied. This question leads to the consideration of time-dependent motions. A simple idealized problem is the sudden application of a constant stress at time $t = 0$. How will a basin of simple geometry respond?

The previous analysis of wind setup suggests that the problem will be simplified further if a narrow rectangular basin is considered first, $fa/\sqrt{gH} << 1$. When a wind parallel to the longer sides of such a narrow rectangular basin is suddenly imposed, it is reasonable to suppose that cross-wind transport remains negligible everywhere, as in the steady-state wind setup:

$$U \cong 0. \tag{2.28}$$

Physically speaking, the two longer shores (where $U = 0$ in virtue of the coastal constraint) are supposed to lie close enough together to suppress everywhere any significant cross-basin transport. Also neglecting bottom friction, the along-wind momentum equation and the continuity equation become

$$\begin{aligned} \frac{\partial V}{\partial t} &= -c^2 \frac{\partial \zeta}{\partial y} + F_y, \\ \frac{\partial V}{\partial y} &= -\frac{\partial \zeta}{\partial t}. \end{aligned} \tag{2.29}$$

The boundary conditions along the longer side are satisfied by hypothesis (Equation (2.28)), while at the end-walls it is necessary that

$$V = 0, \qquad (y = \pm b). \tag{2.30}$$

The transport V is easily eliminated from (2.29) by cross-differentiation, leaving an equation for ζ:

$$\frac{\partial^2 \zeta}{\partial t^2} - c^2 \frac{\partial^2 \zeta}{\partial y^2} = 0. \tag{2.31}$$

This is a simple one-dimensional wave equation. The boundary condition (2.30), expressed as a condition on ζ becomes:

$$\frac{\partial \zeta}{\partial y} = \frac{F_y}{c^2}, \qquad (y = \pm b) \tag{2.32}$$

The forcing will be represented by:

$$\begin{aligned} F_y &= 0, \qquad (t < 0) \\ F_y &= u_*^2, \quad (t \geqslant 0) \end{aligned} \tag{2.33}$$

2.3.1. Rectangular Basin

The properties of the wave equation are discussed in detail in texts on applied mechanics, acoustics and other branches of theoretical physics (see e.g., Courant and Hilbert, 1962; Morse and Feshbach, 1953). Solutions are progressive waves of propagation velocity c or standing waves of a wave length and frequency determined by the boundary conditions. For the initial and boundary conditions specified above a solution may be found e.g., by using Laplace transforms. Alternatively, one may note that wind setup (Equation (2.3)) is a particular solution and add a series of standing waves in such a way that the initial conditions are satisfied. The result is

$$\zeta = \frac{u_*^2}{c^2} y - \frac{8bu_*^2}{\pi^2 c^2} \left\{ \cos \frac{\pi ct}{2b} \sin \frac{\pi y}{2b} - \right.$$
$$\left. - \frac{1}{9} \cos \frac{3\pi ct}{2b} \sin \frac{3\pi y}{2b} + \frac{1}{25} \cos \frac{5\pi ct}{2b} \sin \frac{5\pi y}{2b} - \ldots \right\}. \tag{2.34}$$

The first term on the right is the now familiar wind setup. The other terms may be looked upon as a series of standing oscillations of sinusoidal amplitude distributions along the length of the basin, known as 'seiches'. This is the traditional point of view usually adopted in oceanographic texts: each component seiche is deemed to exist independently of the others; each has its own amplitude distribution $\sin n\pi y/2b$, having n nodes. The frequency of each component seiche is

$$\sigma_n = \frac{n\pi c}{2b}. \tag{2.35}$$

Note that only odd modes $n = 1, 3, 5, \ldots$ are excited by a sudden uniform wind. The period of the fundamental seiche mode is

$$T = \frac{2\pi}{\sigma_1} = \frac{4b}{c}. \tag{2.36}$$

This is equal to the time a progressive wave would take to propagate from one end of the basin to the other, and back. Equation (2.36) has been known for about 150 years and is often called Merian's formula.

At least from some points ov view the traditional interpretation of regarding the terms in Equation (2.34) as independently existing standing waves is not the most convenient. Either at given points within the basin (y fixed), or at given instants (t fixed) the trigonometric series sum to a periodic, but not to a sinusoidal function. For example, at time $t = 0$ all cosine terms in Equation (2.34) are equal to unity, and the remaining sine series sums to $(u_*^2/c^2)y$, cancelling out the wind setup solution. At $t = T/2$, half the fundamental seiche period, the cosine terms are all equal to minus one, the sum of the sine series is the same as before but with a negative sign, and surface elevations are again linearly distributed along the length of the basin, but now at double the amplitude of wind setup.

Other interesting results are obtained by calculating the along-basin transport. From Equations (2.29) and (2.34) one finds without difficulty:

$$V = \frac{8bu_*^2}{\pi^2 c} \left\{ \sin \frac{\pi ct}{2b} \cos \frac{\pi y}{2b} - \frac{1}{9} \sin \frac{3\pi ct}{2b} \cos \frac{3\pi y}{2b} + \right.$$
$$\left. + \frac{1}{25} \sin \frac{5\pi ct}{2b} \cos \frac{5\pi y}{2b} - \ldots + \ldots \right\}. \tag{2.37}$$

At the center of the basin, $y = 0$, V is proportional to the same sine series that follows from (2.34) for ζ at $t = 0$, but now with the argument proportional to time, so that here

$$V = u_*^2 t, \quad (y = 0) \quad (t \leqslant b/c). \tag{2.37a}$$

The restriction to $t < b/c = T/4$, the quarter of the seiche period, arises in analogy with the linear ζ-distribution cancelling the wind setup at $t = 0$, the sine expansion of which is valid to $y = b$ only.

2.3.2. Progressive Wave Interpretation

An illuminating alternative interpretation of the above results arises from a representation of the standing waves as a sum of progressive waves. It is shown in standard texts on theoretical physics that the general solution of the wave equation (2.31) is of the form

$$\zeta = \phi_1(y - ct) + \phi_2(y + ct), \tag{2.38}$$

where $\phi_1(\eta)$ and $\phi_2(\eta)$ are arbitrary functions, representing waves propagating to positive and negative y respectively, without change of wave-form.

Equation (2.34) is brought to the canonical form (2.38) by observing that the sine-cosine products may be written, using a simple trigonometric identity:

$$\cos \frac{n\pi ct}{2b} \sin \frac{n\pi y}{2b} = \frac{1}{2} \sin \frac{n\pi(y+ct)}{2b} + \frac{1}{2} \sin \frac{n\pi(y-ct)}{2b}.$$

Consequently, (2.34) is also expressible as

$$\zeta = \frac{u_*^2}{c^2} y - \frac{4bu_*^2}{\pi^2 c^2} \left\{ \sin \frac{\pi(y+ct)}{2b} - \frac{1}{9} \sin \frac{3\pi(y+ct)}{2b} + \frac{1}{25} \sin \frac{5\pi(y+ct)}{2b} - + \dots + \frac{\sin \pi(y-ct)}{2b} - \frac{1}{9} \sin \frac{3\pi(y-ct)}{2b} + \frac{1}{25} \sin \frac{5\pi(y-ct)}{2b} - + \dots \right\}. \tag{2.39}$$

The brackets now contain *two* sine series, the first one a function of $(y+ct)$, the second of $(y-ct)$, in accordance with (2.38). Each of these sums to a sharp crested wave function illustrated in Figure 2.4, extended beyond the original domain of the solution to $\pm\infty$. The corresponding transport distribution may be represented similarly, in virtue of a general relationship which follows from the second of Equations (2.29) for the canonical solution (2.38):

$$V = c\phi_1(y-ct) - c\phi_2(y+ct). \tag{2.40}$$

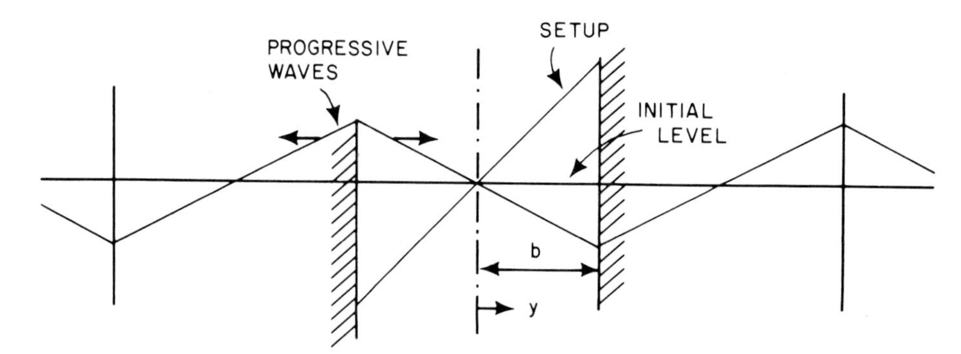

Fig. 2.4. A combination of seiches represented as forward and backward progressive waves of saw-tooth shape, showing surface level amplitudes.

Thus transport is forward directed in a forward wave (ϕ_1) where the elevation is positive, oppositely directed under positive elevation in a backward wave (ϕ_2), and in both cases it is proportional to the elevation.

The solution (2.39) may be regarded as consisting of wind setup, a forward wave of wave form illustrated in Figure 2.4. and a similar backward wave. The amplitudes of the forward and backward waves are the same, and such that at $t = 0$ each cancels one half of the wind setup in the original domain $-b \leqslant y \leqslant b$. Because the forward and backward waves start in an overlapping position at $t = 0$, they are always symmetrically placed relative to the points $y = \pm b$, and have the same amplitude there. Since the transports induced by these waves are oppositely directed, they always cancel out at these points, thus satisfying the boundary conditions.

The progressive wave interpretation of the solution (2.33) enables one to analyze the response of a basin to sudden wind stress with minimum labor, avoiding the need to sum the series in Equations (2.34) and (2.37). Figure 2.5 illustrates the elevation distribution at $t = T/8$, one eighth of a fundamental seiche period. The peaks of the progressive waves

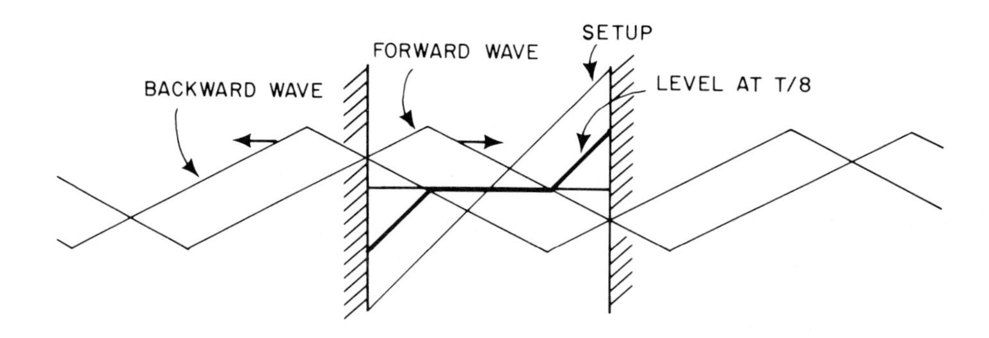

Fig. 2.5. Summation of setup and progressive waves at $t = T/8$ (T = fundamental seiche period). The resulting surface level distribution (thick line) shows the developing setup near the ends of the basin, the level near the center being as yet undisturbed.

have moved a quarter of the way into the basin from both ends. Behind the peaks the surface has moved upwards or downwards, and an elevation gradient has been established equal to the equilibrium value. Ahead of the peaks, however, the water surface is undisturbed. Physically one may interpret the peaks as the signal which announces the presence of end-walls. Once the signal has passed, a surface level gradient is present to balance the applied wind stress. The surface elevations change only in response to the influence of the end-walls, which 'dam up' the downwind transport, and signal their presence to distant points by a wave.

Corresponding to the absence of a pressure gradient ahead of the peaks of the progressive waves, the windward transport before the arrival of the pulse increases as

$$V = u_*^2 t, \quad (|y| \leqslant b - ct). \tag{2.41}$$

In other words, the impulse of the wind stress is absorbed simply by the momentum of the water column. This result generalizes Equation (2.37a), which was found for $y = 0$ by determining the sum of the sine series in (2.37).

In the region behind the peaks the transport remains constant, at the value reached before the peak passed because the wind stress and pressure gradient exactly balance.

At $t = T/4$ the two peaks meet in the center, Equation (2.41) holds at the single point $y = 0$, two component waves exactly cancel and the static elevation distribution prevails everywhere.

At $t > T/4$ the component waves continue to increase elevations and elevation gradients, thereby also beginning to reduce the transport. At half cycle, $t = T/2$, the surface level overshoots by 100% and the transport becomes zero everywhere. In the second half of the seiche cycle return transports are generated by the excessive pressure gradient and the surface returns to its equilibrium position ($\zeta = 0$) at $t = T$. The reader may easily verify these statements by drawing simple diagrams similar to Figure 2.5.

It is also of interest to follow the history of elevation and transport at a fixed point. For $y = -b/2$ this is shown in Figure 2.6. Cycles of acceleration give way to periods of constant transport, then deceleration, again constant transport and so on.

The abrupt changes in the time or space derivatives of surface elevation and transport are, of course, physically unrealistic. These are the consequence of the hydrostatic approximation which disregards vertical accelerations $\partial^2\zeta/\partial t^2$, and of the idealized sudden application of the forcing. With all its blemishes, however, the simple theoretical model places in perspective the relationship of setup and seiche, and vividly illustrates how boundary effects propagate into the interior of a shallow sea. In a basin of depth $H = 100$ m the pressure signal travels 100 km in approximately one hour, so that setup is rapidly establishes everywhere in a typical enclosed shallow sea.

2.4. EVOLUTION OF SETUP ALONG A STRAIGHT OPEN COAST

The results of the previous section were obtained for a suitably narrow basin, wherein the postulate (2.28) eliminated earth rotation effects. An opposite idealization is to consider a semi-infinite sea, bounded by a straight, infinitely long coast. An investigation of this model should reveal how earth rotation affects the propagation of a

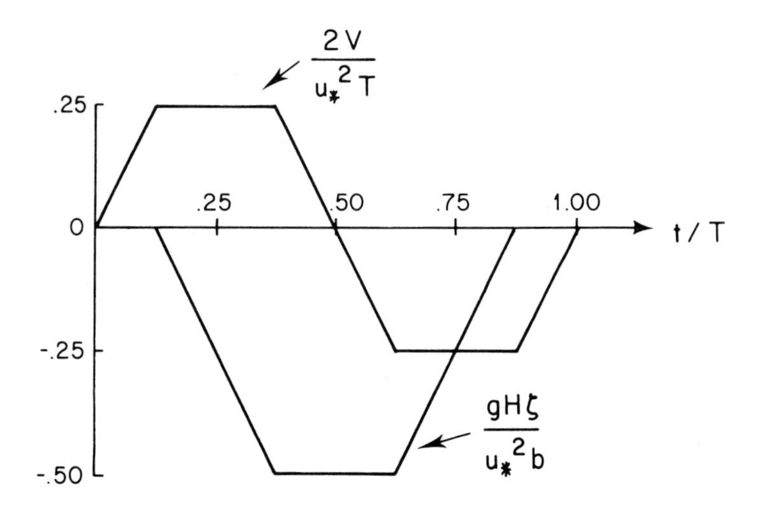

Fig. 2.6. Time history of surface elevation ζ and axial transport V at $y = -b/2$. In the nondimensional units used the setup level is -0.25.

pressure signal into the interior, without complications due to the presence of other shores.

Consider therefore a shallow sea $x \leqslant 0$ bounded by a coast coincident with the y axis. Let a wind perpendicular to the coast suddenly arise at $t = 0$,

$$\begin{aligned} F_x &= u_*^2, \quad (t \geqslant 0) \\ F_x &= 0, \quad (t < 0) \\ F_y &= 0, \quad (\text{all } t). \end{aligned} \tag{2.42}$$

With bottom friction neglected, Equations (2.11) and the no normal transport boundary condition yield:

$$\begin{aligned} &\frac{\partial^2 \zeta}{\partial t^2} - c^2 \frac{\partial^2 \zeta}{\partial x^2} + f^2 \zeta = 0, \\ &\frac{\partial \zeta}{\partial x} = \frac{u_*^2}{c^2}, \quad (x = 0). \end{aligned} \tag{2.43}$$

The y-gradient term in the wave equation has been dropped, since there is no reason why any differences in elevation along y should arise. Once Equations (2.43) are solved, transports may be found from Equations (2.8), dropping bottom stresses, longshore wind stress and pressure gradient:

$$\begin{aligned} &\frac{\partial U}{\partial t} - fV = -c^2 \frac{\partial \zeta}{\partial x} + F_x, \\ &\frac{\partial V}{\partial t} + fU = 0. \end{aligned} \tag{2.44}$$

The second of these equations has a simple content: longshore transport is generated only to the extent that cross-shore transport is present.

A time independent particular solution of Equation (2.43) is easily found, and the corresponding transport calculated from the first of Equations (2.44):

$$\zeta = \frac{u_*^2}{fc}\, e^{x/R}, \qquad V = -\frac{u_*^2}{f}\,(1 - e^{x/R}), \qquad \left(R = \frac{c}{f}\right). \tag{2.45}$$

The cross-shore transport is zero, as may be seen from the second of (2.44). The solution is characterized by zero longshore transport and the no-rotation value of the elevation gradient at the coast, and zero elevation gradient and Ekman transport far away. The transition takes place on the scale of the radius of deformation, $R = c/f$.

How the pressure field becomes established is shown by the full solution of the impulsively forced problem (Equation (2.42)) which may be found e.g., by Laplace transform methods (Crépon, 1967). The inversion of the transform leads to some inconvenient algebra:

$$\zeta = \frac{u_*^2}{c}\int_0^t J_0\left[f\sqrt{t^2 - \frac{x^2}{c^2}}\right] Y\left(t + \frac{x}{c}\right) \mathrm{d}t, \tag{2.46}$$

where J_0 is a Bessel function and $Y(\lambda)$ is the Heaviside function ($Y = 1$, $\lambda \geqslant 0$, $Y = 0$, $\lambda < 0$). At a long time after the imposition of the wind stress this expression may be approximated by

$$\zeta = \frac{u_*^2}{fc}\left[e^{x/R} + \sqrt{\frac{2}{\pi}}\,\frac{\sin(ft - \pi/4)}{\sqrt{ft}} - \ldots\right] \quad (ft >> 1). \tag{2.47}$$

This asymptotic solution is seen to consist of wind setup and inertial oscillations of slowly decaying amplitude. As in similar infinite ocean wave generation problems (Lamb, 1932), the decay comes about through a dispersal of the waves to infinity. Figure 2.7 shows the development of the pressure gradient and of the oscillations at different distances from the coast. The distribution of level at different times is shown in Figure 2.8. A pulse is seen to propagate into the fluid at the wave propagation velocity c, as in the problem without rotation. A wake of near-inertial oscillations is left behind the pulse. Once this wake decays, the wind setup remains, but only at distances of order R from the coast. Further away, the continued action of wind stress is balanced by the Coriolis force of Ekman transport parallel to the coast.

According to the second of Equations (2.44), longshore transport V is established only after some cross-shore transport U has developed. Right at the coast U is zero at all times and hence no longshore transport can develop. At distances of order R and greater, transient cross-shore transport appears at times of order f^{-1}, allowing the establishment of a steady longshore transport pattern.

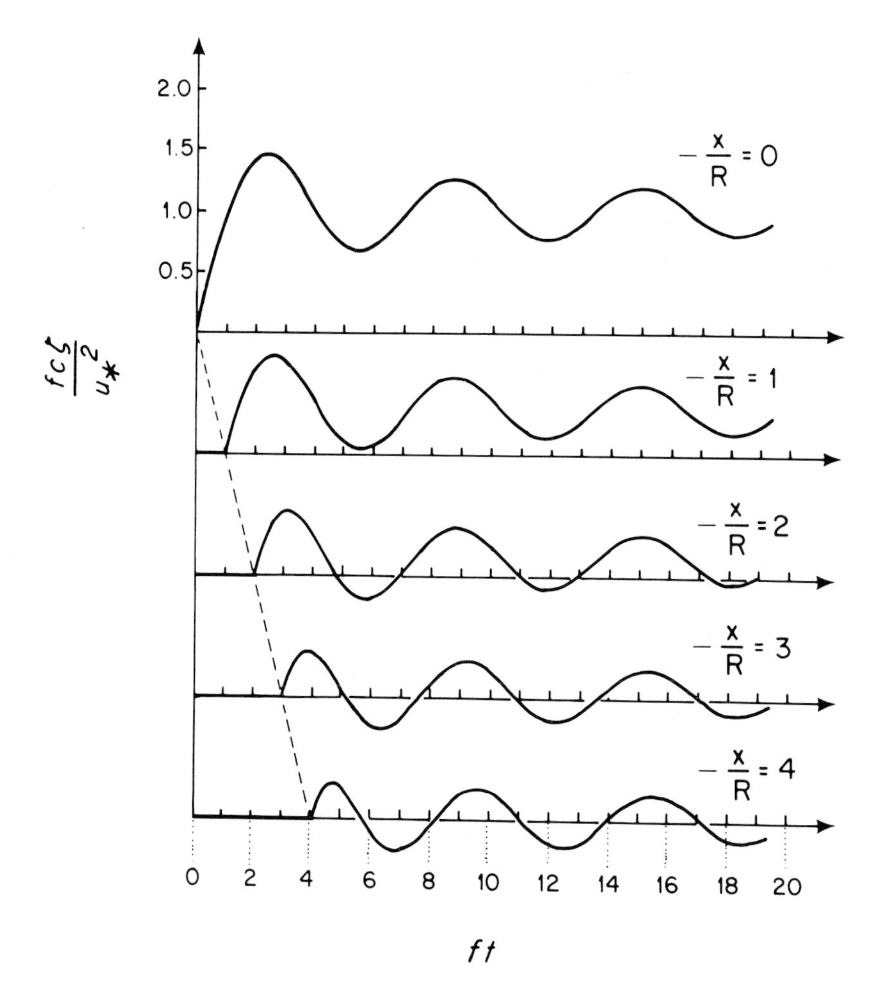

Fig. 2.7. Development of setup and oscillations due to cross-shore wind against an infinite coast. Beyond a few radii of deformation only oscillations arise. As in a closed basin, the level remains undisturbed until a pressure pulse arrives, travelling offshore at the speed $c = fR$. (Redrawn from Crépon, 1967).

2.5. LONGSHORE WIND AND SEA LEVEL

The case of a wind parallel to the coast, imposed suddenly at $t = 0$, leads to completely new insights, particularly important in the dynamics of coastal currents. Without earth rotation, this case is uninteresting: the wind stress accelerates the water downwind and no pressure field is generated, there being no coast to dam up the fluid. The effect of earth rotation is, however, to generate a cross-stream Coriolis force which comes to be balanced by an appropriate pressure field, and also leads to coastal sea level changes.

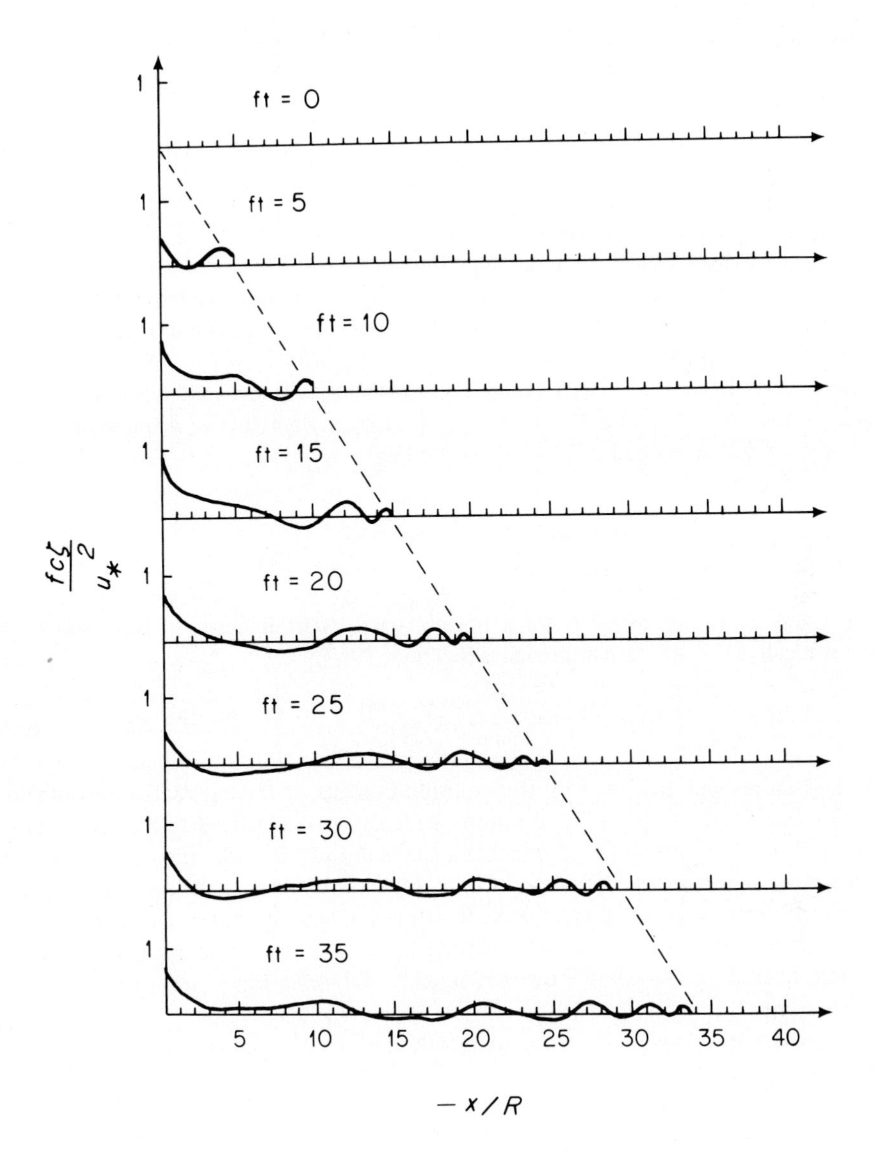

Fig. 2.8. Same case as the previous figure, showing the sea surface shape at successive times (Crépon, 1967).

On the same semi-infinite shallow sea as in the previous section, $x \leqslant 0$, bounded by a coast parallel to the y-axis, a longshore wind along positive y is imposed at $t = 0$:

$$F_x = 0, \qquad F_y = u_*^2, \quad (t \geqslant 0). \tag{2.48}$$

The motion starts from rest, so that the boundary condition at the coast, $U = 0$, leads at once to, from Equations (2.8):

$$V = u_*^2 t, \qquad \frac{\partial \zeta}{\partial x} = ft \frac{u_*^2}{c^2}, \qquad (x = 0). \tag{2.49}$$

The first of these equations is the same that applies without rotation, the fluid simply accelerating downwind, but here this is a boundary condition, applying only at $x = 0$. The second of (2.49) shows that the developing coastal current is in geostrophic balance with a developing pressure field.

The first of Equations (2.43) remains valid in this case and may again be solved by Laplace transform methods. Alternatively, one may notice that (2.43) and the present boundary condition (second of 2.49) is satisfied by

$$\zeta = f \int_0^t \zeta_N(x, t)\, dt, \tag{2.50}$$

where $\zeta_N(x, t)$ is the solution for a normal wind, written down in Equation (2.46). Asymptotically, this solution approaches (Crépon, 1967):

$$\zeta = \frac{u_*^2}{fc}\left[(ft)e^{x/R} - \sqrt{\frac{2}{\pi}}\,\frac{\cos(ft - \pi/4)}{\sqrt{ft}} + \ldots\right], \qquad (ft >> 1). \tag{2.51}$$

The pressure field described by this solution consists of inertial oscillations of slowly decaying amplitude, as before, and a non-oscillatory part confined to a coastal band of scale width R, the amplitude of which increases linearly in time. Figure 2.9 shows the development of this field in function of time, Figure 2.10 its spatial structure at different instants. A pressure pulse is again seen to penetrate into the fluid, its steepness at the outer edge decreasing in time instead of sharpening as in the previous case. The pressure field left behind at the coast, however, quickly outgrows the setup associated with a cross-shore wind.

The transport components associated with the non-oscillatory part of the solution (2.51) are:

$$U = \frac{u_*^2}{f}(1 - e^{x/R}), \qquad V = u_*^2 t e^{x/R}. \tag{2.52}$$

The longshore transport V is, in this non-oscillatory part of the solution, in geostrophic balance with the pressure field, not only at the coast, as given by (2.49), but everywhere in the coastal band of scale width R within which the non-oscillatory solution is significant. At the same time, there is significant cross-shore transport U over most of this band, except very near the coast, ($|x/R| << 1$). The cross-shore transport builds up on the time scale f^{-1}, as the pressure wave propagates out from the coast. The cross-shore transport

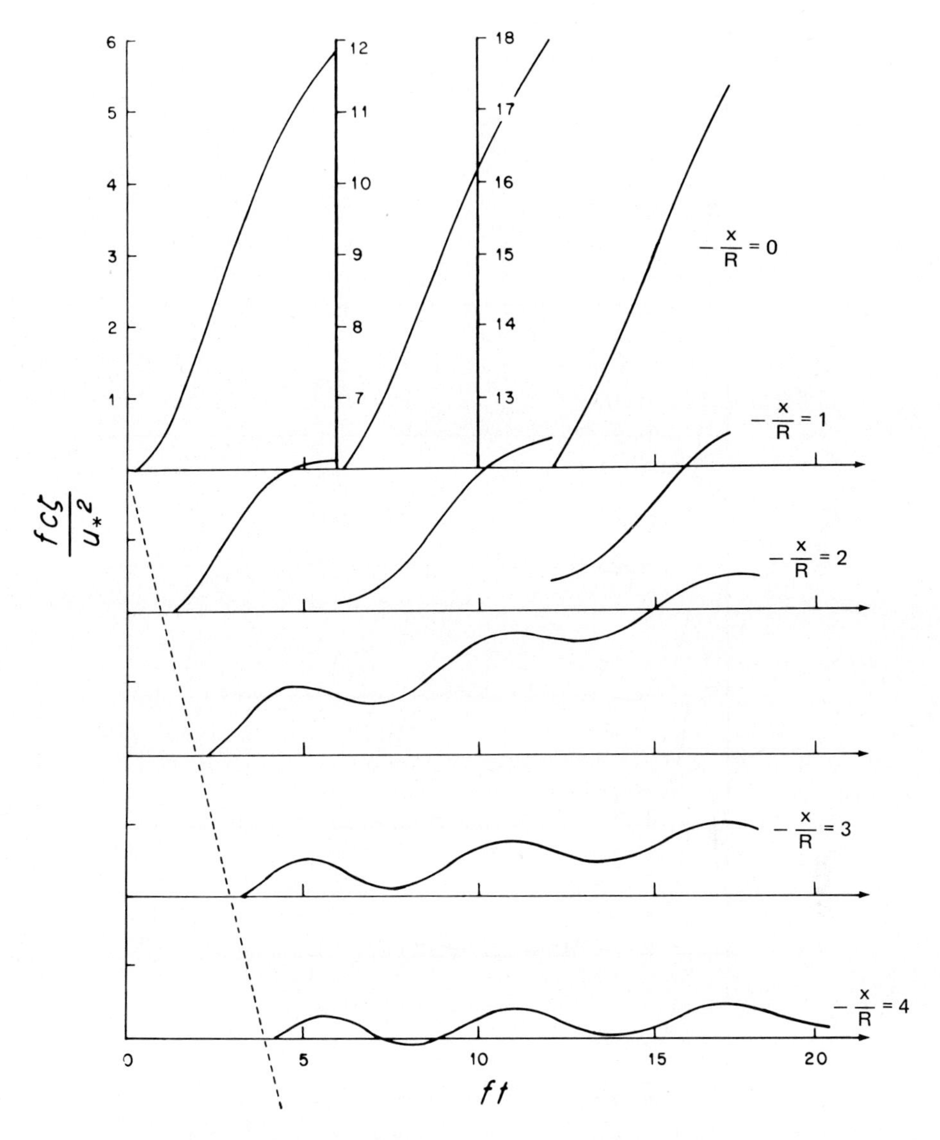

Fig. 2.9. Development of setup and oscillations due to longshore wind along an infinite coast. Close to the coast the level keeps rising (redrawn from Crépon, 1967).

is *not* in geostrophic balance, there being no longshore pressure gradient. The dynamical role of its non-oscillatory component is highlighted by the second of Equation (2.8), which for this case is of the form

$$\frac{\partial V}{\partial t} = u_*^2 - fU. \tag{2.52a}$$

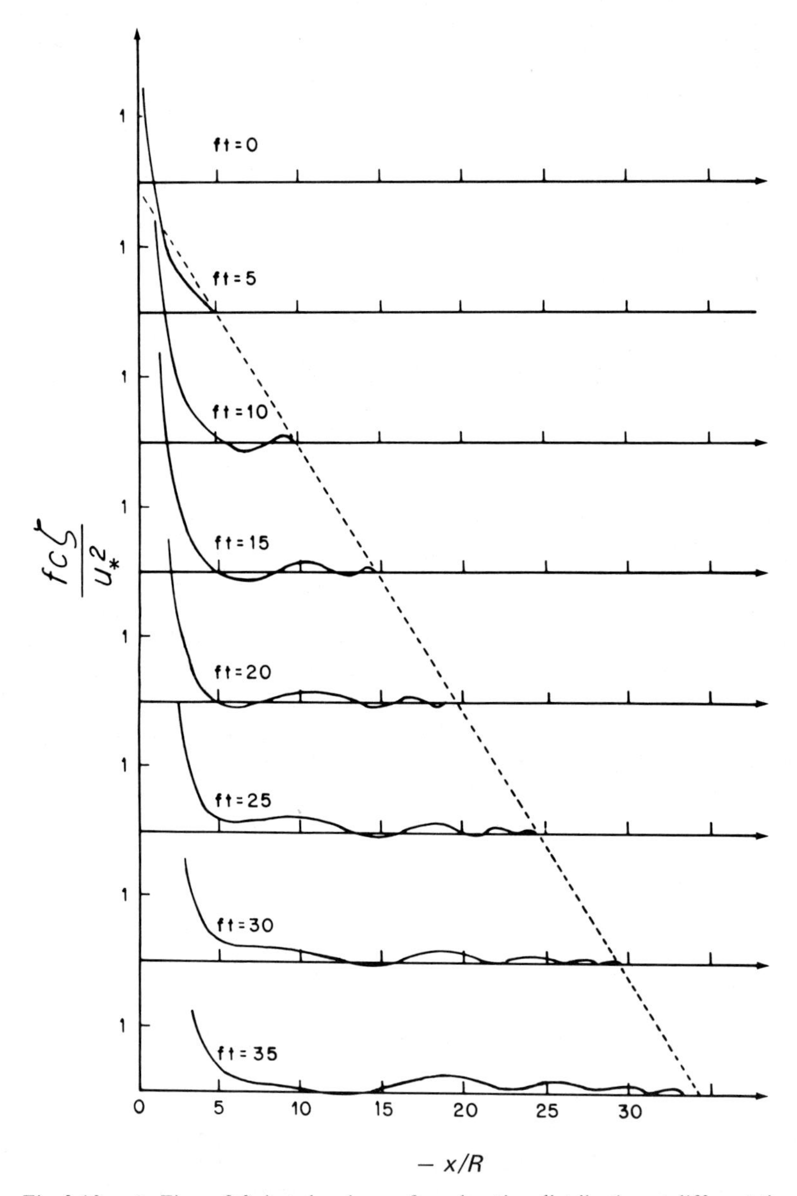

Fig. 2.10. As Figure 2.9, but showing surface elevation distribution at different times (Crépon, 1967).

Near the coast, where $U \cong 0$, the wind stress simply accelerates the fluid. Far from the coast, where $U \cong u_*^2/f$ (the Ekman transport associated with the longshore wind) the Coriolis force associated with Ekman drift balances the wind stress and no longshore transport develops. In between, the gradual change of U provides the transition.

A common notion is that Ekman transport associated with longshore wind in some sense "causes" coastal sea level rise or fall. This is true in the sense that according to the third of (2.8)

$$\frac{\partial U}{\partial x} = -\frac{\partial \zeta}{\partial t}. \tag{2.53}$$

The non-oscillatory part of a semi-infinite sea's response to longshore wind stress is characterized by $U = 0$ at the coast, $U = u_*^2/f$ (Ekman transport) at $|x/R| >> 1$, and the implied changes $\partial U/\partial x$ in between require sea level rise or fall according to (2.53). In order to calculate coastal sea level changes, it is necessary to know not only the offshore value of U, but also the spatial scale of its changes, R.

The mechanism of sea level change due to longshore wind is more directly described by the boundary condition written down in Equation (2.49). Net cross-shore transport must be zero within some neighborhood of the coast, so that the longshore force balance is simply between wind stress and acceleration. It is precisely the *absence* of Ekman transport (or its neutralization by opposite cross-shore flow, see Section 2.7 below) near the coast which leads to the simple impulse equals momentum change relationship embodied in the first of (2.49). Sea level changes result as the pressure field adjusts to geostrophic equilibrium with the developing longshore current. In this light, it is simpler and more direct to regard the Coriolis force as the dynamical cause of coastal sea level changes, associated with longshore currents. This view is all the more useful as longshore winds are not the only possible cause of longshore currents. Coastal sea level changes accompany the development of longshore currents due to longshore pressure gradients or to freshwater influx at the shore in exactly the same way as they accompany wind driven longshore flow.

2.6. WIND ACTING FOR A LIMITED PERIOD

Further interesting results are obtained by comparing the responses to cross-shore and to longshore winds, when the duration of the forcing is limited. The two cases in question are described by

$$\begin{array}{llll} (1) & F_x = u_*^2, \quad (0 \leqslant t \leqslant T) & (2) & F_y = u_*^2, \quad (0 \leqslant t \leqslant T) \\ & F_x = 0 \text{ otherwise,} & & F_x = 0 \text{ otherwise,} \\ & F_y = 0 \text{ at all } t. & & F_y = 0 \text{ at all } t. \end{array} \tag{2.54}$$

As before, the coast coincides with the y-axis, and the domain of interest is $x \leqslant 0$.

In case (1) of cross-shore wind the removal of wind stress at $t = T$ excites inertial oscillations anew, which combine with those started at $t = 0$. These are of no particular interest in the context of the circulation problem. The non-oscillatory part of the solution, which during the period of wind action is the wind setup (limited to a coastal band of scale width R), drops abruptly to zero with the cessation of the wind. The oscillations decay slowly and all motion ceases at $ft >> 1$.

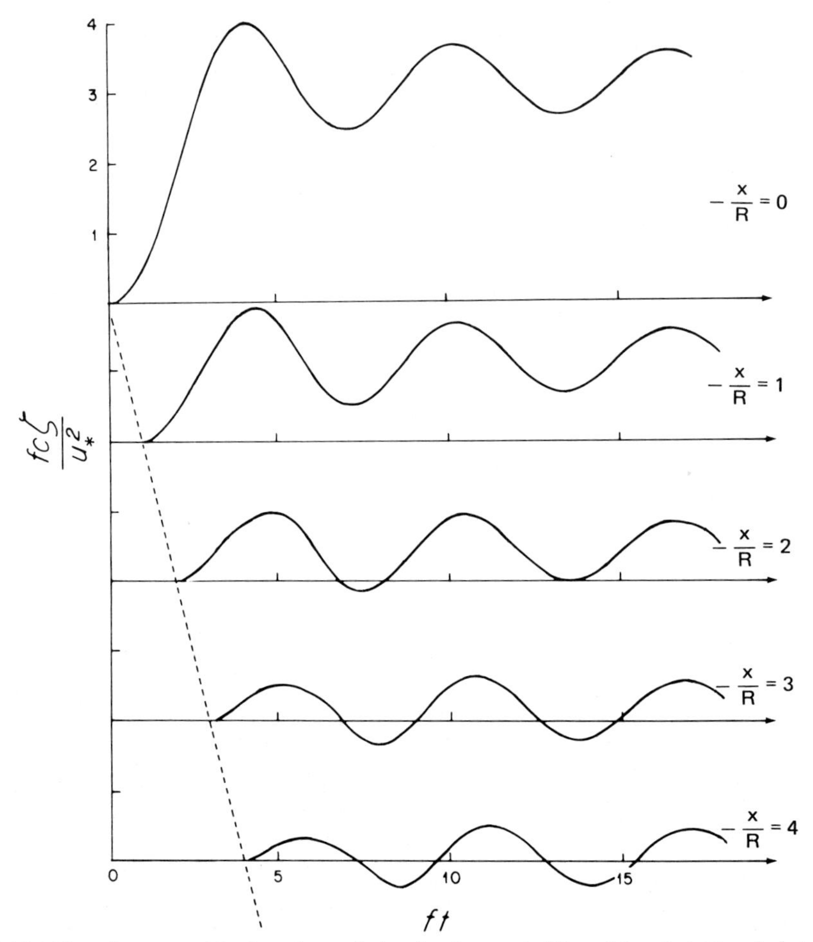

Fig. 2.11. Elevations caused by longshore wind acting for one half inertial period, $T = \pi f^{-1}$ (redrawn from Crépon, 1967).

In case (2) of longshore wind the result is completely different. The nonoscillatory part of the solution, linear in time for $0 \leqslant t \leqslant T$, simply stops developing at $t = T$, but the pressure field and the longshore current in geostrophic balance with it remain. The oscillatory components behave in much the same way as in case (1). Figure 2.11 shows the development of the pressure field at different distances from the coast for an impulse lasting for a period $T = \pi f^{-1}$. The non-oscillatory part of the field at $t > T$, and the associated transport distribution is:

$$\begin{aligned} \zeta &= \frac{u_*^2 T}{c} e^{x/R}, \\ U &= 0, \\ V &= u_*^2 T e^{x/R}. \end{aligned} \tag{2.55}$$

In this case the cross-shore transport within the coastal band drops abruptly (i.e., on a time scale f^{-1}) to zero at $t = T$, but the coastal current retains the intensity it reached by that time, (indefinitely, if bottom friction is absent as supposed in this simple model).

2.7. PRESSURE FIELD-INDUCED AND FRICTIONAL INTERIOR VELOCITIES

In the last few sections the focus was on solutions to the global problem and various results have been derived regarding the development of the pressure field under forcing by wind. To put these results in perspective, the interior velocity distribution needs to be considered at least for some of the global solutions found.

The linearized equations of motion (Equation (1.32)) will be taken to apply:

$$\frac{\partial u}{\partial t} - fv = -g\frac{\partial \zeta}{\partial x} + \frac{\partial}{\partial z}\left(K\frac{\partial u}{\partial z}\right),$$
$$\frac{\partial v}{\partial t} + fu = -g\frac{\partial \zeta}{\partial y} + \frac{\partial}{\partial z}\left(K\frac{\partial v}{\partial z}\right) \tag{2.56}$$

with K = constant below and above surface and bottom wall layers, which will be considered to be of negligible thickness. The solution of the global problem is supposed to have yielded the local sea level gradients $\partial\zeta/\partial x$, $\partial\zeta/\partial y$, which are depth-independent forcing terms in (2.56). Correspondingly the local solutions $u(z, t)$, $v(z, t)$ of Equations (2.56) (depending only parametrically on x, y) may be written as

$$u = u_1(t) + u_2(z, t),$$
$$v = v_1(t) + v_2(z, t), \tag{2.57}$$

where $u_1(t)$, $v_1(t)$ is a solution of (2.56) with the stress-gradient terms deleted:

$$\frac{du_1}{dt} - fv_1 = -g\frac{\partial \zeta}{\partial x},$$
$$\frac{dv_1}{dt} + fu_1 = -g\frac{\partial \zeta}{\partial y}. \tag{2.58}$$

With the surface slopes considered given in function of time, these are ordinary differential equations for u_1, v_1. For the impulsive problems considered in the past few sections they may, in principle, be solved with the initial conditions $u_1 = v_1 = 0$ at $t = 0$. This solution will be considered to describe the 'pressure-field induced' component of the flow.

Subtracting (2.58) from (2.56) one finds for the depth-dependent or frictionally induced velocity components:

$$\frac{\partial u_2}{\partial t} - fv_2 = \frac{\partial}{\partial z}\left(K\frac{\partial u_2}{\partial z}\right),$$
$$\frac{\partial v_2}{\partial t} + fu_2 = \frac{\partial}{\partial z}\left(K\frac{\partial v_2}{\partial z}\right). \tag{2.59}$$

These equations may be solved with the initial conditions $u_2 = v_2 = 0$ at $t = 0$ and stress boundary conditions at surface and bottom:

$$
\begin{aligned}
K\frac{\partial u_2}{\partial z} &= F_x, \quad & K\frac{\partial v_2}{\partial z} &= F_y, \quad & (z = 0)\\
K\frac{\partial u_2}{\partial z} &= B_x, \quad & K\frac{\partial v_2}{\partial z} &= B_y, \quad & (z = -H).
\end{aligned}
\tag{2.60}
$$

Equations (2.59) and (2.60) with K = constant specify a time-dependent Ekman layer evolution problem. Problems of this type have been treated by Ekman (1905), Gonella (1971) and others. The solutions consist of steady state Ekman layers, more or less as already discussed in Chapter 1, and inertial oscillations, which propagate from the surface downward. To consider first the global properties of the frictional currents, Equations (2.59) are integrated with respect to depth, yielding the following degenerate form of the transport equations:

$$
\begin{aligned}
\frac{\partial U_2}{\partial t} - fV_2 &= F_x - B_x,\\
\frac{\partial V_2}{\partial t} + fU_2 &= F_y - B_y,
\end{aligned}
\tag{2.61}
$$

where U_2, V_2 are depth-integrated velocities u_2, v_2. For the simple global problems discussed above these equations may be solved easily. For example, with bottom stresses neglected, and a longshore stress $F_y = u_*^2$ suddenly imposed at $t = 0$ one finds

$$
\begin{aligned}
U_2 &= \frac{u_*^2}{f}(1 - \cos ft),\\
V_2 &= \frac{u_*^2}{f}\sin ft.
\end{aligned}
\tag{2.62}
$$

In terms of the simple conceptual models discussed at the end of Chapter 1 this may be regarded as a superposition of Ekman transport (Equation (1.47a)) and inertial oscillations (Equation (1.45)). Note, however, that the equivalence to the simple models applies to the frictional *transport* $\mathbf{U}_2(U_2, V_2)$, not necessarily the interior distribution of frictional velocity $u_2(z, t)$, $v_2(z, t)$. It is also of interest to observe that the inertial oscillations do not decay (in the sense that the modulus of $\mathbf{U}_2$ remains constant) as long as bottom friction is negligible. The hodograph of the transport vector $\mathbf{U}_2$ moves along a circle of radius u_*^2/f, centered at $U_2 = u_*^2/f$ (Figure 2.12).

The local solution of Equations (2.59), with suddenly imposed wind stress at the surface was first given by Fredholm (Ekman, 1905) in an integral form:

$$
\begin{aligned}
u_2 &= \frac{u_*^2}{\pi^{1/2}}\int_0^t \frac{\sin fs}{(Ks)^{1/2}}\exp\left(-\frac{z^2}{4Ks}\right)\mathrm{d}s,\\
v_2 &= \frac{u_*^2}{\pi^{1/2}}\int_0^t \frac{\cos fs}{(Ks)^{1/2}}\exp\left(-\frac{z^2}{4Ks}\right)\mathrm{d}s.
\end{aligned}
\tag{2.62a}
$$

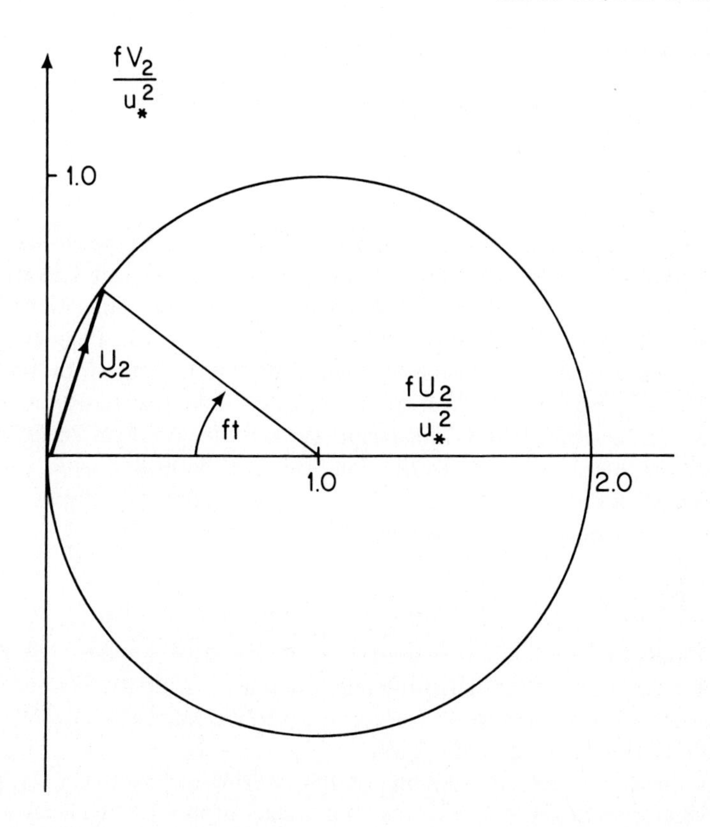

Fig. 2.12. Hodograph of frictional contribution U_2 to transport vector following sudden application of wind stress toward positive y. Inertial frequency oscillations cause directional changes between $+\pi/2$ and $-\pi/2$, magnitude changes between zero and twice the Ekman transport.

It may be shown (for a detailed discussion see Gonella, 1971) that this solution consists of a steady state Ekman spiral and inertial oscillations. The former is exactly as given by the steady state solution (Equations (1.51)), i.e., it extends to a layer of limited depth, scaled by $D = (2Kf^{-1})^{1/2}$. The inertial oscillations, however, penetrate to gradually increasing depth, at time t to a depth of order $(2Kt)^{1/2}$, which is the depth to which momentum diffusion penetrates in the absence of rotation. The oscillating transport retains a constant amplitude because the oscillating velocity amplitude decreases as $(u_*{}^2 f^{-1})(2Kt)^{-1/2}$. Thus after a long period the inertial oscillations become imperceptible as their (constant) momentum becomes distributed over a very deep water column. Because inertial oscillations do not contribute to 'circulation' in the sense discussed earlier, their detailed structure is not analyzed further below.

Returning now to the pressure-field induced interior velocities, one recalls that the solution of the global problem has yielded the *total* depth-integrated transports U, V. The resolution according to (2.57) also applies to the transports, so that with u_1, v_1

being independent of depth one has at once:

$$u_1 = \frac{1}{H}(U - U_2),$$
$$v_1 = \frac{1}{H}(V - V_2). \tag{2.63}$$

The calculation of the interior velocity distribution may thus be carried out in two steps: first determine the frictionally induced, depth-dependent velocities (u_2, v_2), by solving an Ekman layer evolution problem. With inertial oscillations neglected, this reduces to the steady state surface Ekman layer discussed in Chapter 1. The pressure field induced velocities are calculated then from (2.63). Again, with inertial oscillations neglected, it is simply the Ekman transport, which is subtracted from the total transport according to Equation (2.63). This recipe was already put into effect in Section 2.1.1, to calculate the interior velocity distribution associated with wind setup, postulating steady state and vanishing transport along x and y. The above discussion demonstrates the validity of the recipe for more complex cases.

2.7.1. Local Problem for Impulsive Onshore Wind

The determination of interior velocities for the impulsive problems discussed in Sections 2.4–2.6 now becomes particularly simple. Consider first the case (1) of Section 2.6 for $0 \leqslant t \leqslant T$. With bottom stress neglected, a surface Ekman layer evolution problem is encountered in solving Equation (2.59).

Both the global problem solution and the evolving surface Ekman layer solution contain oscillatory components, the oscillations being of inertial frequency and therefore of no particular interest in the circulation problem. The calaulations below will be restricted to the non-oscillatory components of the flow. The global solution gave for these components:

$$\zeta = \frac{u_*^2}{fc} e^{x/R},$$
$$U = 0, \tag{2.64}$$
$$V = -\frac{u_*^2}{f}(1 - e^{x/R}).$$

The frictionally induced flow has a depth integrated non-oscillatory transport component according to (2.61) of:

$$V_2 = -\frac{u_*^2}{f}, \qquad U_2 = 0. \tag{2.65}$$

Therefore the non-oscillatory pressure field-induced velocities are, according to (2.63):

$$u_1 = 0,$$
$$v_1 = \frac{u_*^2}{fH} e^{x/R}. \tag{2.66}$$

This velocity is seen to be in geostrophic balance with the pressure field, see the first of (2.64), as it should by the first of (2.58) for $u_1 = 0$ (note especially that this applies to the non-oscillatory part of the velocities only).

The frictionally induced velocities, neglecting inertial oscillations, are distributed according to a classical steady-state, surface Ekman layer profile, much as in Section 1.11.3, but transposing x and y axes:

$$\frac{u_2}{u_*} = \frac{u_*}{fD}\left(\cos\frac{z}{D} + \sin\frac{z}{D}\right),$$
$$\frac{v_2}{u_*} = -\frac{u_*}{fD}\left(\cos\frac{z}{D} - \sin\frac{z}{D}\right), \tag{2.67}$$

where $D = \sqrt{2K/f}$. The combined non-oscillatory velocity distribution is illustrated in Figure 2.13. At large distances from the coast velocities are only significant in the Ekman layer. Close to the coast ($|x/R| << 1$) the geostrophic longshore transport exactly balances the longshore Ekman transport, so that the net cross-shore Coriolis force is zero. This corresponds to the exact balance of wind stress and pressure gradient in the cross-shore direction. As pointed out in Section 2.1.1, the Coriolis force associated with longshore flow may be thought to transfer the wind force downward. This effect is brought about by the pressure field-induced velocity v_1.

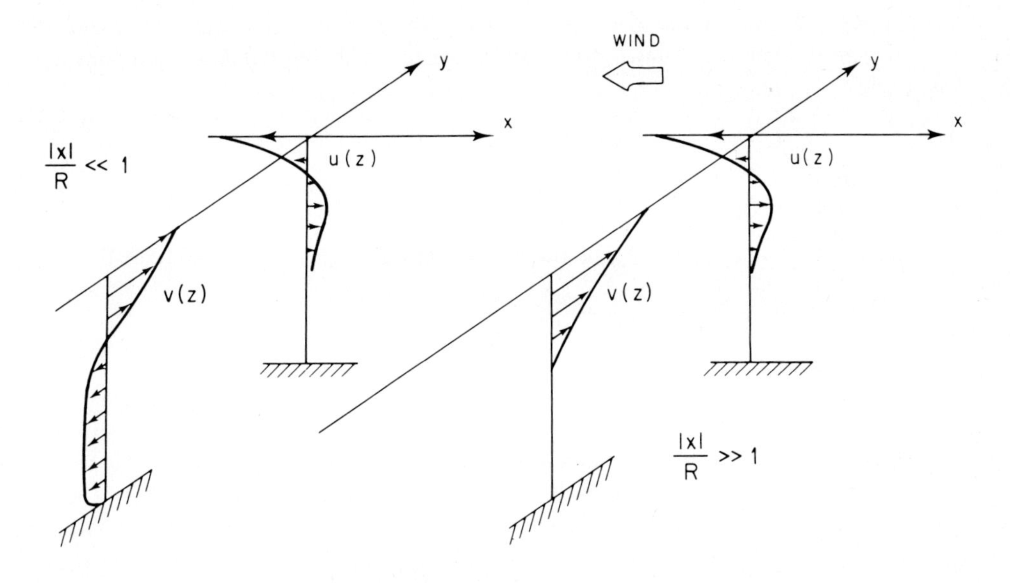

Fig. 2.13. Interior velocities caused by cross-shore wind close to the coast ($|x|/R << 1$) and far from it. The difference is that longshore transport in the Ekman layer is cancelled by reverse geostrophic transport close to shore.

The validity of the no bottom stress hypothesis is restricted to a water column so deep that v_1 at the coast is small enough to give rise to only negligible bottom stress. The neglected inertial oscillations may also induce bottom stress, although this, too, is small for a sufficiently deep water column.

In focussing on the non-oscillatory flow components only, the longshore flow $v_1(x)$ near the coast appears to be generated abruptly at $t = 0$. Reference back to the second of Equations (2.58) reveals that, with $\partial\zeta/\partial y = 0$, v_1 must be generated by appropriate cross-shore motion in an initial period. This occurs on a time scale of f^{-1}, more or less as a fallout of the developing inertial oscillations.

2.7.2. The Case of Longshore Wind

For the case of impulsive longshore wind treated in Section 2.5, or in Section 2.6 for the interval $0 \leqslant t \leqslant T$, again restricting considerations to the non-oscillatory components of the motion, the global solution gave

$$\begin{aligned} \zeta &= \frac{u_*^2 t}{c} e^{x/R}, \\ U &= \frac{u_*^2}{f}(1 - e^{x/R}), \\ V &= u_*^2 t e^{x/R}. \end{aligned} \tag{2.68}$$

Resolving the velocity into a pressure field induced and a frictional component, the nonoscillatory depth integrated transport associated with the frictional component is, see (2.62):

$$\begin{aligned} U_2 &= \frac{u_*^2}{f}, \\ V_2 &= 0. \end{aligned} \tag{2.69}$$

The non-oscillatory pressure field-induced velocities are therefore, from (2.63):

$$\begin{aligned} u_1 &= -\frac{u_*^2}{fH} e^{x/R}, \\ v_1 &= \frac{u_*^2 t}{H} e^{x/R}. \end{aligned} \tag{2.70}$$

The evolving non-oscillatory longshore velocity v_1 is seen to be in geostrophic balance with the pressure field at each instant, see the first of Equation (2.68). This is in accord with the first of Equations (2.58), given $\mathrm{d}u_1/\mathrm{d}t = 0$. The second of these equations, for $\partial\zeta/\partial y = 0$ yields on integration:

$$v_1 = -f \int_0^t u_1 \, \mathrm{d}t = -f\zeta_1, \tag{2.71}$$

where ζ_1 is fluid particle displacement during the period t. Equations (2.70) of course satisfy this requirement, which means in physical terms that the longshore velocity component v_1 is generated by the Coriolis force associated with the offshore velocity component u_1. The pressure field is seen to induce longshore flow indirectly by establishing offshore motion. Again as in the previous section, when only the non-oscillatory parts of the motion are considered, a pattern of (in this case offshore) flow seems to arise abruptly. Actually, this develops on the time scale f^{-1} along with the inertial period motions.

The non-oscillatory, frictionally induced velocities, for negligible bottom friction, are confined to a surface Ekman layer, the distribution being exactly that given in Equation (1.51). The negligible bottom friction hypothesis is valid only as long as v_1 is suitably small, which is true only for some initial period.

On the cessation of the longshore wind at $t = T$ a longshore current remains in geostrophic balance with the surface elevation gradient, see equations (2.55). There is now no frictionally induced motion (if bottom friction is negligible) and the pressure field induced motion is simply

$$v_1 = \frac{V_1}{H} = \frac{u_*^2 T}{H} e^{x/R}. \tag{2.72}$$

This motion was built up during the period of action of the wind according to Equation (2.71). At the end of that period all that remains is longshore flow in geostrophic equilibrium with an appropriate pressure field (and some slowly decaying inertial oscillations). Similar results have been arrived at by Rossby (1938), Cahn (1945), Charney (1955), and others, in investigations relating to the adjustment of the flow and pressure fields to geostrophic equilibrium. In all such problems (whatever the initial conditions or forcing which ultimately produces geostrophic flow) the geostrophic current v_1 is established via cross-stream velocities, or in a time-integrated form, see Equation (2.71), cross-stream displacement of fluid particles, giving rise to a Coriolis force which establishes the current. It will be convenient in the discussion of similar phenomena later in this text to have a descriptive term for such a pressure field-induced flow component *not* itself in geostrophic equilibrium, the principal dynamical role of which is to effect adjustment of the flow and pressure fields to geostrophic equilibrium. Similar motions will be referred to therefore as 'adjustment drift', the term drift being chosen to indicate that in practical cases the velocity magnitude u_1 is relatively low, although when acting for a period long compared to f^{-1}, it may produce a substantial geostrophic current, see Equation (2.71) again. Figure 2.14 illustrates the contribution of Ekman drift and of adjustment drift to an evolving longshore geostrophic current, generated by longshore wind.

The considerations in this section are among the most important in coastal oceanography. Failure to recognize the importance of adjustment drift in generating longshore currents is the cause of many a misleading remark found in the literature.

2.8. RESPONSE OF A CLOSED BASIN TO SUDDEN WIND

The separate treatment of cross-shore and longshore wind effects along a straight infinite coast has yielded much insight, but these over-idealized case need to be connected to

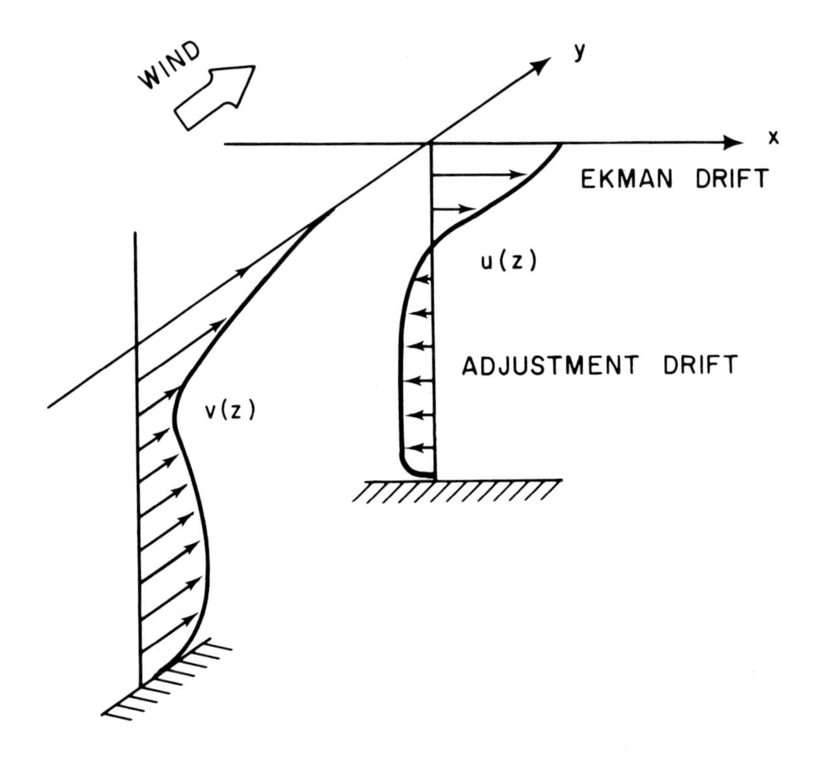

Fig. 2.14. Longshore and cross-shore velocities nearshore ($|x| << R$) accompanying generation of longshore current.

more realistic geometry. Along a natural coast even a spatially uniform wind has variable cross-shore and longshore ˏomponents owing to the change of orientation of the coastline. In a closed basin subject to uniform wind some of the coastline is nearly transverse to the wind, some of it nearly parallel, with a gradual transition in between. Without further investigation it is not clear whether or how the previous results relate to such more realistic cases.

In view of earlier discussion in this chapter a rectangular basin would suggest itself as a suitable enclosed sea model. This, however, has some unrealistic singularities at the corners and also leads to clumsy algebra. The transition from a parallel to a transverse portion of the coast in a rectangular model is not gradual, but abrupt. A circular basin is easier to treat algebraically and is particularly suitable because in the very large basin limit its shores become nearly straight, in appropriate locations respectively parallel and perpendicular to a uniform wind. Thus the investigation of a circular basin's response to suddenly imposed uniform wind should reveal any essential differences between finite and infinite sea models, while it should also readily connect in the large basin limit to results found for straight coasts.

2.8.1. Circular Basin Model

Figure 2.15 illustrates the geometry of a circular basin and of the polar coordinate system which is conveniently introduced to deal with it. The transport equations (1.35) written in terms of radial and tangential transport components (V_r, V_ϕ) become:

$$\frac{\partial V_r}{\partial t} - fV_\phi = -c^2 \frac{\partial \zeta}{\partial r} + F_x \cos\phi + F_y \sin\phi,$$

$$\frac{\partial V_\phi}{\partial t} + fV_r = -c^2 \frac{\partial \zeta}{r\,\partial \phi} - F_x \sin\phi + F_y \cos\phi, \qquad (2.73)$$

$$\frac{1}{r}\frac{\partial (rV_r)}{\partial r} + \frac{1}{r}\frac{\partial V_\phi}{\partial \phi} = -\frac{\partial \zeta}{\partial t},$$

where F_x is the wind stress component along $\phi = 0$, F_y along $\phi = \pi/2$. The impulsive forcing is represented by

$$F_x = u_*^2, \quad (t \geqslant 0)$$
$$F_y = 0. \qquad (2.74)$$

The no-normal transport boundary condition at the shore is

$$V_r = 0, \quad (r = r_0). \qquad (2.75)$$

These equations may be solved for the surface elevation distribution by Laplace transform methods (for example) with the result (Csanady, 1968; Birchfield, 1969):

$$\zeta = \frac{u_*^2}{c^2} r_0 \left[A_0(r) \cos\phi + \sum_n a_n A_n(r) \cos(\phi - \sigma_n t) \right]. \qquad (2.76)$$

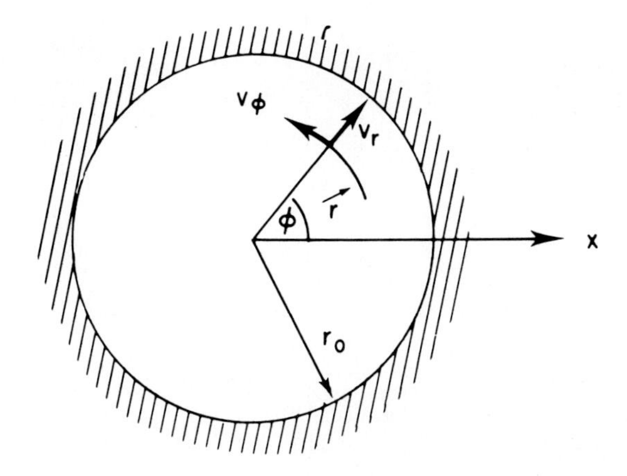

Fig. 2.15. Polar coordinate system used in calculations for circular basin.

The solution consists of wind setup and a series of azimuthal waves with radial amplitude distributions of $A_0(r)$ and $A_n(r)$ respectively, $n = 1, 2, 3 \ldots$, and is in this respect analogous to the solution found for a narrow or non-rotating rectangular basin acted upon by sudden wind, Equation (2.34).

The wind setup in a circular basin is described by the amplitude distribution

$$A_0(r) = \frac{I_1(r/R)}{I_1(r_0/R)}, \tag{2.77}$$

where $I_1(\ )$ is a modified Bessel funciton. The properties of this elevation distribution are analogous to what was discussed in detail in Section 2.2.1 for a rectangular basin. Figure 2.16 illustrates transport streamlines for a large ($r_0/R = 15$) circular basin. In a small basin the surface is an inclined plane, the pressure gradient (nearly) balancing the wind stress everywhere, and there is (almost) no net transport anywhere in the basin. In a large basin surface level gradients are much larger at the ends of the diameter parallel to the wind than in a small basin. Along these end-walls in a large basin considerable longshore transport takes place within a coastal boundary layer. This end-wall boundary layer transport constitutes the return flow for the Ekman transport collected by the side-wall boundary layers. Over most of the basin, outside the coastal boundary layers, the surface is undisturbed and the transport is Ekman transport, confined to a surface Ekman layer if the basin is much deeper than the Ekman depth.

The azimuthal waves in a rotating circular basin excited by uniform wind are all of wavenumber one, i.e., have a wavelength at the shore equal to the basin perimeter $2\pi r_0$.

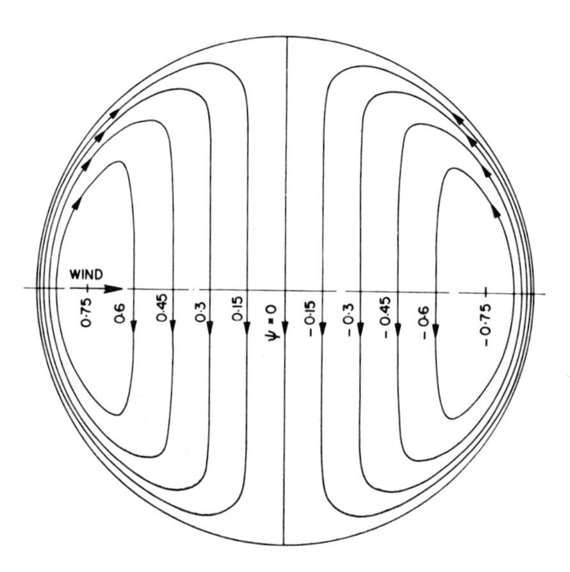

Fig. 2.16. Calculated distribution of transport streamlines associated with wind setup in very large circular basin ($r_0/R = 15$), from Csanady (1968), Copyright American Geophysical Union.

They propagate around the basin with angular speed σ_n, counterclockwise should σ_n be positive, clockwise for negative σ_n. The (circular) frequency of oscillations observable at a fixed point is also σ_n.

The frequencies σ_n are the natural frequencies of free oscillations in a rotating circular basin of constant depth and have been discussed in detail by Lamb (1932). The frequencies of wavenumber one azimuthal waves may be calculated from the equation.

$$\gamma \frac{r_0}{R} J_1' \left(\gamma \frac{r_0}{R}\right) + \frac{f}{\sigma} J_1 \left(\gamma \frac{r_0}{R}\right) = 0, \tag{2.78}$$

where $J_1(\;)$ is a Bessel function, $J_1'(\;)$ its derivative and

$$\gamma = \sqrt{\frac{\sigma_2}{f_2} - 1}. \tag{2.79}$$

The distribution of amplitude over radius in the azimuthal waves is

$$A_n(r) = \frac{J_1(\gamma r/R)}{J_1(\gamma r_0/R)}. \tag{2.80}$$

The constants a_n in the solution (2.76) determine to what extent each individual azimuthal wave is excited. They are given by

$$a_n = \frac{\sigma_n/f - 1}{\sigma_n/f + 1 - (\sigma_n/f)^3 \; (r_0/R)^2}. \tag{2.81}$$

In a detailed discussion of the natural frequencies of a circular basin, Lamb (1932) shows that, if the basin is sufficiently large, the lowest frequency azimuthal wave, for given wavenumber, comes to have a structure unlike that of the higher frequency waves. For aximuthal wavenumber one this occurs when

$$\frac{r_0^2}{R^2} \geqslant 2. \tag{2.82}$$

For such large basins the characteristic equation (2.78) has one imaginary root γ_1, the corresponding frequency σ_1 being thus less than the inertial (by Equation (2.79)). Writing for this case $\gamma_1 = i\gamma_*$, the associated amplitude distribution becomes

$$A_1(r) = \frac{I_1(\gamma_* r/R)}{I_1(\gamma_* r_0/R)}, \tag{2.83}$$

where $\gamma_* = (1 - \sigma_1^2/f^2)^{1/2}$, a real quantity, and $I_1(\;)$ is a modified Bessel function.

All other roots of Equation (2.78) give frequencies higher than the inertial, although in a very large basin many of them differ only slightly from f. Oscillations of inertial frequency or higher are of no direct interest in the circulation problem. In any case, these higher frequency oscillations of wavenumber one in a circular basin possess some special properties on account of axial symmetry which makes them a poor model of high frequency standing waves in rotating basins of natural shape. Of further interest here is the static (wind setup) solution, already discussed above and in Section 2.2.1, and, in the large basin limit, the lowest frequency azimuthal wave, which then becomes a slowly developing, i.e., quasi-stationary mode of motion.

2.8.2. The Large Basin Limit

For a very large basin $r_0/R >> 1$, the lowest root of (2.76) is approximately (Lamb, 1932):

$$\frac{\sigma_1}{f} \cong \frac{R}{r_0}, \quad \left(\frac{R}{r_0} << 1\right) \tag{2.84}$$

So that also $\sigma_1/f << 1$. The corresponding low frequency wave travels along the perimeter of the basin with linear speed $\sigma_1 r_0$, which is, according to the last result

$$\sigma_1 r_0 \cong fR = c. \tag{2.84a}$$

In this limit $\gamma_* = (1 - \sigma_1^2/f^2)^{1/2}$ is nearly unity and the amplitude distributions of wind setup and low frequency wave, $A_0(r)$ and $A_1(r)$, Equations (2.80) and (2.83), become nearly the same. Both become vanishingly small over most of a very large basin, except for a band of scale-width R near the perimeter. This is so because the Bessel functions $I_1(\ )$ behave for large argument as exponentials, and provide very large denominators in (2.80) and (2.83). The distributions $A_0(r)$ and $A_1(r)$ may be approximated for large arguments in the large basin limit, by

$$A_0(r) \cong A_1(r) \cong \exp\left(\frac{r-r_0}{R}\right), \quad (r_0/R >> 1). \tag{2.85}$$

The steady and low-frequency parts of the solution (2.76) may now be combined into the following approximate expression, valid in the large basin limit to order R/r_0:

$$\zeta \cong \frac{u_*^2}{c^2}\, r_0 \left[\cos\phi - \left(1 - \frac{R}{r_0}\right)\cos(\phi - \sigma_1 t)\right] \exp\left(\frac{r-r_0}{R}\right), \quad (r_0/R >> 1). \tag{2.86}$$

In addition to this quasi-stationary elevation field there are oscillations of inertial frequency and higher, which are neglected here, as in the previous sections dealing with straight infinite shore models.

In the solution (2.86), at time $t = 0$, the wind setup is cancelled to zeroth order in R/r_0, leaving a first order residue:

$$\zeta = \frac{u_*^2}{fc} \cos\phi \exp\left(\frac{r-r_0}{R}\right), \quad (t = 0). \tag{2.87}$$

The radial pressure gradient is thus at the coast $r = r_0$:

$$\frac{\partial \zeta}{\partial r} = \frac{u_*^2 \cos\phi}{c^2}, \quad (r = r_0, t = 0) \tag{2.88}$$

which exactly balances the applied wind stress, and is much less than the level gradient in the wind setup solution (Figure 2.16) at $\phi = 0, \pi$. In the neighborhood of the point $\phi = 0$ or $\phi = \pi$ along the coast (where the applied wind stress is perpendicular to the coast) the quasi-steady solution (2.87) at $t = 0$ is thus exactly the same as found for the non-oscillatory pressure field along an infinite coast transverse to the wind, see Section 2.4, Equation (2.46).

As the low frequency wave propagates away from its initial position at $t \neq 0$, it no longer cancels the wind setup entirely, and the elevation field becomes to first order in R/r_0:

$$\zeta = \frac{u_*^2}{fc} \exp\left(\frac{r - r_0}{R}\right)(\cos\phi - ft \sin\phi), \quad (t \ll \sigma^{-1}). \tag{2.89}$$

The part proportional to $\cos\phi$ is the same as (2.86), but there is now also a component which peaks at the ends of the diameter *perpendicular* to the wind, and which evolves linearly in time. At $\phi = \pi/2$, where the shore is parallel to the wind and lies to the left of it (looking along wind) the level begins to drop, at $\phi = -\pi/2$, to the right of the wind, the level rises. This pressure field is thus exactly the same as found in Section 2.5 for the case of wind blowing along a straight, infinite shore, neglecting the oscillatory components.

The transport components corresponding to the quasi-static elevation field of Equation (2.86) may be determined from the transport equations (2.73). For $t \ll \sigma_1^{-1}$, one finds for the tangential transport, for example, to first order in R/r_0:

$$\begin{aligned} V_\phi = {} & -u_*^2 t \sin\phi \exp\left(\frac{r - r_0}{R}\right) - \\ & - \frac{u_*^2}{f} \cos\phi \left[1 - \exp\left(\frac{r - r_0}{R}\right)\right], \quad (r_0/R \gg 1,\ t \ll \sigma_1^{-1}). \end{aligned} \tag{2.90}$$

This corresponds exactly to the results found for a straight infinite coast, transverse and parallel to the wind at $\phi = 0$ and $\pi/2$ respectively, with a smooth variation in between.

At times of order σ_1^{-1} and longer the two components of the quasi-static solution, the wind setup and the slow aximuthal wave, combine in more complex ways. Intuition is perhaps best served then by viewing the two as independent components of the solution which sometimes reinforce each other, sometimes nearly cancel, as at $t = 0$. The properties of the wind setup on its own were discussed above: it remains to analyze the slow aximuthal wave in a little more detail.

2.9. THE KELVIN WAVE

The slow azimuthal wave of the quasi-steady solution (2.86) corresponds to a well-known classical model called the 'Kelvin wave', see e.g., Proudman (1953). In the large basin limit $r_0/R \to \infty$ one may put $r - r_0 = x$, $y = r_0\phi$, the domain of interest being the interior,

$x \leqslant 0$. Noting that in this limit $\sigma_1 = fR/r_0 = c/r_0$ (Equation (2.84)) one may write for the low frequency wave component of (2.86)

$$\zeta = ae^{x/R} \cos\left(\frac{y-ct}{r_0}\right) \tag{2.91}$$

with $a = (u_*^2/c^2)(r_0 - R)$, the shore amplitude. One easily verifies that this expression satisfies the homogeneous differential equation (2.11). From Equations (2.8) with forcing neglected, an expression for the transport component U normal to the coast may be derived:

$$\frac{\partial^2 U}{\partial t^2} + f^2 U = -c^2\left(\frac{\partial^2 \zeta}{\partial x \partial t} + f\frac{\partial \zeta}{\partial y}\right). \tag{2.92}$$

Substitution of (2.91) shows that the right-hand side vanishes, so that U is zero everywhere at all t, hence the boundary condition of $U = 0$ at the coast $x = 0$ is automatically satisfied. Far from the coast, $x \to -\infty$ the elevation ζ tends to zero, thus satisfying a realistic boundary condition at infinity. Physically this means that the wave is 'trapped' at the coast. One should observe also that if the direction of propagation were changed (i.e., $-ct$ replaced by $+ct$ in (2.91)) the boundary condition $U = 0$ could be satisfied only by also changing the sign of the exponent x/R. This would describe a Kelvin wave in the positive x half-plane, propagating to negative y, i.e., so as to leave the coast to the right, much as the Kelvin wave of Equation (2.91), only turned 180° around. Directions here refer to the northern hemisphere: in the southern hemisphere they are opposite. In a closed basin a Kelvin wave propagates around the perimeter always in the cyclonic direction.

The alongshore transport component associated with the Kelvin wave may readily be calculated from geostrophic balance, i.e., the first of (2.8), with $U = 0$, $F_x = B_x = 0$:

$$V = ace^{x/R} \cos\left(\frac{y-ct}{r_0}\right). \tag{2.93}$$

This contains the first contribution to (2.90) at $t \ll \sigma_1^{-1}$, as well as the part that cancels the transport associated with the wind setup. At times of order σ_1^{-1} and longer the cancellation between the wind setup and the Kelvin wave is generally ineffective (until a full cycle is completed by the wave around the perimeter) so that the longshore transports are of order $u_*^2\, r_0/c$, or quite large, just as in the flow pattern accompanying the wind setup (Figure 2.16).

To sum up the discussion of the last few sections, the pressure field in a finite basin under impulsive wind has many of the characteristics revealed by infinite coast models. Elevation and transport along portions of the coast perpendicular to the wind have a structure initially identical with what was found in a straight coast model with transverse wind. Similarly, the initial development of longshore flow along portions of the coast parallel to the wind is much as discussed earlier using the straight coast model. However, this can only be said to be true at times $t \ll \sigma_1^{-1}$: at larger times a more complex behavior ensues.

The practical significance of these results is again undermined by the fact that most closed shallow seas are smaller than the radius of deformation. In that case, of course, earth rotation effects are unimportant: Section 2.3 to 2.3.2 dealt with no-rotation dynamics in detail. Rotational effects and the Kelvin wave become of considerable importance in the circulation in a stratified shallow sea, the behavior of which is discussed in the next chapter.

CHAPTER 3

The Behavior of a Stratified Sea

3.0. GENERAL REMARKS

As discussed at some length in Chapter 1, an important factor in the pressure distribution within a shallow sea is the stratification of the water column. Whether over continental shelves or in enclosed basins, the temperature and salinity distribution gives rise to density differences of the order of 1‰. The lighter water generally overlies heavier layers, but the surfaces of constant density are only approximately horizontal. Figure 3.1 shows an observed distribution of temperature in a cross section of Lake Michigan, which in a freshwater lake is also a direct indication of density, with the vertical scale exaggerated by a factor of about 3000. The density of the top layers compared to those at the bottom (18 °C versus 4 °C in freshwater) was about 1.4‰ on this occasion. Most of the density change is seen to take place in a relatively thin layer, the thermocline, between about the 5° and 15 °C isotherms.

In this example, the dynamic height of the lake surface ζ_d (Equation (1.12), taking the density of 4 °C water as reference, $\rho_0 = 1.0$), relative to the deep layers, where the water is more or less homogeneous, is of the order of 4 cm. This quantity varies, however, with the depth of the thermocline and reaches a value close to 7 cm near the western shore, where the thermocline becomes much deeper. If the pressure is constant along level surfaces in the deep water, the surface layers are subject in some places to sea level changes of 3 cm in 6 km, or a surface slope of 5×10^{-6}. If, on the other hand, the sea surface were entirely flat, the bottom layers would be subject to pressure gradients of the same magnitude. Horizontal pressure gradients this large are certainly of first order importance in the dynamics of shallow seas (as pointed out before, surface slopes of 10^{-7} are significant). Clearly the vertical position of constant density surfaces (isopycnals) is an essential part of the response of a shallow sea to forcing by wind, or by whatever other physical factor.

The isopycnal surfaces not only depart significantly from their horizontal position of hydrostatic equilibrium in a shallow sea, but also move about in a most marked manner. Figure 3.2 shows the same cross section as the previous illustration, retaining only the 10 °C isotherm (the center of the thermocline), but showing several transects taken in the course of four consecutive days (from a car-ferry criss-crossing the lake). This picture suggests that large standing waves occur on the thermocline in the center of the lake, while the nearshore temperature distribution is more nearly persistent. Cross sections taken at other times show that also the nearshore isopycnals move, and that the sharp upslope to the eastern shore, downslope to the western shore picture shown in Figure 3.1 is sometimes reversed or replaced by some other pattern (Mortimer, 1968).

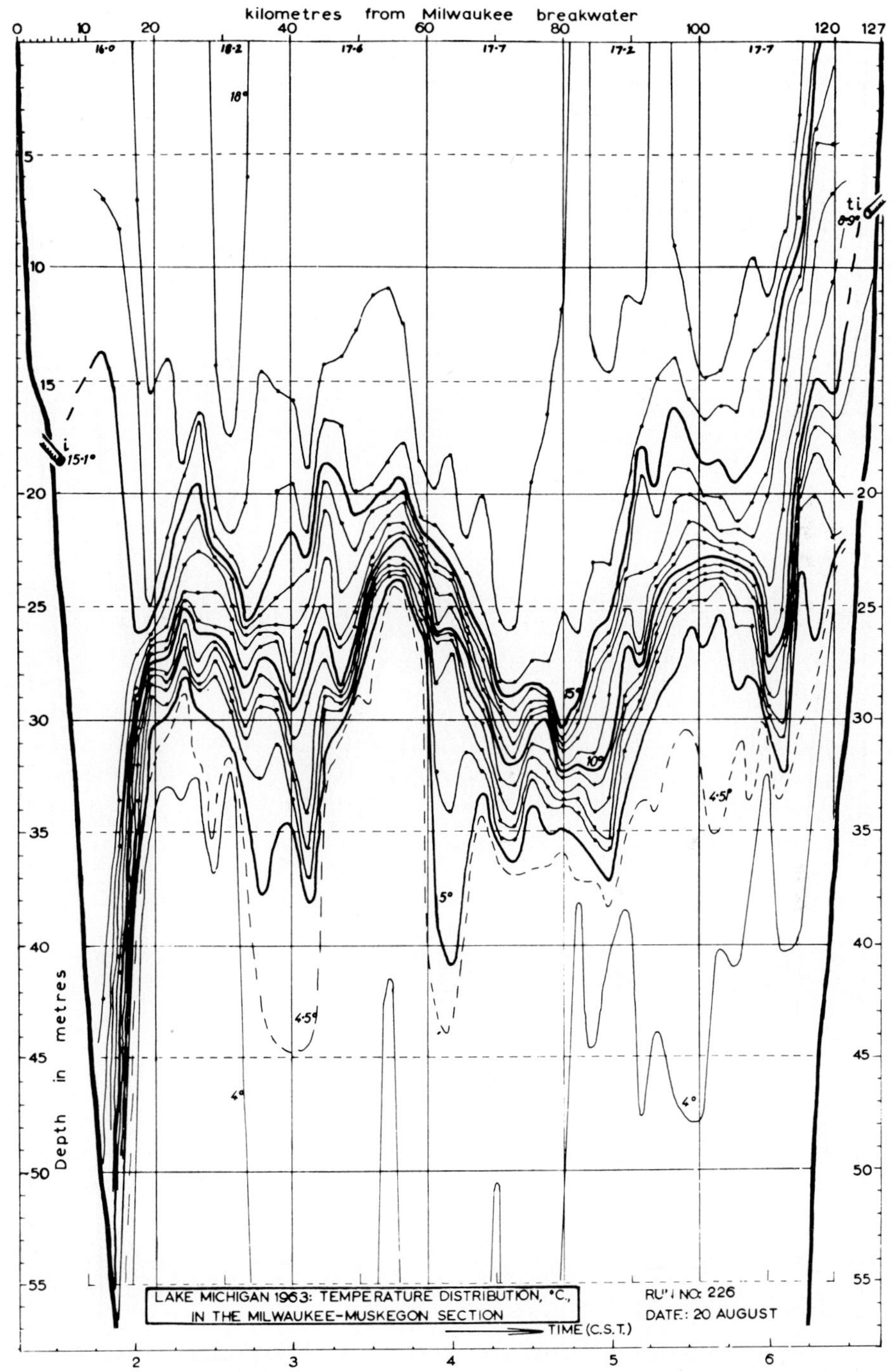

Fig. 3.1. Distribution of temperature (°C) in a transect of Lake Michigan between Milwaukee and Muskegon, observed on a car-ferry run (Mortimer, 1968).

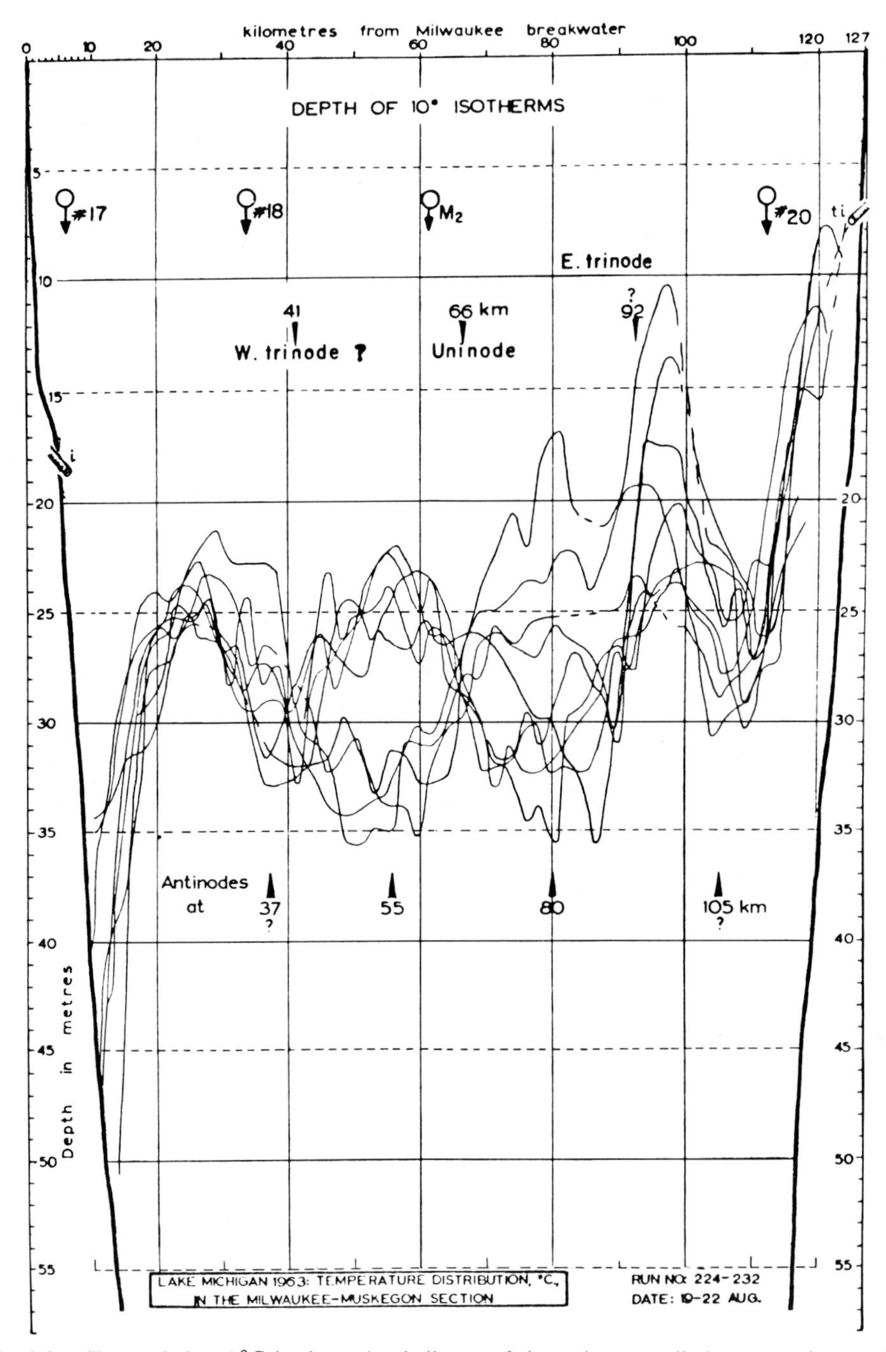

Fig. 3.2. Shape of the 10°C isotherm (an indicator of the main pycnocline) on several successive transects of Lake Michigan, observed as Figure 3.1 (Mortimer, 1968).

To gain physical insight into similar phenomena, it is useful to analyze the behavior of very simple 'model basins' of constant depth and simple horizontal geometry (e.g., circular shape) containing a stratified fluid and subject to simple distributions of surface stress, e.g., spatially uniform wind stress imposed suddenly at time zero and then remaining constant. In order to obtain explicit mathematical solutions, the displacements of the isopycnals will in much of this chapter be supposed small compared to their equilibrium depth. This hypothesis linearizes the pressure term in the equations of motion and renders these equations tractable. However, the hypothesis is restrictive: as may be seen from the illustrations, it does not strictly apply to the commonly observed cases. Yet, the linearized theory yields the basic modes of motion, time and length scales, with adequate accuracy. The theoretical framework for treating the linearized equations of stratified flow is well developed – as Eckart (1960) points out, it is 19th century fluid mechanics. The complete mathematical solutions obtainable from the linearized theory provide the basis for choosing theoretical approximations for the further investigation of motions in which pycnocline displacements do *not* remain small.

Another approximation made in this chapter is to neglect mixing and internal and bottom friction, except in a surface layer of homogeneous density. These approximations are reasonable in the stratified interior in view of the well-known turbulence-suppressing property of stratification.

3.1. PERTURBATION THEORY

According to classical theory (see Eckart, 1960; or Krauss, 1966) the state of motion in a stratified basin of constant depth is usefully considered a 'perturbation' of a state of rest relative to the rotating earth, the equilibrium density distribution in the zero order state being $\rho(z)$. The distribution may contain one or more mixed layers of constant density. Small vertical displacements of particles (which are supposed to conserve their density) lead to local density perturbations ρ' small compared to a reference density ρ_0. These also imply pressure perturbations p' through the hydrostatic equation. The appropriate linearized equations of momentum balance, density conservation and continuity are, from Equations (1.1), (1.6), (1.16), and (1.18), but in xyz notation and after the introduction of the hydrostatic approximation:

$$\begin{aligned}
&\frac{\partial u}{\partial t} - fv = -\frac{1}{\rho}\frac{\partial p'}{\partial x} + \frac{1}{\rho}\frac{\partial \tau_x}{\partial z},\\
&\frac{\partial v}{\partial t} + fu = -\frac{1}{\rho}\frac{\partial p'}{\partial y} + \frac{1}{\rho}\frac{\partial \tau_y}{\partial z},\\
&\frac{\partial p'}{\partial z} = -g\rho',\\
&\frac{\partial \rho'}{\partial t} + w\frac{\mathrm{d}\rho}{\mathrm{d}z} = 0,\\
&\frac{\partial u}{\partial x} + \frac{\partial v}{\partial y} + \frac{\partial w}{\partial z} = 0.
\end{aligned} \tag{3.1}$$

The only equation requiring comment here is the one expressing conservation of density of fluid particles (fourth of 3.1): both ρ' and w are small (perturbation) quantities, $\mathrm{d}\rho/\mathrm{d}z$ is the vertical density gradient in the undisturbed mean state, and can be relatively large.

It is convenient to work in terms of the small vertical displacements of particles from their equilibrium level, $\zeta(x, y, z, t)$ (note that ζ now refers to particles at all levels, not only at the surface as in previous chapters):

$$\zeta = \int_0^t w \, \mathrm{d}t. \tag{3.2}$$

Eliminating the density perturbation from the third and fourth of Equations (3.1) an equation connecting p' to ζ is obtained:

$$\frac{\partial p'}{\partial z} = g\zeta \frac{\mathrm{d}\rho}{\mathrm{d}z}. \tag{3.3}$$

The boundary condition at the free surface is, on the assumption that the atmospheric pressure is constant:

$$\frac{\partial p'}{\partial t} + w \frac{\mathrm{d}p}{\mathrm{d}z} = 0 \quad (z = 0)$$

or (3.4)

$$\frac{\partial p'}{\partial t} = g\rho w = g\rho \frac{\partial \zeta}{\partial t} \quad (z = 0).$$

Equation (3.3) may now be integrated and yields

$$p' = g\rho_s\zeta_s - g \int_z^0 \frac{\mathrm{d}\rho}{\mathrm{d}z'} \zeta(z') \, \mathrm{d}z', \tag{3.5}$$

where ρ_s and ζ_s are values at the free surface, $z = 0$. The boundary condition at the bottom is

$$w = 0 \quad \text{at} \quad z = -H \quad (H = \text{constant}). \tag{3.6}$$

Boundary conditions at the shores are that normal velocities vanish.

3.2. NORMAL MODE EQUATIONS

The solution of (3.1) subject to the prescribed boundary conditions can be expressed as a combination of normal modes, each characterized by a certain vertical distribution of horizontal velocity and vertical displacement:

$$u = \sum_k U_k(x, y, t) \frac{dQ_k}{dz},$$

$$v = \sum_k V_k(x, y, t) \frac{dQ_k}{dz}, \tag{3.7}$$

$$\zeta = \sum_k Z_k(x, y, t) Q_k, \qquad (k = 1, 2, \ldots),$$

where $Q_k = Q_k(z)$. Setting vertical displacements proportional to Q_k and horizontal velocities to dQ_k/dz ensures that this formulation is consistent with the continuity equation, the last of (3.1). The distribution function $Q_k(z)$ will be taken to be non-dimensional, so that U_k, V_k have physical dimensions of transport (velocity × length) while Z_k is a length.

On substitution of (3.7) into (3.1) the terms containing the velocities are seen to be proportional to dQ_k/dz, while the pressure terms come to contain the depth integral of $Q_k(z)$ (see Equation (3.5)). The two are consistent, provided that

$$\frac{d}{dz}\left(\rho \frac{dQ_k}{dz}\right) = \frac{g}{c_k^2} \frac{d\rho}{dz} Q_k(z), \tag{3.8}$$

where c_k is an arbitrary constant, of the physical dimensions of velocity. Equation (3.8) may be written as

$$\frac{d^2 Q_k}{dz^2} - \frac{N^2}{g} \frac{dQ_k}{dz} + \frac{N^2}{c_k^2} Q_k = 0, \tag{3.9}$$

where $N^2 = -(g/\rho)(d\rho/dz)$, N being known as the Väisälä frequency. Boundary conditions on Q_k are, from (3.4) and (3.6):

$$\frac{dQ_k}{dz} - \frac{g}{c_k^2} Q_k = 0, \qquad (z = 0)$$

$$Q_k = 0, \qquad (z = -H). \tag{3.10}$$

Equations (3.9) and (3.10) pose an eigenvalue problem, discussed in detail by Krauss (1966), the solution of which yields a set of eigenfunctions $Q_k(z)$ or dQ_k/dz, and eigenvalues c_k, given a continuous distribution of density, i.e., smooth $N(z)$. The eigenfunctions dQ_k/dz form a complete orthogonal set: on multiplying Equation (3.9) by Q_m and integrating with respect to z and then interchanging the indices k and m, it may be shown directly that

$$\int_{-H}^{0} \rho \frac{dQ_k}{dz} \frac{dQ_m}{dz} dz = 0 \quad \text{for} \quad k \neq m. \tag{3.11}$$

With some qualifications, see Section 3.4 below, a friction force arbitrarily distributed in the vertical may be expressed in terms of the eigenfunctions dQ_k/dz (Lighthill, 1969):

$$\frac{1}{\rho}\frac{\partial \tau_x}{\partial z} = gH \sum_k \phi_{xk} \frac{dQ_k}{dz},$$
$$\frac{1}{\rho}\frac{\partial \tau_y}{\partial z} = gH \sum_k \phi_{yk} \frac{dQ_k}{dz}, \tag{3.12}$$

where the coefficients ϕ_{xk}, ϕ_{yk} may be functions of (x, y, t). Multiplying this expansion by $\rho\partial Q_m/\partial z$ and integrating from $z = -H$ to 0 one finds:

$$\phi_{xk} = \frac{\int_{-H}^{0} \frac{\partial \tau_x}{\partial z}\frac{dQ_k}{dz}\,dz}{gH\int_{-H}^{0} \rho \left(\frac{dQ_k}{dz}\right)^2 dz} \tag{3.13}$$

with an analogous expression for ϕ_{yk}.

Substituting now back into the equations of motion and continuity (3.1), and taking into account also (3.7) and (3.8), one finds the equations satisfied if the coefficients of dQ_k/dz balance for each k, which is the case if:

$$\frac{\partial U_k}{\partial t} - fV_k = -c_k^2\frac{\partial Z_k}{\partial x} + gH\phi_{xk},$$
$$\frac{\partial V_k}{\partial t} + fU_k = -c_k^2\frac{\partial Z_k}{\partial y} + gH\phi_{yk}, \qquad (k = 1, 2, \ldots). \tag{3.14}$$
$$\frac{\partial U_k}{\partial x} + \frac{\partial V_k}{\partial y} = -\frac{\partial Z_k}{\partial t},$$

These equations are of the same form as the depth integrated or transport equations for a homogeneous fluid (1.41), subject to wind stress components $\rho gH(\phi_{xk}, \phi_{yk})$, in a basin of a depth $c_k^2/g = H_k$, which one might term the 'equivalent' depth corresponding to the kth normal mode. In this manner, the problem of determining the response of a stratified sea to arbitrary forcing has been reduced to solving homogeneous fluid equations. The reduction is possible for a constant depth, linearized model, and is thus restrictive, but it makes a whole array of theoretical solutions available, which turn out to be very instructive, if interpreted with due caution.

As usual in such problems, the amplitude of the eigenfunctions dQ_k/dz is arbitrary and needs to be normalized in some manner. Here it turns out to be most convenient to take the nondimensional surface layer velocity as unity in all natural modes, i.e.,

$$H\left.\frac{dQ_k}{dz}\right|_{z=0} = 1, \qquad k = 1, 2, \ldots. \tag{3.15}$$

The resultant response of a stratified basin to a given distribution of interior stress may now be determined by solving (3.14) for $k = 1, 2, \ldots$, and summing over all k. The resultant surface velocity, for example, will have an x component:

$$u = \frac{1}{H} \sum_k U_k (x, y, t), \quad (z = 0) \tag{3.15a}$$

as a direct consequence of the normalization (3.15).

3.3. STRATIFICATION MODEL

As already pointed out in connection with Figure 3.1, observed vertical distributions of density in shallow seas are usually characterized by sharp density changes in a 'pycnocline' region. Close to the free surface the density is nearly homogeneous on account of the stirring effect of wind stress, and one often speaks of the surface 'mixed layer'. Below the pycnocline density changes are also relatively small, although it is not always realistic to describe the bottom layer as homogeneous. Figure 3.3 shows one typical observed density profile.

In order to illustrate the properties of normal modes discussed in the previous section a concrete typical example will be analyzed. A reasonably simple set of calculations result if one supposes

$$\begin{aligned}
\frac{d\rho}{dz} &= 0 \quad &&\text{for} \quad 0 \geqslant z \geqslant -\alpha H,\\
\frac{d\rho}{dz} &= \rho\Gamma \quad &&\text{for} \quad -\alpha H \geqslant z \geqslant -\beta H,\\
& &&(\Gamma = \text{constant})\\
\frac{d\rho}{dz} &= 0 \quad &&\text{for} \quad -\beta H > z > -H.
\end{aligned} \tag{3.16}$$

In a typical shallow sea one might have $\alpha = 0.2$, $\beta = 0.4$, $H = 75$ m, $\Gamma H = 10^{-2}$, corresponding to a Väisälä period of about 3 min in the pycnocline, with mixed layers above and below. These quantitative data are used in the calculated results given below.

Within the mixed layers Equation (3.9) yields at once

$$H \frac{dQ_k}{dz} = \text{constant}, \quad (k = 1, 2, 3, \ldots) \tag{3.17}$$

so that the velocity amplitudes are independent of depth. Within the pycnocline layer with linear stratification the solution of Equation (3.9) is of the form

$$Q_k = e^{-\Gamma z/2} (a_k \cos q_k z + b_k \sin q_k z) \tag{3.18}$$

with a_k, b_k = constants and

$$q_k = \sqrt{-\frac{g\Gamma}{c_k{}^2} - \frac{\Gamma^2}{4}}. \tag{3.18a}$$

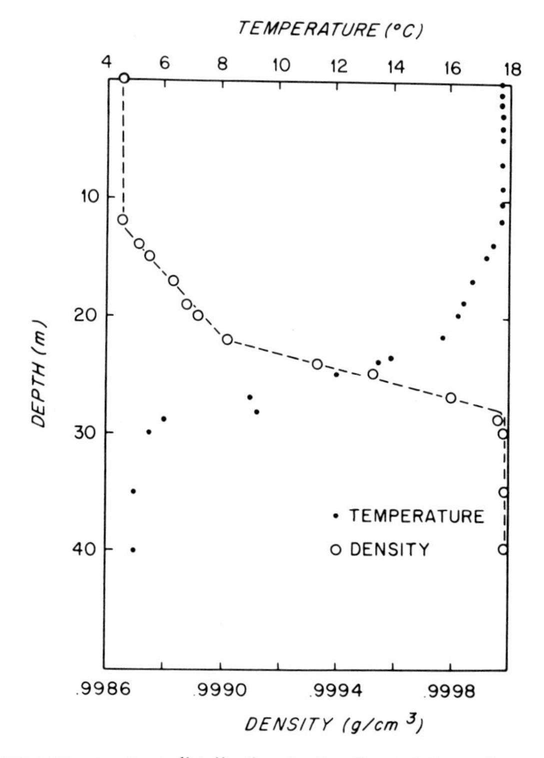

Fig. 3.3. Typical summer temperature distribution in the Great Lakes, observed off Douglas Point, Lake Huron.

The boundary conditions (3.10) are supplemented in the 3-piece stratification model by matching conditions at the interfaces, requiring continuity of Q_k and its vertical derivative. The eigenvalue problem may then be reduced to the following characteristic equation (Csanady, 1972):

$$\tan[(\alpha-\beta)q_k H] = q_k H \frac{1+\alpha-\beta-\dfrac{c_k^2}{gH}}{1+\Gamma H\left[\alpha(1-\beta)\dfrac{gH}{c_k^2}+\dfrac{1}{2}(\beta-\alpha-1)\right]+\dfrac{\Gamma H}{2}\dfrac{c_k^2}{gH}}. \tag{3.19}$$

Together with (3.18a) this equation allows the calculation of c_k and of the eigenfunctions. The latter are characterized by a vertical wavenumber q_k in the pycnocline region and a nondimensional bottom layer velocity amplitude B_k.

For the usual small density variation between top and bottom layers in a stratified shallow sea, the equivalent depth in the lowest mode, $k = 1$, is to a good approximation always $H_1 \cong H$, i.e.,

$$c_1 \cong \sqrt{gH}. \tag{3.20}$$

The corresponding velocity distribution is uniform, as if density variations were completely absent:

$$H \frac{dQ_1}{dz} = 1, \quad (0 \geqslant z \geqslant -H). \tag{3.20a}$$

This mode is legitimately referred to as 'barotropic' because the inclination of the density surfaces plays no role in it. The higher modes are often described as 'baroclinic', but it is perhaps appropriate to emphasize again that the easy separability of motion into barotropic and baroclinic modes is the property of a constant depth model. Loose usage of the term 'baroclinic' to describe nonuniform velocity distributions actually observed in the ocean can be very misleading.

For the example specified by Equations (3.16), with the 'typical' quantities α, β, etc. as quoted, characteristic data of the first four modes are:

$$\begin{aligned}
c_1 &= 27.0 \text{ m s}^{-1}, & q_1 H &= 0.1, & B_1 &= 1.0,\\
c_2 &= 0.51 \text{ m s}^{-1}, & q_2 H &= 5.3, & B_2 &= -0.42,\\
c_3 &= 0.156 \text{ m s}^{-1}, & q_3 H &= 17.3, & B_3 &= 0.35,\\
c_4 &= 0.0832 \text{ m s}^{-1}, & q_4 H &= 32.4, & B_4 &= -0.335.
\end{aligned}$$

These data reveal that within the pycnocline region, the eigenfunction $\partial Q_k/\partial z$, which is proportional to the velocity U_k/H, has exactly $(k-1)$ velocity-nodes. Bottom velocities correspondingly alternate between positive and negative values, as mode number increases.

3.4. MODEL OF FORCING

The wind generates motions in the water by exerting a high horizontal stress at the free surface. In the interior of a stratified water column, however, the stress is usually small, so that significant stress gradients $\partial\tau_x/\partial z$, $\partial\tau_y/\partial z$ (which are the forcing terms in Equations (3.1)) remain confined to a relatively thin layer at the surface. At the bottom of the mixed layer, in particular, it is quite realistic to suppose vanishing stress.

Within the framework of the stratification model adopted these physical facts may be represented in the following manner. The evaluation of the force coefficients becomes simple, because the integral in the numerator of Equation (3.13) reduces to:

$$\int_{-H}^{0} \frac{\partial\tau_x}{\partial z}\frac{dQ_k}{dz}\,dz = \tau_x(0)\left.\frac{dQ_k}{dz}\right|_0 = \frac{\tau_x(0)}{H}, \tag{3.21}$$

where $\tau_x(0)$ is the x-component of the surface (wind) stress and $H\;dQ_k/dz|_0$ is the homogeneous layer value of the nondimensional velocity amplitude, already chosen to be unity. A similar result applies along the y-axis.

Equation (3.21) constitutes an important simplification but it should be noted also that the expansion of the friction force within the mixed layer according to Equations (3.12) is imperfect. The stress gradient is represented by this expansion as a constant force, equal to surface stress divided by mixed layer depth, which is an average value, not the actual stress gradient within a frictional surface shear layer. Correspondingly, calculated velocities in the mixed layer are to be interpreted as average values over the mixed layer depth. The actual distribution of frictional velocities is discussed further in Section 3.5.3 below.

With the simplification of Equation (3.21), the force coefficients are, simplifying the integral in the denominator in Equation (3.13) by substituting a reference density ρ_0 = constant for ρ with a small error:

$$\begin{aligned} \phi_{xk} &= \frac{\tau_x(0)}{\rho_0 g H}\,\phi_k, \\ \phi_{yk} &= \frac{\tau_y(0)}{\rho_0 g H}\,\phi_k, \end{aligned} \tag{3.22}$$

where

$$\phi_k = \left[H \int_{-H}^{0} \left(\frac{\mathrm{d}Q_k}{\mathrm{d}z} \right)^2 \mathrm{d}z \right]^{-1} \tag{3.22a}$$

this quantity being nondimensional and depending on the density distribution only.

The force coefficients are thus equal to the wind-setup surface slope in a homogeneous basin times the influence coefficient ϕ_k, the latter being a measure of the importance of each mode in the stratified basin's response to wind stress. With the previously specified typical quantities, and stratification as per Equation (3.16), the value of ϕ_k is, in the first four normal modes

$$\begin{aligned} \phi_1 &= 1.0, \\ \phi_2 &= 2.778, \\ \phi_3 &= 0.584, \\ \phi_4 &= 0.224. \end{aligned}$$

As a function of k, these are seen to speak at k = 2, i.e., in the first baroclinic mode. It is therefore usually supposed that the barotropic and first few baroclinic modes are responsible for most of the motions excited by the wind in a stratified sea. However, the peak in ϕ_k at k = 2 is not very sharp and ϕ_k drops ultimately only as k^{-2} at large mode numbers.

3.5. RESPONSE OF CONTINUOUSLY STRATIFIED MODEL TO FORCING

The modal transport equations (3.14) are, as already pointed out, identical in form with the transport equations for a homogeneous fluid of constant depth (Equation (2.8)). In the lowest or barotropic mode, moreover, $c_1^2 \cong gH$ and ϕ_1 = 1.0, so that the constants are also the same, with an error of order $\epsilon = \rho'/\rho_0$, a typical value of the proportionate density defect. The higher or baroclinic mode equations differ mainly in that c_k^2 are small, or order ϵc_0^2 for low k, dropping as k^{-2} for high k, and that the forcing coefficients ϕ_k also drop asymptotically as k^{-2}, a point already made above.

The boundary condition of zero normal transport at coasts has to be satisfied in each normal mode separately. With the equations and boundary conditions the same, homogeneous fluid solutions such as those found in the previous chapter apply, describing the response to forcing in each mode separately in a continuously stratified sea. Each modal solution yields a pressure field characterized by isopycnal displacements $z_k(x, y, z, t)$, proportional to the vertical distribution $Q_k(z)$, and a velocity field proportional to $\mathrm{d}Q_k/\mathrm{d}z$ (Equation (3.7)). The composite field of pressure and velocity then follows by summing over all mode numbers k, provided that the series in Equation (3.7) converge.

It is generally taken for granted that this reduction of stratified fluid behavior to a set of well known equations turns the determination of stratified fluid response to forcing into a perhaps laborious but in principle trivial exercise. The behavior of each mode is often viewed in isolation on the tacit or explicit supposition that a superposition of the first two or three modes is likely to give a sufficiently detailed description of motions in a stratified sea. One should be wary of such an automatic assumption, recalling the results of Section 2.3 on the excitation of seiches: the sum of an infinite series of normal modes may well describe a behavior in many respects quite unlike the fundamental mode or the lowest few harmonics. In the stratified flow case specifically, it has already been pointed out that forcing by sudden wind gives rise to force coefficients ϕ_k dropping relatively slowly with mode number. Without further investigation it is certainly not clear how an infinite series of normal modes would combine into a resultant response, for some specific application of external forcing.

3.5.1. Sudden Longshore Wind

Consider, therefore, one of the simple forced motion problems discussed in the previous chapter, that of a horizontally semi-infinite sea bounded by a straight infinite coast, and acted upon by a sudden longshore wind (Section 2.5). The fluid column is now supposed stratified. Transposing the solution found in Section 2.5, the modal transport equations, (3.14), with forcing modelled as described by Equation (3.22), have solutions containing a non-oscillatory part and inertial oscillations. The non-oscillatory solution for each mode may be written down as:

$$Z_k = \phi_k \frac{u_*^2}{fc_k} (ft)\, e^{x/R_k},$$

$$U_k = \phi_k \frac{u_*^2}{f} \left(1 - e^{x/R_k}\right), \qquad (3.23)$$

$$V_k = \phi_k\, u_*^2 t e^{x/R_k}.$$

It should be recalled that $x \leqslant 0$ is the region of interest, the wind stress is ρu_*^2, applied along positive y at $t = 0$, and $R_k = c_k/f$ is the radius of deformation of each mode. With $f = 10^{-4}\ \mathrm{s}^{-1}$ the values of R_k are, for the first four modes of the example discussed previously (c_k were quoted in Section 3.3):

R_1	270	km
R_2	5.10	km
R_3	1.56	km
R_4	0.832	km

The difference between R_1 and the other radii of deformation is especially large. The resultant response of the stratified water column is given by the sum of all the normal mode contributions, i.e.,

$$\zeta = u_*^2 t \sum_{k=1}^{\infty} \frac{\phi_k}{c_k} Q_k(z)\, e^{x/R_k},$$

$$u = \frac{u_*^2}{fH} \sum_{k=1}^{\infty} \phi_k\, H \frac{dQ_k}{dz} \left(1 - e^{x/R_k}\right), \tag{3.24}$$

$$v = \frac{u_*^2 t}{H} \sum_{k=1}^{\infty} \phi_k\, H \frac{dQ_k}{dz}\, e^{x/R_k}.$$

3.5.2. The Far Field

At distances from the coast large compared to all R_k, including R_1, the exponentials vanish, so that level displacements ζ and longshore velocities v are zero. The cross-shore velocity is, however,

$$u = \frac{u_*^2}{fH} \sum_{k=1}^{\infty} \phi_k\, H \frac{dQ_k}{dz}, \qquad (-x \gg R_1). \tag{3.25}$$

In the absence of pressure gradients, the applied wind stress at these distances must be balanced by the Coriolis force of cross-shore motion. Since the stress vanishes at the base of the mixed layer by hypothesis, this implies

$$\int_{-\alpha H}^{0} u\, dz = \frac{u_*^2}{f}, \qquad (-x \gg R_1). \tag{3.26}$$

$$\int_{-H}^{-\alpha H} u\, dz = 0,$$

The frictionless stratified flow model yields constant velocity in the mixed layer, because for each mode $H(dQ_k/dz) = 1.0$, constant from the surface to $z = \alpha H$. To reconcile this with physical reality, it is necessary to interpret the calculated velocities as depth-averages, as already pointed out. The actual interior velocities may be regarded as composed of pressure field-induced and frictional contributions, on the principle discussed in Section 2.7:

$$u = u_1(t) + u_2(z, t) \tag{3.27}$$

The frictional component is subject to the pair of Equations (2.59), but now with the boundary conditions

$$\begin{aligned} K \frac{\partial u_2}{\partial x} &= 0, \qquad K \frac{\partial v_2}{\partial z} = F_y, \qquad (z = 0) \\ K \frac{\partial u_2}{\partial x} &= 0, \qquad K \frac{\partial v_2}{\partial z} = 0, \qquad (z = -\alpha H). \end{aligned} \tag{3.28}$$

This is a surface Ekman layer problem with the modification that the zero stress condition is imposed at the finite depth $z = -\alpha H$, instead of at infinity. The solution of the time-dependent Ekman layer evolution problem, with this lower boundary condition, has been discussed in detail by Gonella (1971), its application to a turbulent mixed layer by Csanady (1972a). The next section gives details on the distribution of frictional velocities within the mixed layer. Continuing here the general analysis, one observes that the depth integral of the cross-shore frictional velocity is, in virtue of the boundary conditions (3.28):

$$\int_{-\alpha H}^{0} u_2 \, \mathrm{d}z = \frac{u_*^2}{f}. \tag{3.29}$$

This is the same as applies to the total non-oscillatory velocity, Equation (3.26). One concludes that at large distances from the coast u_1 and v_1 both vanish, or more precisely that any pressure-field induced velocities are oscillatory, so that $u = u_2(z)$, $v = v_2(z)$ are the total non-oscillatory velocities. The far-field velocity distribution is thus seen to be almost identical with what was found for a homogeneous fluid column (except for the distribution of frictional velocities near the free surface, see below).

The vanishing of the pressure-field induced (non-oscillatory) velocities at $-x \to \infty$ implies the following values for the sums of the series involved:

$$\begin{aligned} &\sum_{k=1}^{\infty} \phi_k = \alpha^{-1}, \\ &\sum_{k=1}^{\infty} \phi_k \left(H \frac{\mathrm{d}Q_k}{\mathrm{d}z} \right) = 0, \qquad (z < -\beta H). \end{aligned} \tag{3.30}$$

Applied to the bottom mixed layer the latter series is also

$$\sum_{k=1}^{\infty} \phi_k B_k = 0, \qquad (z < -\beta H). \tag{3.30a}$$

It may be verified directly that this sum approaches zero already for the first four modes of the example for which values of ϕ_k and B_k were given above.

3.5.3. The Stratification-limited Ekman Layer

The distribution of the frictional velocity components $u_2(z, t)$, $v_2(z, t)$ in a mixed layer of depth $h = \alpha H$, with a zero stress boundary condition at the top of the pycnocline, and wind stress suddenly applied at the surface, may be calculated by solving Equations (2.59), subject to boundary conditions (3.28). The result consists of oscillatory and non-oscillatory components, the latter being a solution of (2.59) with time dependent terms deleted. This solution is most easily written down as a complex velocity vector:

$$w_2 = u_2 + iv_2 \tag{3.31}$$

which is subject to the equation and boundary conditions

$$\begin{aligned} &\frac{d^2 w_2}{dz^2} - i\frac{f}{K} w_2 = 0, \\ &K\frac{dw_2}{dz} = iu_*^2, \qquad (z = 0) \\ &K\frac{dw_2}{dz} = 0, \qquad (z = -h). \end{aligned} \tag{3.32}$$

After a few simple calculations one finds:

$$\begin{aligned} &\frac{w_2}{u_*} = \frac{u_*}{fD}\left\{A \exp\left[(i+1)\frac{z}{D}\right] + B \exp\left[-(i+1)\frac{z}{D}\right]\right\}, \\ &A = (i+1)\frac{\exp[2(i+1)h/D]}{\exp[2(i+1)h/D] - 1}, \\ &B = (i+1)\frac{1}{\exp[2(i+1)h/D] - 1}, \end{aligned} \tag{3.33}$$

where $D = \sqrt{2K/f}$ is again the Ekman depth. The distribution $w_2(z)$ depends on the parameter h/D. For large h/D, $A = i + 1$, $B = 0$ and the classical Ekman spiral is recovered. For small H/D, $A \cong B \cong D/2h$ and

$$\frac{w_2}{u_*} \cong \frac{u_*^2}{fh} \cong \text{constant}, \quad \left(\frac{h}{D} \ll 1\right). \tag{3.34}$$

This is the limit of very large eddy viscosity, when the mixed layer is so well stirred that velocity gradients vanish and the frictional velocity is equal to its depth-average value, which points 90° to the right of the wind, i.e., is real in the chosen coordinates.

The changes in the velocity distribution with reducing values of h/D are discernible in a hodograph of the velocity vector at the surface and at the top of the thermocline, $w_2(0)$ and $w_2(-h)$. Such a hodograph was given by Gonella (1971), reproduced here in Figure 3.4. The numbers on the hodograph give the endpoints of the nondimensional

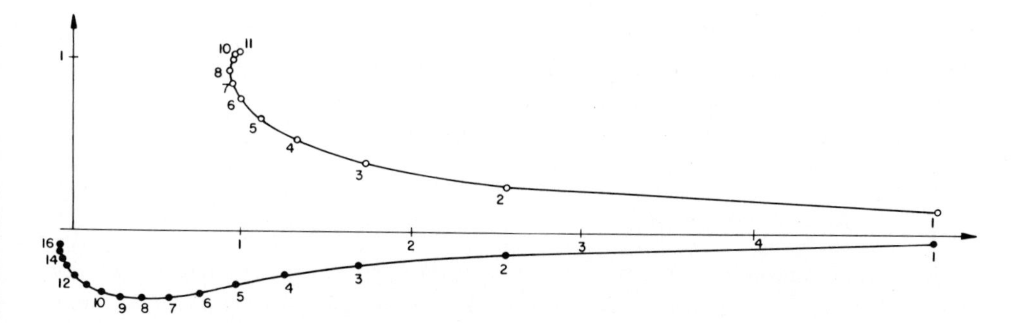

Fig. 3.4. Endpoint of velocity vector at the surface (top curve) and at the pycnocline (bottom curve), for various ratios of mixed layer depth h to Ekman depth D. The numbers shown on the curves are $n = (16/\pi)\ (h/D)$. The velocity for given h/D points from the origin to the marked points (from Gonella, 1971). Units on the axes are u_*^2/fD, so that one unit is typically equal to $7u_*$.

velocity vector at the surface and the top of thermocline (the { } bracketed expression in Equation (3.33)), for given h/D, the number actually shown being $n = (16/\pi)h/D$. At $h/D = 1$, for example, this number is about 5, and one can see that the velocity vector has an x-component u_2 about equal to the depth average value throughout the mixed layer, with a modest shear in the wind direction.

Empirical data on the velocity distribution in such stratification-limited mixed layers have been used to infer the value of the eddy viscosity (Csanady, 1972a). They were found to be given by an expression of the form

$$K = \frac{u_* h}{Re}, \tag{3.35}$$

where Re is an eddy Reynolds number. The value of this was about 12 in Gonella's (1971) observations in the Mediterranean, about 20 in Lake Huron (Csanady, 1972a). The key nondimensional parameters in the above expression become, with K as per Equation (3.35):

$$\frac{h}{D} = \left(\frac{Re}{2}\frac{fh}{u_*}\right)^{1/2}, \qquad \frac{u_*}{fD} = \left(\frac{Re}{2}\frac{u_*}{fh}\right)^{1/2}. \tag{3.36}$$

At such typical values of mixed layer depth and wind stress as $h = 20$ m, $u_* = 1$ cm s^{-1}, and at mid-latitudes, $f = 10^{-4}$ s^{-1}, these parameters have values of about $h/D = 1.4$ and $u_*/fD = 7$. The Ekman spiral corresponding to these numbers may be visualized by filling in between points labeled $n = 8$ in Gonella's diagram. Near the free surface, there is not much difference compared to the classical Ekman solution, but the velocity at the top of the thermocline is relatively large, about $3u_*$.

After this detour it is appropriate to return to other aspects of the response of a stratified fluid to forcing.

3.4.5. Intermediate Distances

The modal solution (3.23) shows that the pressure field associated with the barotropic mode $k = 0$ is trapped within a band of scale width R_1, which is much larger than the trapping width of any of the other modes. Thus there is a range of 'intermediate' distances from the coast characterized by

$$R_2 \ll -x \ll R_1.$$

At such intermediate distances from the coast the exponential containing R_1 is unity, while the other exponentials vanish as before. The barotropic cross-shore velocity amplitude is uniform, $H(\mathrm{d}Q_1/\mathrm{d}z) \cong 1.0$ at all z, while the force coefficient is also unity, $\phi_1 = 1.0$. The surface level displacements and velocities due to all modes summed are thus

$$\left.\begin{aligned} \zeta &= \frac{u_*^2 t}{c_1}\left(1 + \frac{z}{h}\right), \\ u &= \frac{u_*^2}{fH} \sum_{k=2}^{\infty} \phi_k \left(H \frac{\mathrm{d}Q_k}{\mathrm{d}z}\right), \\ v &= \frac{u_*^2 t}{H}, \end{aligned}\right\} \quad (R_2 \ll -x \ll R_1). \tag{3.37}$$

Observe that the sum in the second equation is over all baroclinic modes only. Using the value of the infinite sums determined before (Equation (3.30)), the cross-shore velocity is also

$$\begin{aligned} u &= (\alpha^{-1} - 1)\frac{u_*^2}{fH}, \quad (z \geqslant -\alpha H) \\ u &= -\frac{u_*^2}{fH}, \qquad (z < -\alpha H). \end{aligned} \tag{3.37a}$$

Again, as before, the mixed layer velocity is to be interpreted as an average value. The mixed-layer average of the frictional cross-shore velocity, u_2, is $u_*^2/f\alpha H$, so that the pressure-field indiced cross-shore velocity is seen to be

$$u_1 = -\frac{u_*^2}{fH} \tag{3.38}$$

at all levels. This is a constant adjustment drift, continuously generating longshore velocity $v_1 = u_*^2 t/H$, according to Equation (2.58). In the mixed layer a frictional longshore velocity $v_2(z)$ is superimposed, the average value of which is zero, consistent with Equation (3.37): the average total velocity in the mixed layer is the same as v_1. The surface level rise is what is needed for geostrophic balance. These results are essentially identical with those found in a homogeneous fluid at distances $-x \ll R_1$, i.e., 'close' to the coast on the distance scale of the radius of deformation R_1. Any differences lie again in the precise structure of the frictional velocities, which are the same as at large distances, discussed in the previous section. Stratification at these distances has no effect on the response of the fluid column to suddenly applied wind.

3.5.5. Conditions at the Coast

At the coast, $x = 0$, all exponentials are unity, and one finds on summing all modes $k = 1$ to ∞:

$$\left.\begin{aligned} \zeta &= u_*^2 t \sum_{k=1}^{\infty} \frac{\phi_k}{c_k} Q_k(z), \\ u &= 0, \\ v &= \frac{u_*^2 t}{H} \sum_{k=1}^{\infty} \phi_k \left(H \frac{\mathrm{d}Q_k}{\mathrm{d}z} \right), \end{aligned}\right\} \quad (x = 0). \tag{3.39}$$

The sum giving v at $x = 0$ is the same as found above for u at $x \to \infty$. From the above results therefore

$$\left.\begin{aligned} v &= \frac{u_*^2 t}{\alpha H}, & &(0 \geqslant z \geqslant -\alpha H) \\ v &= 0, & &(-\alpha H > 0) \end{aligned}\right\} \quad (x = 0). \tag{3.39a}$$

In other words, the impulse of the wind is distributed over the mixed layer only, the deeper levels remaining stangnant. The physical explanation for the vanishing of longshore velocity right at the coast, below the mixed layer, is that in the absence of cross-shore velocity there is no longshore force to accelerate the fluid, the Coriolis force fu vanishing on account of the coastal constraint at all depths.

The abrupt change in longshore velocity across the bottom of the mixed layer leads to some difficulties: cross-shore momentum balance should be geostrophic, so that $\partial p'/\partial x$ should also change abruptly between the mixed layer and slightly below. A reference back to Equation (3.5) reveals at once that this is not possible for a realistic distribution of particle displacements $\zeta(z')$, if the density distribution is continuous.

From the first of Equations (3.10) the surface amplitude of the $Q_k(z)$ distribution is:

$$Q_k(0) = \frac{c_k^2}{gH}. \tag{3.40}$$

Correspondingly, the surface level perturbation is:

$$\zeta(0) = \frac{u_*^2 t}{gH} \sum_k \phi_k c_k \, e^{x/R_k}. \tag{3.41}$$

At the coast, $x = 0$, the infinite sum is only slightly greater than c_1, being about 28.6 m s^{-1} in the example, compared with $c_1 = 27.0$ m s^{-1}. Thus the level at the coast rises only a little more than in the case of homogeneous fluid (by an order $\epsilon^{1/2}$ quantity).

The difficulties come to light on calculating interior fluid displacements. Within the surface mixed layer, given unit nondimensional velocity amplitude and the surface displacement (3.40) the displacement amplitude distribution is:

$$Q_k(z) = \frac{c_k^2}{gH} + \frac{z}{H}. \tag{3.42}$$

The displacement of the isopycnal surface forming the bottom of the mixed layer is therefore, from (3.39) with $z/H = -\alpha$:

$$\zeta = \zeta(0) - \alpha u_*^2 t \sum_{k=1}^{\infty} \frac{\phi_k}{c_k} e^{x/R_k}. \tag{3.43}$$

The infinite sum

$$\sum_{k=1}^{\infty} \frac{\phi_k}{c_k}$$

diverges for the stratification-forcing model adopted above. From the algebraic relationships in Sections 3.3 and 3.4 it may be shown that, for large k, ϕ_k varies as k^{-2}, c_k as k^{-1}, hence ϕ_k/c_k also as k^{-1}. Since the terms of the series are all positive, the series diverges, giving unbounded isopycnal displacements at $x = 0$.

At finite negative x, however small, the exponential term e^{x/R_k} suppresses the terms of order greater than some k for which $x \cong R_k$. Thus at $x \neq 0$ the series (3.43) may be summed, although it yields very large displacements at small x. At the coast a singularity arises, the top of the pycnocline moving upward through a very large distance even for a small wind impulse $u_*^2 t$.

At the bottom of the pycnocline, the value of the displacement function $Q_k(-\beta H)$ is most easily calculated from the bottom velocity amplitude B_k, and the boundary condition $Q_k(-H) = 0$:

$$Q_k(-\beta H) = (1 - \beta) B_k. \tag{3.44}$$

Therefore the isopycnal displacement is here

$$\zeta = (1 - \beta) u_*^2 t \sum_k \frac{\phi_k}{c_k} B_k e^{x/R_k}. \tag{3.45}$$

Because B_k alternate between positive and negative values, the infinite series in (3.45) now converges even at $x = 0$. It may be shown that B_k remains of order one for all k, tending to $B = \pm \frac{1}{3}$ at large k. For the example above the series in (3.45) at $x = 0$ sums to about -1.5 m^{-1} s, so that a 'typical' wind impulse $I = u_*^2 t = 3$ m^2 s^{-1} causes a rise or fall of about 2.7 m of the isopycnal forming the bottom of the pycnocline (positive wind stress causing a depression, negative wind a rise).

The singularity at the coast in isopycnal behavior is inconvenient, but is of the kind that often arises in frictionless linearized theory. Physically, the cause of this behavior is that the calculated longshore velocity distribution in the vertical is discontinuous on account of the distribution of wind-imparted momentum over the top layer only. The fluid tries to adjust to geostrophic equilibrium, but a discontinuous velocity distribution would require a discontinuity in the internal pressure, only possible if there is a finite density jump at the base of the mixed layer.

One could continue this discussion by writing down the oscillatory components of the solution, and the comprehensive expressions containing both oscillatory and non-oscillatory parts found in the previous chapter, and summing over all normal modes. The oscillatory solutions are models of the wave-like isopycnal perturbations shown in Figures 3.1 and 3.2 above. However, these oscillations are not directly relevant to the circulation problem and their detailed discussion is outside the scope of this monograph.

Summing up the results of the above calculations, the most outstanding finding is perhaps that the non-oscillatory part of the solution is almost the same as without stratification over most of the semi-infinite domain, excepting only a narrow nearshore band of scale width R_2. Within this band, longshore velocity is generated by a longshore wind only in the surface mixed layer, rather than in the entire fluid column. Adjustment to geostrophic balance, given this velocity distribution, brings about the rise or fall of the pycnocline, commonly known as 'upwelling' or 'downwelling'. When the wind impulse is such that the coast is to the left in the Northern Hemisphere (i.e., when the wind is directed toward negative y, over the semi-infinite field $x \leqslant 0$), an upwelling develops, otherwise a downwelling. The top of the pycnocline moves up or down much more readily than its bottom, showing a tendency to open up the isopycnals during upwelling, to squeeze them together during downwelling. However, the linear model with a mixed layer gives unrealistically large isopycnal movements at the coast, being somewhat over-idealized.

Returning to a question raised earlier, to view the resultant solution as a superposition of a few normal modes is definitely unhelpful, at least as far as the non-oscillatory part of the solution is concerned. If one conceptualized the motions generated by longshore wind as the sum of a barotropic and one or two baroclinic components, all rather different in character, one might very well miss some of the important physics which emerges only on summing all normal mode contributions. One is also left with the feeling that the resolution into an infinite series of normal modes is a rather laborious procedure to arrive at relatively simple results in the end.

3.6. THE TWO-LAYER MODEL

Much of the complexity of linear stratified fluid theory may be avoided by a frequently used further idealization: this is to represent the pycnocline as a sharp interface separating two fluids of slightly different density. A density distribution discontinuous at the pycnocline is a not unreasonable representation of observed density profiles, such as the example shown in Figure 3.3. The two-layer model is illustrated in Figure 3.5. In the next few sections a linearized theory of this model will be developed, on the hypothesis that both surface and interface displacements, ζ and ζ' are small compared to the equilibrium

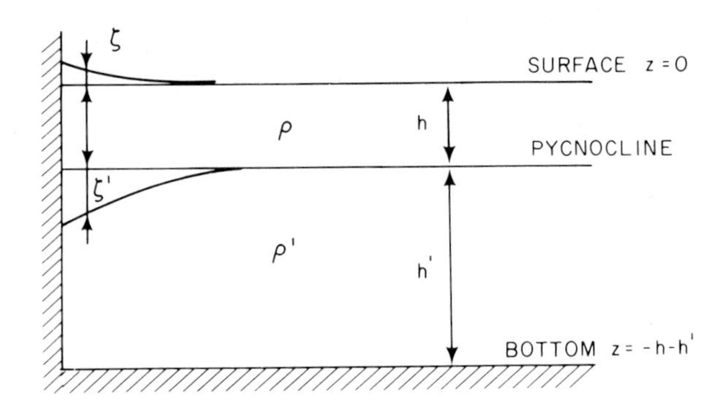

Fig. 3.5. Definition sketch for two-layer model showing displacements from equilibrium level of surface, ζ, and pycnocline, ζ'.

depth of either layer, h or h'. The proportionate density defect of the top layer:

$$\epsilon = \frac{\rho' - \rho}{\rho'}$$

is in practice small, of order 10^{-3}, a fact which will be used to simplify some algebraic results.

The development of linearized theory for the two-layer model follows ideas familiar from earlier sections and needs little detailed discussion. Depth-integrated velocities, i.e., transport components are defined for the two layers separately:

$$U = \int_{-h+\zeta'}^{\zeta} u \, dz, \qquad V = \int_{-h+\zeta'}^{\zeta} v \, dz,$$

$$U' = \int_{-H}^{-h+\zeta'} u \, dz, \qquad V' = \int_{-H}^{-h+\zeta'} v \, dz, \tag{3.46}$$

where $H = h + h'$. The hydrostatic approximation gives the pressure in the top and bottom layers respectively as:

$$p = \rho g(\zeta - z),$$

$$p' = \rho g(\zeta - \zeta' + h) + \rho' g(\zeta' - h - z). \tag{3.47}$$

Interface and bottom friction will be supposed to be zero. A depth-integration of the equations of motion and continuity for top and bottom layer separately now leads to the following set of six linearized transport equations:

$$\begin{aligned}
\frac{\partial U}{\partial t} - fV &= -gh\frac{\partial \zeta}{\partial x} + F_x,\\
\frac{\partial V}{\partial t} + fU &= -gh\frac{\partial \zeta}{\partial y} + F_y,\\
\frac{\partial U}{\partial x} + \frac{\partial V}{\partial y} &= -\frac{\partial}{\partial t}(\zeta - \zeta'),\\
\frac{\partial U'}{\partial t} - fV' &= -gh'\frac{\rho}{\rho'}\frac{\partial \zeta}{\partial x} - \epsilon gh'\frac{\partial \zeta'}{\partial x},\\
\frac{\partial V'}{\partial t} + fU' &= -gh'\frac{\rho}{\rho'}\frac{\partial \zeta}{\partial y} - \epsilon gh'\frac{\partial \zeta'}{\partial y},\\
\frac{\partial U'}{\partial t} + \frac{\partial V'}{\partial y} &= -\frac{\partial \zeta'}{\partial t}.
\end{aligned} \tag{3.48}$$

Note that these equations for top and bottom layers are coupled, ζ' appearing in the first set of three, ζ in the second set. In analogy with normal modes in the continuously stratified case, sets of three uncoupled transport equations are sought. A linear combination of the first and second sets of three equations above gives rise to equations of motion and continuity of the following form:

$$\begin{aligned}
&\frac{\partial}{\partial t}(aU + bU') - f(aV + bV') = -gh\frac{\partial}{\partial x}\left(a\zeta + b\frac{h'\rho}{h\rho'}\zeta + b\frac{h'}{h}\epsilon\zeta'\right),\\
&\frac{\partial}{\partial x}(aU + bU') + \frac{\partial}{\partial y}(aV + bV') = -\frac{\partial}{\partial t}(a\zeta - a\zeta' + b\zeta'),
\end{aligned} \tag{3.49}$$

where a, b are arbitrary constants. A set of such equations for the linear transport combinations $(aU + bU')$, $(aV + bV')$ will be of the form of the homogeneous transport equations if the pressure variables on the right of the equations of motion and of the equation of continuity are the same, except for an arbitrary constant factor β:

$$\beta(a\zeta - a\zeta' + b\zeta') = a\zeta + b\frac{h'\rho}{h\rho'}\zeta + b\frac{h'}{h}\epsilon\zeta'. \tag{3.50}$$

This is an identity and the coefficients of both ζ and ζ' have to agree separately:

$$\begin{aligned}
\beta a &= a + \frac{\rho h'}{\rho' h}b,\\
-\beta a + \beta b &= \frac{h'}{h}\epsilon b.
\end{aligned} \tag{3.51}$$

These two homogeneous equations for a, b are compatible only if their determinant vanishes, which is the case if

$$\beta^2 - \beta\left(\frac{h'}{h} + 1\right) + \frac{h'}{h}\epsilon = 0. \tag{3.52}$$

This is called 'Stokes' equation' by Lamb (1932). Its two roots are

$$\beta_{1,2} = \frac{h+h'}{2h}\left[1 \pm \left(1 - \frac{4hh'\epsilon}{(h+h')^2}\right)^{1/2}\right] \tag{3.53}$$

For the small values of ϵ of interest the two roots may be expanded as

$$\begin{aligned}\beta_1 &= 1 + \frac{h'}{h} - \epsilon\frac{h'}{h+h'} + 0(\epsilon^2),\\ \beta_2 &= \epsilon\frac{h'}{h+h'} + 0(\epsilon^2).\end{aligned} \tag{3.53a}$$

One of Equations (3.51) may now be used to determine the ratio of the two constants a and b. This is

$$\frac{a}{b} = 1 - \frac{h'\epsilon}{h\beta}. \tag{3.54}$$

One of the constants a, b may be chosen arbitrarily. The choice amounts to a normalization of the combined transport variables. The ratio a/b has two values corresponding to the two roots β_1, β_2. The product of the two roots is, from Stockes' equation, (3.52):

$$\beta_1\beta_2 = \frac{h'}{h}\epsilon \tag{3.55}$$

so that the amplitude ratios are

$$\frac{a_1}{b_1} = 1 - \beta_2, \qquad \frac{a_2}{b_2} = 1 - \beta_1. \tag{3.56}$$

Substituting the values of the roots, one finds to order ϵ:

$$\frac{a_1}{b_1} = 1 - \epsilon\frac{h'}{h+h'}, \qquad \frac{a_2}{b_2} = -\frac{h'}{h} + \epsilon\frac{h'}{h+h'}.$$

The two sets of independent transport equations corresponding to the solutions β_1 and β_2 are

$$\begin{aligned}\frac{\partial U_k}{\partial t} - fV_k &= -g\beta_k h\frac{\partial\zeta_k}{\partial x} + F_{xk},\\ \frac{\partial V_k}{\partial t} + fU_k &= -g\beta_k h\frac{\partial\zeta_k}{\partial y} + F_{yk}, \qquad (k = 1, 2)\\ \frac{\partial U_k}{\partial x} + \frac{\partial V_k}{\partial y} &= -\frac{\partial\zeta_k}{\partial t},\end{aligned} \tag{3.58}$$

where

$$\begin{aligned}U_k &= a_k U + b_k U',\\ V_k &= a_k V + b_k V',\end{aligned}$$

$$\zeta_k = a_k\zeta + (b_k - a_k)\zeta',$$
$$F_{xk} = a_k F_x,$$
$$F_{yk} = a_k F_y.$$

Equations (3.58) are two normal mode equations in every way analogous to the normal modes of a continuously stratified fluid, with the simplification that there are now only two of them.

3.6.1. The Surface or Barotropic Mode

The structure of either one of the modes may be elucidated easily from the results already obtained. A convenient choice for the first mode is $b_1 = 1$, which gives with (3.57)

$$\begin{aligned} U_1 &= U + U', \\ V_1 &= V + V', \\ \zeta_1 &= \zeta, \\ F_{x1} &= F_x, \\ F_{y1} &= F_y, \end{aligned} \qquad (3.59)$$

neglecting quantities of order ϵ where they occur together with quantities of order unity, supposing that ζ' is of the same order as ζ, and that transports in both layers are also of the same order. These suppositions may be verified by noting that if the motion of the two-layer model is entirely in the first mode, the second-mode transports and elevations vanish. Setting $U_2 = V_2 = 0$ and $\zeta_2 = 0$, and using Equation (3.57) again one easily finds, for motions in mode $k = 1$ only, neglecting $0(\epsilon)$ quantities:

$$U' = \frac{h'}{h} U, \quad V' = \frac{h'}{h} V, \quad \zeta' = \frac{h'}{h+h'}\zeta, \quad (U_2 = V_2 = 0,\ \zeta_2 = 0). \qquad (3.60)$$

Thus the velocities U/h and U'/h', V/h and V'/h' are the same in this mode, and the interface moves a little less than the surface, consistent with a linear distribution of particle displacements throughout the water column. This is the same as in a homogeneous fluid (true now to order ϵ). The equations describing this mode are, substituting the variables (3.59) into the modal equations (3.58):

$$\begin{aligned} \frac{\partial}{\partial t}(U+U') - f(V+V') &= -\,gH\frac{\partial\zeta}{\partial x} + F_x, \\ \frac{\partial}{\partial t}(V+V') + f(U+U') &= -\,gH\frac{\partial\zeta}{\partial y} + F_y, \\ \frac{\partial}{\partial x}(U+U') + \frac{\partial}{\partial y}(V+V') &= -\frac{\partial\zeta}{\partial t}, \end{aligned} \qquad (3.61)$$

which are indeed identical with the homogeneous transport equations, recalling that $U+U'$ and $V+V'$ are now total transports.

3.6.2. The Internal or Baroclinic Mode

To elucidate the structure of the second mode, suppose that motions in the first mode vanish. This gives

$$U' = -U, \qquad V' = -V$$

$$\zeta = -\epsilon \frac{h+h'}{h'} \zeta', \tag{3.62}$$

$$(U_1 = V_1 = 0, \; \zeta_1 = 0)$$

to the lowest order of ϵ in each case. The surface and interface elevations are now of a different order, while transports in the two layers are equal and opposite. A convenient choice of the constant b_2 is

$$b_2 = \frac{h'}{h+h'} \tag{3.63}$$

which yields the second mode variables, using the relationships (3.62):

$$U_2 = U',$$

$$V_2 = V',$$

$$\zeta_2 = \zeta',$$

$$F_{x2} = -\frac{h'}{h+h'} F_x,$$

$$F_{y2} = -\frac{h'}{h+h'} F_y.$$

The equations governing motions in this mode are, substituting the chosen variables into the modal equations (3.58):

$$\frac{\partial U'}{\partial t} - fV' = -\epsilon g \frac{hh'}{h+h'} \frac{\partial \zeta'}{\partial x} - \frac{h'}{h+h'} F_x,$$

$$\frac{\partial V'}{\partial t} + fU' = -\epsilon g \frac{hh'}{h+h'} \frac{\partial \zeta'}{\partial y} - \frac{h'}{h+h'} F_y, \tag{3.65}$$

$$\frac{\partial U'}{\partial x} + \frac{\partial V'}{\partial y} = -\frac{\partial \zeta'}{\partial t}.$$

Physically, these equations may be regarded almost as equations for the bottom layer alone, the interface being the effective upper boundary. The pressure term and the forcing term are, however, not quite the same as would follow from such a simple notion. The presence of the small quantity ϵ in the pressure term means that large interface displacements ζ' are necessary if this term is to play a significant role. The negative sign of the forcing term means that the interface tends to set up in a direction opposite to the free surface.

It should be remembered, of course, that Equations (3.61) and (3.65) describe the excitation of motions in the surface and internal modes separately, supposing in both cases the other mode dormant. Transports U or U' etc. as determined from these two sets of equations in any forced motion problem have to be added in the end. Equation (3.61) yields, for example, total transports $U+U'$, $V+V'$, while Equation (3.60) specifies how this is distributed in the water column. From (3.60) and (3.61) one obtains therefore the surface mode contributions to U and U'. Similarly, (3.65) yields the internal mode contributions to U' and V' directly, while (3.62) shows that surface layer transports are equal and opposite, specifying U and V as arising from the excitation of the internal mode. The total transports are obtained by simple algebraic addition, the original equations (3.48) being linear.

3.7. IMPULSIVE LONGSHORE WIND

To illustrate the strengths and weaknesses of the two layer model the same problem will be discussed next that was analyzed earlier with the aid of the continuously stratified fluid model. A wind stress ρu_*^2 along positive y is suddenly imposed at $t = 0$, over a semi-infinite ocean occupying $x < 0$. The excitation of the barotropic mode is governed by Equation (3.61) and the non-oscillatory part of the solution is the same as discussed in Chapter 2, noting that total transports are now $U + U'$, $V + V'$:

$$\begin{aligned} \zeta &= \frac{u_*^2 t}{c_1} e^{x/R_1}, \\ U + U' &= \frac{u_*^2}{f} (1 - e^{x/R_1}), \\ V + V' &= u_*^2 t e^{x/R_1}, \end{aligned} \tag{3.66}$$

with $c_1 = (gH)^{1/2}$, $R_1 = c_1/f$. The baroclinic mode responds in accordance with Equation (3.65) and yields interface displacements and bottom layer transports as follows (again, the non-oscillatory contributions to these quantities only being considered):

$$\begin{aligned} \zeta' &= -\frac{u_*^2 t}{c_1}\left(\frac{h'}{h+h'}\right) e^{x/R_2}, \\ U' &= -\frac{u_*^2 h'}{f(h+h')}\left(1 - e^{x/R_2}\right), \\ V' &= -u_*^2 t \frac{h'}{h+h'} e^{x/R_2}, \end{aligned} \tag{3.67}$$

with $c_2 = \left(\epsilon g h h'/(h+h')\right)^{1/2}$, $R_2 = c_2/f$. Noting the earlier rules on the relationship of top and bottom layer transports, surface and interface displacements, in the two modes, the combined response is easily written down:

$$\begin{aligned}
\zeta &= \frac{u_*^2 t}{c_1}\left[e^{x/R_1} + \left(\frac{\epsilon H^2}{hh'}\right)^{1/2} e^{x/R_2}\right],\\
U &= \frac{u_*^2}{f}\left[1 - \frac{h}{h+h'}\, e^{x/R_1} - \frac{h'}{h+h'}\, e^{x/R_2}\right],\\
V &= \frac{h}{h+h'}\, u_*^2 t\left[e^{x/R_1} + \frac{h'}{h}\, e^{x/R_2}\right],\\
\zeta' &= -\frac{u_*^2 t}{c_2}\left(\frac{h'}{h+h'}\right)\left[e^{x/R_2} + \left(\frac{\epsilon h'}{h}\right)^{1/2} e^{x/R_1}\right],\\
U' &= \frac{u_*^2 h'}{f(h+h')}\left[e^{x/R_2} - e^{x/R_1}\right],\\
V' &= u_*^2 t\,\frac{h'}{h+h'}\left[e^{x/R_1} - e^{x/R_2}\right].
\end{aligned} \tag{3.68}$$

Far from the coast, $-x \gg R_1$, five of the six variables above vanish, while the top layer transport U becomes equal to the Ekman transport. At intermediate distances, $R_2 \ll -x \ll R_1$, one finds only barotropic mode contributions to ζ, ζ', V, and V', and cross-shore transports of

$$\begin{aligned}
U &= \frac{u_*^2}{f}\left(\frac{h'}{h+h'}\right),\\
U' &= -\frac{u_*^2}{f}\left(\frac{h'}{h+h'}\right),
\end{aligned} \qquad (R_2 \ll -x \ll R_1). \tag{3.69}$$

This may be viewed as Ekman transport u_*^2/f through frictional velocities in the surface layer, compensated for exactly by adjustment drift at constant velocity $u_*^2/f(h+h')$, throughout the water column, just as discussed in the homogeneous fluid case close to the coast, $-x \ll R_1$. The distribution of frictional velocities in the surface layer may be modified by a shallow pycnocline as discussed in Section 3.5.3, but otherwise the structure of the flow is the same as without stratification.

Sufficiently close to the coast, $-x \ll R_2$, cross-shore transports vanish in both layers, as does longshore transport in the bottom layer. Surface layer transport is

$$V = u_*^2 t, \qquad (-x \ll R_2) \tag{3.70}$$

all of the wind-imported momentum being used to accelerate the surface layer alongshore. The pycnocline displacement is, neglecting an order $\epsilon^{1/2}$ quantity:

$$\zeta' = -\frac{u_*^2 t}{c_2}\left(\frac{h'}{h+h'}\right), \qquad (-x \ll R_2). \tag{3.70a}$$

Given a typical wind-stress impulse $I = u_*^2 t = 3\ \mathrm{m^2\ s^{-1}}$ (0.1 Pa wind acting for about 10 hr) and $c_2 = 0.5\ \mathrm{m\ s^{-1}}$, the typical pycnocline displacement is of the order of 5 m, downward for positive F_y, upward for negative F_y. The surface level displacement is what is required for geostrophic balance, and differs from the homogeneous fluid value $u_*^2 t/c_1$ by a quantity of order $\epsilon^{1/2}$, which is only a few percent.

The two layer model clearly yields most of the physical characteristics of the response to wind of a stratified fluid column already determined from the continuously stratified model. In one technical respect the two layer model is superior: it predicts finite pycnocline displacements at the coast. Its major deficiency is that it does not come to grips with motions within the pycnocline, the opening-up of the isopycnals in an upwelling, their squeezing together in a downwelling. This deficiency is largely removed by a three-layer model, which may be developed on lines similar to those above, at the cost of some increase in complexity.

Unless motions of intermediate density fluid are of particular interest in a given problem, the simplicity of the two-layer model proves to be a decisive advantage. This model is used in the remaining sections of this chapter to discuss further the response of a stratified sea to forcing by wind.

3.8. CROSS-SHORE WIND

The case of a wind blowing perpendicular to a straight coast over a two-layer ocean may be treated similarly. Transposing the solution found in Section 2.4, the non-oscillatory motions are described by, in the barotropic mode

$$\begin{aligned} \zeta &= \frac{u_*^2}{fc_1}\, e^{x/R_1}, \\ U + U' &= 0, \\ V + V' &= -\frac{u_*^2}{f}\,(1 - e^{x/R_1}), \end{aligned} \tag{3.71}$$

with $c_1 = [g(h+h')]^{1/2}, R_1 = c_1 f^{-1}$. The baroclinic mode contributions are:

$$\begin{aligned} \zeta' &= -\frac{h'}{h+h'}\frac{u_*^2}{fc_2}\, e^{x/R_2}, \\ U' &= 0, \\ V' &= \frac{h'}{h+h'}\frac{u_*^2}{f}\,(1 - e^{x/R_2}), \end{aligned} \tag{3.72}$$

again with $c_2 = [\epsilon g h h'/(h+h')]^{1/2}, R_2 = c_2 f^{-1}$. The combined solutions are:

$$\begin{aligned} \zeta &= \frac{u_*^2}{fc_1}\left[e^{x/R_1} + \frac{h'}{h+h'}\left(\frac{\epsilon h'}{h}\right)^{1/2} e^{x/R_2}\right], \\ U &= 0, \\ V &= -\frac{u_*^2}{f} + \frac{u_*^2 h}{f(h+h')}\, e^{x/R_1} + \frac{u_*^2 h'}{f(h+h')}\, e^{x/R_2}, \end{aligned}$$

(3.73)

$$\zeta' = -\frac{h'}{h+h'} \frac{u_*^2}{fc_2} \left[e^{x/R_2} - \frac{c_2}{c_1} e^{x/R_1} \right],$$

$$U' = 0,$$

$$V' = \frac{u_*^2 h'}{f(h+h')} \left[e^{x/R_1} - e^{x/R_2} \right].$$

In the far field, $-x >> R_1$, surface and pycnocline elevations and bottom transport vanish. Surface layer transport is Ekman transport only, carried by the frictional contribution to interior velocities. At intermediate distances, $R_2 << -x << R_1$, a constant longshore velocity, $u_*^2/f(h+h')$ is added in both layers, the depth-integrated transport associated with which exactly cancels Ekman transport. The velocity distribution is thus exactly the same as close to shore without stratification, except possibly for a different distribution of frictional velocities in the surface mixed layer. Very close to the coast, $-x << R_2$, the longshore Ekman transport is cancelled by a pressure-field induced transport in the surface layer only. The surface elevation rises by a little more (by an order $\epsilon^{1/2}$ quantity) than in a homogeneous fluid. The pycnocline is depressed at the coast under an onshore wind, raised under an offshore wind, by an amount of order u_*^2/fc_2, which is typically several meters. The pycnocline tilt reduces to negligible values (compared to its nearshore inclination) at distances of order R_2. However, the surface elevation gradient becomes negligible compared to its nearshore value only at distances of order R_1 (*small* pycnocline displacements, of the order of the surface displacement also reach to such larger distances, but they are dynamically unimportant).

The depth-integrated balance of forces at the coast in this case is between wind stress and pressure gradient, the latter acting in the surface layer only. The pressure gradient in the bottom layer vanishes. The depth-integrated Croiolis force also vanishes in each layer separately, there being no net transport in either layer, cross-shore or alongshore. On the cessation of the wind, therefore, the surface and the pycnocline return to their equilibrium position, and no longshore current in geostrophic equilibrium is left over, unlike the case of the longshore wind-stress.

3.9. TWO-LAYER CLOSED BASIN

As in the previous chapter, it is necessary to connect the findings for infinite coast models to somewhat more realistic geometry in order to place those findings in appropriate perspective and discover their limitations. For reasons already given in Section 2.8, a circular basin model is most suitable for this purpose.

A typical two layer circular basin model might have the following physical characteristics:

$$\epsilon = 2 \times 10^{-3},$$

$$h = 15 \text{ m},$$

$$h' = 60 \text{ m},$$

$$f = 10^{-4} \text{ s}^{-1},$$

$$r_0 = 75 \text{ km}.$$

These data will be used in numerical examples below. The two-layer model chosen here is very similar to the 'typical' continuously stratified model used in earlier calculations in this chapter. The quantitative characteristics of the two normal modes in the two-layer model are:

$$c_1 = 27.1 \text{ m s}^{-1}, \qquad R_1 = 271 \text{ km},$$

$$c_2 = 0.49 \text{ m s}^{-1}, \qquad R_2 = 4.9 \text{ km}.$$

These closely parallel the barotropic and first baroclinic mode characteristics in the continuous density model. The typical basin radius r_0 is thus considerably smaller than R_1, but much larger than R_2.

Motions impulsively forced by a suddenly imposed wind, $F_x = u_*^2$ = constant for $t \geqslant 0$ may be described using the results of Section 2.8.1, applied to the barotropic and baroclinic modes separately, Equations (3.61) and (3.65). As in that section, polar coordinates will be used. The excitation of the barotropic mode closely conforms to the behavior of a basin small compared to the radius of deformation, $r_0/R_1 \ll 1$. The only non-oscillatory solution is wind setup, described approximately by the radial amplitude distribution (Equation (2.77), for small r_0/R):

$$A_0(r) \cong \frac{r}{r_0}, \qquad (r_0/R_1 \ll 1). \tag{3.74}$$

The barotropic mode contribution to surface elevations is therefore approximately, from Equation (2.76)

$$\zeta = \frac{u_*^2}{c_1^2}\, r \cos\phi = \frac{u_*^2}{c_1^2}\, x. \tag{3.75}$$

This is the same wind setup as occurs in a rectangular or any other shape basin. The corresponding transports vanish:

$$\begin{aligned} U + U' &= 0, \\ V + V' &= 0. \end{aligned} \tag{3.76}$$

The excitation of the baroclinic mode is very different, and follows the properties of 'large' basins, $r_0/R_2 \gg 1$. Equations (3.65) yield for the non-oscillatory part of the elevation, to first order in R_2/r_0, from Equation (2.86):

$$\zeta' = -\frac{h'}{h+h'}\frac{u_*^2}{c_2^2}\, r_0 \left[\cos\phi - \left(1 - \frac{R_2}{r_0}\right) \times \right.$$
$$\left. \times \cos(\phi - \sigma_1 t)\right] \exp\left(\frac{r-r_0}{R_2}\right) \tag{3.77}$$

with $\sigma_1 \cong f(R_2/r_0)$.

As may be seen from this result, there is no baroclinic contribution to the pycnocline elevation field outside coastal boundary layers of scale width R_2. Equations (3.65) yield then for bottom layer transports near the center of the basin:

$$V_r' = \frac{h'}{h+h'} \frac{u_*^2}{f} \sin \phi, \qquad (r_0 - r >> R_2) \qquad (3.78)$$
$$V_\phi' = \frac{h'}{h+h'} \frac{u_*^2}{f} \cos \phi,$$

which in Cartesian coordinates is simply

$$U' = 0, \qquad (2.78a)$$
$$V' = \frac{h'}{h+h'} \frac{u_*^2}{f}$$

or bottom layer transport to the *left* of the wind stress (for positive f, i.e., in the Northern Hemisphere), and of a magnitude produced by a pressure gradient $\partial\zeta/\partial x = u_*^2/g(h+h')$. Surface layer transports are, outside the coastal boundary layers:

$$U = 0, \qquad (3.79)$$
$$V = -\frac{h'}{h+h'} \frac{u_*^2}{f}.$$

The results in (3.78a) and (3.79) may be interpreted as Ekman transport to the *right* of the wind in the surface layer, exactly compensated for by transport due to the (constant) geostrophic velocity associated with the surface level gradient. Observe that the *combined* flow and pressure fields outside the coastal boundary layers consist of the barotropic mode contribution to surface elevation, and of the baroclinic mode contribution to transports: the barotropic mode has no transport associated with it, the baroclinic mode no pressure gradients. It is only in combination that the results make sense physically.

The pycnocline elevation within the coastal boundary layer given by (3.77), may be regarded as the superposition of a static solution and an 'internal' (baroclinic mode) Kelvin wave. At $t << \sigma_1^{-1}$ the two nearly cancel, except for the residue

$$\zeta' = -\frac{h'}{h+h'} \frac{u_*^2}{fc_2} \left[\cos \phi - ft \sin \phi \right] \exp\left(\frac{r-r_0}{R_2}\right), \qquad (t << \sigma_1^{-1}) \qquad (3.80)$$

Along the diameter parallel to the wind, $\phi = 0$ there is thus a pycnocline elevation and bottom layer transport distribution:

$$\zeta' = -\frac{h'}{h+h'} \frac{u_*^2}{fc_2} \exp\left(\frac{r-r_0}{R_2}\right), \qquad (3.80a)$$
$$V_\phi' = \frac{h'}{h+h'} \frac{u_*^2}{f} \left[1 - \exp\left(\frac{r-r_0}{R_2}\right)\right], \qquad (t << \sigma_1^{-1}, \phi = 0).$$

At distances from the boundary large compared to R_2 the transport V_ϕ' merges into the center value, Equation (3.78). Along the diameter perpendicular to the wind, $\phi = \pi/2$, one finds

$$\zeta' = \frac{h'}{h+h'}\,\frac{u_*^2 t}{c_2}\exp\left(\frac{r-r_0}{R_2}\right),$$
$$V_\phi' = \frac{h'}{h+h'}\,u_*^2 t\exp\left(\frac{r-r_0}{R_2}\right), \tag{3.80b}$$

having neglected quantities of order R_2/r_0 in comparison with unity. Surface layer longshore transports, V_ϕ are equal and opposite to V_ϕ', noting again that the barotropic contributions to transport vanish.

The combined barotropic and baroclinic solutions along the diameter $\phi = 0$, at $t \ll \sigma_1^{-1}$ are thus within the approximations $r_0/R_1 \ll 1$, $r_0/R_2 \gg 1$, identical with the solution found for an infinite coast, subject to a transverse wind, at 'intermediate' distances, $R_2 \ll -x \ll R_1$, and closer to the coast, see Section 3.8. Similarly, along the diameter $\phi = \pi/2$ the solutions are the same as long an infinite coast subject to longshore wind, again at intermediate distances and closer (Section 3.7).

3.9.1. The Internal Kelvin Wave

Differences develop on the time scale $t = \sigma_1^{-1}$. The internal Kelvin wave propagates by this time significantly away from the position where it nearly cancels the wind setup, in somewhat the same way as seiches propagate away and fail to cancel the wind setup in a small basin, see Section 2.3 et seq. The flow and pressure fields then consist of the wind-setup solution and the internal Kelvin wave existing more or less independently. Before analyzing this case quantitatively, it is desirable to discuss the general properties of internal Kelvin waves. When basin radius is large compared to the internal radius of deformation R_2 a straight-coast model is appropriate. The internal Kelvin wave component of the solution (3.77) transcribed to Cartesian coordinates (replacing the angle ϕ by $ky = y/r_0$, k = horizontal wave number, $2\pi k^{-1}$ = wavelength) becomes:

$$\zeta' = a\cos(ky - \sigma_1 t)\exp\left(\frac{x}{R_2}\right), \tag{3.81}$$

where a is the shore amplitude of pycnocline displacement and the domain of interest is $x < 0$. Previous discussion in Section 2.9 has shown that the cross-shore transport vanishes identically in a barotropic Kelvin wave. From Equations (3.65) the same result follows for the internal Kelvin wave. The longshore transport in the bottom layer may be shown to be:

$$V' = c_2\zeta'. \tag{3.82}$$

Thus where the pycnocline rises, V' is positive or points in the cyclonic direction, which is also the direction of Kelvin wave propagation. The top layer transport V is equal in magnitude but opposite in direction.

On differentiating V' one obtains the vorticity tendency balance

$$\frac{\partial V'}{\partial x} = f\zeta' \tag{3.83}$$

which is the same as Equation (2.10), but for the bottom layer of a two-layer basin. It shows that a rising pycnocline generates vorticity by stretching the vortex lines on a rotating earth. Similarly, top layer vorticity is generated by the squashing of vortex lines by a rising pycnocline:

$$\frac{\partial V}{\partial x} = -f\zeta' \tag{3.83a}$$

to order $\epsilon^{1/2}$, neglecting surface elevation changes in the internal mode.

The important difference compared to the homogeneous fluid case is that ζ' is typically much larger than surface elevation changes. For typical values of the parameters in Equation (3.77), $u_*^2 = 1$ cm s^{-1}, $c_2 = 50$ cm s^{-1}, $r_0 = 50$ km, the amplitude of pycnocline displacements is about the same as the typical top layer depth, h. Although the validity of the linear theory becomes doubtful under these circumstances, the results suggest that the vorticity $(\partial/\partial x)\,(V/H)$ is large, of order f in the baroclinic coastal currents accompanying such relatively large amplitude internal Kelvin waves. Peak longshore velocities are of order c_2, or typically 50 cm s^{-1}, decaying to a negligible value within a coastal trapping width of order 5 km. Figure 3.6 illustrates schematically the flow structure of an internal Kelvin wave.

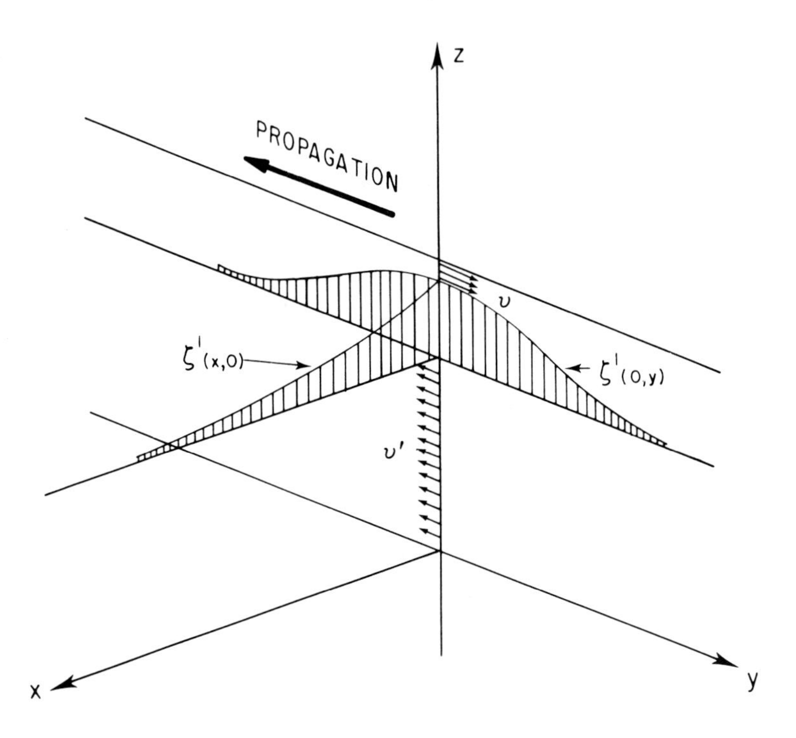

Fig. 3.6. Schematic illustration of internal Kelvin wave along a straight coast. Velocities in top and bottom layer are opposite, and both are in phase with the interface elevation.

3.9.2. Internal Kelvin Wave Propagation

To consider further the development of the flow at times of order σ_1^{-1} after the imposition of wind stress it is possible simply to analyze the expression (3.77). However, because winds do not usually last that long, it is more realistic to suppose that at time $t = T << \sigma_1^{-1}$ the wind suddenly stops acting. Solving the impulsive problem associated with the removal of the wind stress one obtains the same expression as (3.77) (for the quasi-steady flow components), with the opposite sign, and with $(t-T)$ replacing t. Combining the two solutions, due respectively to the imposition and the removal of the wind stress, one arrives at a result which may be written as

$$\zeta' = -\frac{2h'}{h+h}\frac{u_*^2}{c_2^2}(r_0 - R_2)\sin\frac{\sigma_1 T}{2}\sin\left(\phi - \sigma_1 t - \frac{\sigma_1 T}{2}\right)\exp\left(\frac{r-r_0}{R_2}\right). \tag{3.84}$$

Because $\sigma_1 T$ is usually small, and $\sigma_1 \cong fR_2/r_0 = c_2/r_0$, this may be simplified to

$$\zeta' = -\frac{h'}{h+h}\frac{u_*^2 T}{c_2}\left(1 - \frac{\sigma_1}{f}\right)\sin\left(\phi - \sigma_1 t - \frac{\sigma_1 T}{2}\right)\exp\left(\frac{r-r_0}{R_2}\right). \tag{3.84a}$$

The product $u_*^2 T$ may be recognized as the impulse of the wind stress. The typical magnitude of this is 10 m^2 s^{-1}, a 7 m s^{-1} wind blowing for about a day. The amplitude of ζ' according to (3.84a) is then 15 m for a basin of characteristics listed earlier in Section 3.9. After the cessation of the wind, an internal Kelvin wave of large amplitude is thus left over, of a wavelength equal to basin perimeter, propagating cyclonically with speed c_2 around the basin. Initially, maximum downward pycnocline displacement is found to the right of the wind, upward displacement to the left. Corresponding longshore transports are directed downwind in the surface layer along both coastal sections at the end of the diameter perpendicular to the wind. Within a half cycle, $t = \pi\sigma_1^{-1}$, the position of the pycnocline elevation and depression reverses, and longshore transports also reverse direction at fixed points along the coast. The slow development of a coastally trapped pressure field a long time after the application of an impulse is a surprising property of a stratified sea on a rotating earth. Its observational manifestation may well be described as a spectacular event, as will be discussed further in Chapter 5.

Similar results may be obtained for a rectangular basin, with the shorter sides supposed much shorter than the length of the basin, using Taylor's (1920) Kelvin wave reflection model (Csanady and Scott, 1974). When a wind blows along the long axis of the basin, the static setdown of the pycnocline is cancelled initially by a sine series of Kelvin waves, much as wind setup is cancelled by seiches, see Figure 2.4 and discussion in Chapter 2. As the Kelvin waves propagate away from their original position, the pycnocline begins to rise or fall according to location along the coast. On the cessation of the wind a second set of Kelvin waves is started in a somewhat different phase and a combined flat-topped wave remains, propagating back and forth along the basin (see (Figures 3.7 and 3.8). While along the longer coasts this gives much the same picture as the exact solution for the circular basin, the end walls respond in the same phase in the model, as a result of the narrow basin approximation. The exact solution of the impulsively forced problem for a rectangular basin does not seem to have been given in the literature, presumably on account of the clumsiness of the algebra involved.

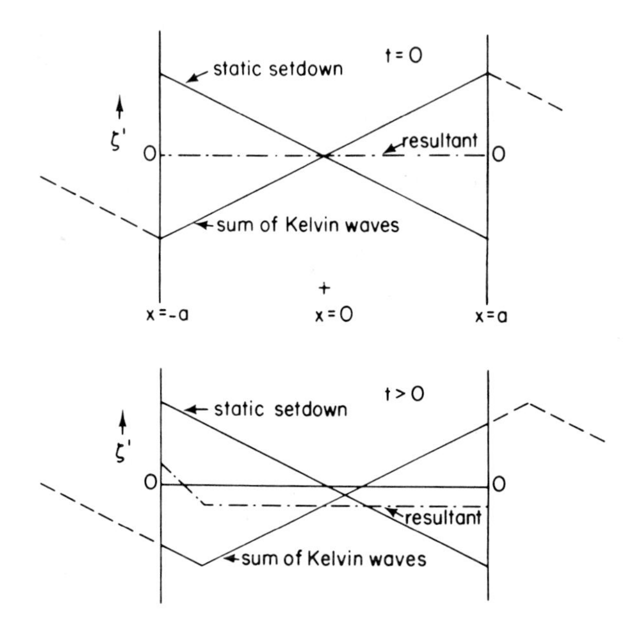

Fig. 3.7. Superposition of wind setup (marked 'static setdown') and sawtooth shape internal Kelvin wave (marked 'sum of [sinusoidal] Kelvin waves') in a long rectangular basin. At the start of the wind impulse (t = 0) the two exactly cancel, leaving a flat thermocline as the resultant (top). At a later time, after the Kelvin wave migrated some distance into the basin, the cancellation is incomplete and the thermocline dips along most of the coast to the right of the wind (bottom). Along most of the opposite shore the thermocline rises, the resultant distribution being similar, except that the Kelvin wave propagates in the opposite (upwind) direction. (From Csanady and Scott (1974).)

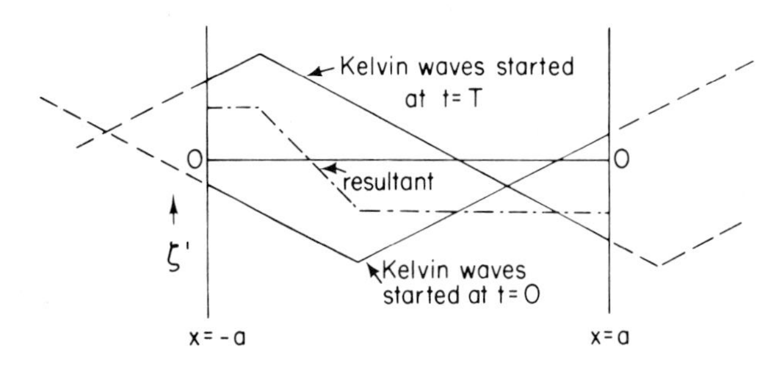

Fig. 3.8. Internal Kelvin waves left over in a rectangular basin after cessation of wind at time T, on the shore to the right of the wind. (From Csanady and Scott (1974).)

3.10. SURFACING OF THE PYCNOCLINE

As is evident from order of magnitude estimates given above, one disturbing weakness remains in the linear theory calculations: the pycnocline displacement induced by longshore wind along a coast quickly outgrows the linearizing assumption $\zeta' \ll h$. In the example given, for $I = u_*^2 T = 3 \text{ m}^2 \text{ s}^{-1}$, $c_2 = 0.5 \text{ m s}^{-1}$, ζ' is already a substantial fraction of h, and a three times stronger wind impulse, no rarity in practice, certainly invalidates the linearized approach, because the calculated ζ' becomes as large as h. Observations following such strong longshore wind impulses show that the pycnocline in fact becomes displaced through vertical distances of the order calculated: with the coast to the left of the wind the developing upwelling results in a surface outcropping and offshore displacement of the pycnocline, Figure 3.9. However, one cannot very well trust the linear theory to show how strong an impulse is required to bring about the surfacing of the pycnocline, nor does one gain any guidance from the theory in regard to the question, how far this surface pycnocline intersection will be displaced from shore.

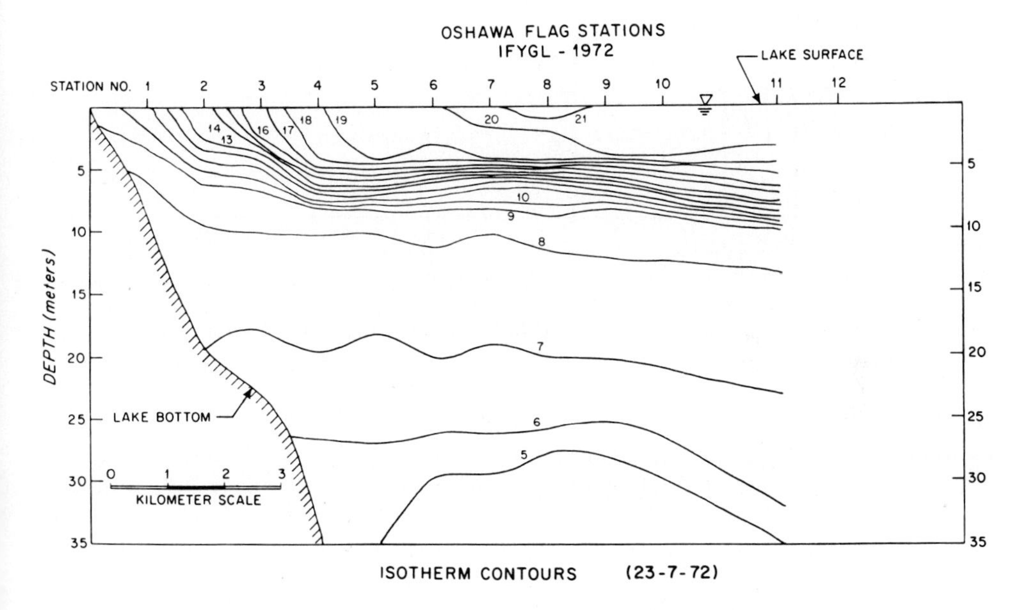

Fig. 3.9. Temperature distribution in coastal zone transect on north shore of Lake Ontario, on an occasion when main pycnocline intersected the surface, following a strong eastward wind stress impulse. (From Csanady (1977b).)

An alternative to the linear theory is the use of the potential vorticity equation derived in Section 1.10, Equation (1.40). In the present application this equation is applied separately to top and bottom layers. Neglecting interface and bottom friction, one supposes the wind stress distributed evenly over the top layer. Furthermore, in order to keep calculations reasonably simple, it is necessary to hypothesize that the wind stress is exerted as a short burst, before the top layer has had time to change depth. A spatially uniform wind and constant top layer depth during the wind impulse results in zero curl of the wind stress force, i.e., a vanishing potential vorticity tendency on the right of Equation (1.40). An impulsively exerted wind stress is clearly an overidealization, but this assumption only affects relatively unimportant details of the solution, as more complex calculations show (Csanady, 1977b). Following the wind stress impulse the two-layer system is allowed to adjust to geostrophic equilibrium: this final state is calculated below, ignoring any inertial oscillations which presumably accompany the adjustment process. The results correspond to those calculated earlier from linear theory for a wind stress acting for a finite period, the final state depending on the impulse (time integral) of the stress only.

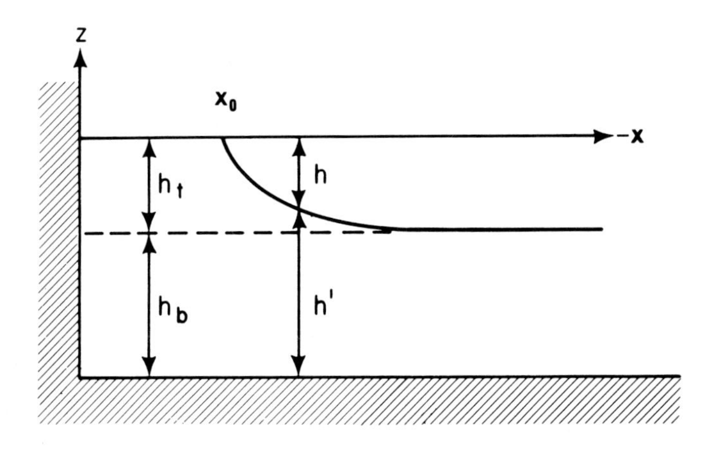

Fig. 3.10. Idealization of fully upwelled pycnocline, using a two-layer model.

An infinite coast, two layer model is considered again, with layer depth changes now comparable to equilibrium depths $h = h_t$ and $h' = h_b$. Figure 3.10 illustrates a two-dimensional, two-layer idealization of the observed transect of Figure 3.9, with the domain of interest being again $x \leqslant 0$ as in previous sections of this chapter for ease of comparison. A sufficiently strong wind impulse exerted within a period so short that top layer depth does not have time to change is supposed to be the cause of the surfaced pycnocline and associated flow structure. During the application of the wind stress the longshore force balance is taken to be

$$\begin{aligned} \frac{\mathrm{d}v}{\mathrm{d}t} + fu &= \frac{u_*^2}{h_t}, \\ \frac{\mathrm{d}v'}{\mathrm{d}t} + fu' &= 0, \end{aligned} \tag{3.85}$$

i.e., the force of the wind is supposed evenly distributed over the top layer only. The time derivatives are total, following the motion of fluid columns in surface and bottom layers.

After one time integration Equation (3.85) become

$$\begin{aligned} v + f\xi &= \frac{I}{h_t}, \\ v' + f\xi' &= 0 \end{aligned} \tag{3.86}$$

with

$$I = u_*^2 t = \int \frac{\tau_y}{\rho}\, dt,$$

$$\xi = \int_0^t u\, dt, \qquad \xi' = \int_0^t u'\, dt.$$

Following the application of the impulsive wind, the fluid columns adjust to geostrophic equilibrium. Inertial oscillations which inevitably attend the adjustment process are of no interest here and only the final state of geostrophic equilibrium will be investigated. In the process of impulsive wind application and subsequent adjustment potential vorticity is conserved in each layer, the wind-stress curl vanishing for the hypothesized spatially uniform wind. Applying Equation (1.40) to the two-layer model, the initial potential vorticities $f/h_t, f/h_b$ equal the potential vorticities after adjustment, or:

$$\begin{aligned} f + \frac{dv}{dx} &= f\,\frac{h}{h_t}, \\ f + \frac{dv'}{dx} &= f\,\frac{h'}{h_b}, \end{aligned} \tag{3.87}$$

where dv/dx, dv'/dx are velocity gradients, *after* adjustment. Geostrophic equilibrium of the two-layer flow field after adjustment is expressed by

$$\left.\begin{aligned} v &= \frac{g}{f}\frac{d}{dx}(h + h'), \\ v' &= \frac{g}{f}\frac{d}{dx}(h + h' - \epsilon h), \end{aligned}\right\} \begin{aligned} &\left(\epsilon = \frac{\rho' - \rho}{\rho'}\right), \\ &(x \leqslant x_0). \end{aligned} \tag{3.88}$$

(3.86)–(3.88) are six linear equations for six unknowns h, h', v, v', ξ, and ξ', being four equations of motion and two vorticity equations, the latter replacing equations of continuity. Equations (3.88) apply in the two-layer portion only. Between the coast and $x = x_0$ only bottom layer fluid is present after adjustment and

$$v' = \frac{g}{f}\frac{dh'}{dx}, \qquad (x \geqslant x_0). \tag{3.89}$$

The second of (3.87) still applies in this region and may be used to eliminate v' to arrive at a single equation for h':

$$\frac{d^2 h'}{dx^2} - \frac{f^2}{g h_b}\, h' = -\frac{f^2}{g}, \qquad (x \geqslant x_0). \tag{3.90}$$

Similarly, from (3.87) and (3.88) three of the four variables may be eliminated to arrive at a single equation for top layer depth h at $x < x_0$:

$$\epsilon \frac{f^4 h}{dx^4} - \frac{f^2}{g}\left(\frac{1}{h_t} + \frac{1}{h_b}\right)\frac{d^2 h}{dx^2} + \frac{f^4}{g^2 h_t h_b} h = \frac{f^4}{g^2 h_b}, \quad (x < x_0). \tag{3.91}$$

Six integration constants appear in the solutions of (3.90) and (3.91). For the case where the pycnocline intersects the free surface some unknown distance x_0 offshore seven boundary conditions are needed to determine these constants plus x_0.

At the coast, where the bottom layer remains in continuous contact with the coast the cross-shore displacement ξ' vanishes:

$$\xi' = 0, \quad (x = 0). \tag{3.92}$$

At the free surface-pycnocline intersection the top layer water column originally at $x = 0$ is found: its total cross-shore displacement is exactly x_0. The same column has also been squashed to vanishing depth. The bottom layer water column has unknown total depth h' or horizontal displacement ξ' here, but both these quantities are continuous.

$$\left.\begin{aligned} \xi &= x_0, \\ h &= 0, \\ h'_- &= h'_+, \\ \xi'_- &= \xi'_+, \end{aligned}\right\} \quad (x = x_0). \tag{3.93}$$

Finally, at large enough distances from the coast both layer depths revert to their equilibrium value:

$$\left.\begin{aligned} h &\to h_t \\ h &\to h_b \end{aligned}\right\} \quad -x \to \infty. \tag{3.94}$$

Three length scales arise in the solution of Equations (3.90) and (3.91). Between the pycnocline intersection and the coast a modified surface-mode radius of deformation scales the solution:

$$R_s = f^{-1}(gh_b)^{1/2}, \quad (x \geqslant x_0). \tag{3.95}$$

In the two-layer portion the radii of deformation already familiar from linear theory appear again:

$$R_1 = f^{-1}[g(h_t + h_b)]^{1/2}, \quad R_2 = f^{-1}[\epsilon g h_t h_b/(h_t + h_b)]^{1/2}, \quad (x \leqslant x_0) \tag{3.96}$$

The calculations are a little tedious but straightforward. For the offshore displacement x_0 of the pycnocline one finds (Csanady, 1977b):

$$x_0 = \frac{I}{fh_t}\frac{h_b}{h_t + h_b} + R_2. \tag{3.97}$$

Since $x \leqslant 0$ is the domain of interest, a negative I is required to result in offshore pycnocline displacement, and one large enough in magnitude: $x_0 \leqslant 0$ provided that

$$-I \geqslant \frac{h_t(h_t+h_b)}{h_b} c_2, \tag{3.97a}$$

where $c_2 = fR_2$, as before. To a satisfactory approximation the pycnocline shape is given by

$$h = h_t \left[1 - \exp\left(\frac{x-x_0}{R_2}\right)\right], \qquad (x \leqslant x_0). \tag{3.98}$$

Correspondingly, the velocity difference between the two layers is

$$v - v' = -\frac{\epsilon g h_b}{c_2} \exp\left(\frac{x-x_0}{R_2}\right), \qquad (x \leqslant x_0). \tag{3.99}$$

Note that both thermocline shape and velocity difference remain the same for any value of I, once a pycnocline surfaces. The only effect of a larger (negative) impulse on the pycnocline is to push it further from the coast.

Longshore velocity in the layer of heavier fluid, which is insulated from the direct wind impulse, is generated through adjustment drift, according to the second of (3.86). Just underneath the surfaced pycnocline, $x = x_0$, a simple mass balance yields neglecting surface level changes of order ϵ,

$$v' = -fx_0 \frac{h_t}{h_b} = -\frac{I}{h_t+h_b} - c_2 \frac{h_t}{h_b}. \tag{3.100}$$

This steadily increases with $(-I)$ and, in virtue of the constancy of $v - v'$, so does the top layer velocity. The extra momentum along negative y is distributed evenly over the entire water column. These results are valid at $-x \ll R_1$, and are identical with those found for a homogeneous fluid. The physical reason is that once the pycnocline surfaces, no further change in the nearshore arrangement of warm and cold layers takes place, the inclined portion of the pycnocline moving offshore as a rigid scoop.

The fluid displacements ξ suffered by the surface layer are again best interpreted as average values over the surface mixed layer. The details are that a surface Ekman layer conducts away the fluid 'scooped up' by the offshore moving front, as well as the adjustment drift arriving above the pycnocline. The fluid participating in the shoreward adjustment drift above and below the pycnocline is supplied from a region of scale width R_1, due to slow surface and pycnocline depression (both of the same order, ϵh).

Further details of this model and its extensions may be found in Csanady (1977b). One point to emphasize is that a similar theory of downwelling can be developed. The principal outcome of such investigations is again that except for some narrowly circumscribed details, the response of the stratified fluid column to impulsive wind is very much as in the absence of stratification. This specifically applies to coastal sea levels and longshore flow velocities over most of a shallow sea. The exceptions to this rule are confined to a narrow band of scale width R_2 and are advantageously viewed as a separate, smaller scale phenomenon associated with stratification. To put the general conclusion another way, the (frictionless) response of a body of stratified fluid to wind does not lead to the development of important horizontal density gradients except in narrow boundary layers. Elsewhere isopycnal surfaces remain horizontal and the presence of stratification is of no consequence. All this is to be understood as relating to the circulation problem, excluding specifically inertial oscillations.

CHAPTER 4

The Subtle Effects of Topography

4.0. INTRODUCTION

Constant depth models of shallow seas discussed in the previous two chapters have elucidated many interesting dynamical phenomena in particularly simple ways, but of course these models are overidealized. Before attempting any comparison between the theoretical models and observation, the dynamical effects of variable depth on basin response have to be investigated. All natural basins possess a complex depth distribution, characterized in particular by a depth reducing *gradually* to zero at the shores. Figure 4.1 shows the depth distribution of Lake Ontario by a contour map; a cross section near the western end is shown later (Figure 4.5). It should be remembered that the vertical scale on such cross sections is exaggerated by a large factor (200 in Figure 4.5): without such an exaggeration the depth distribution would simply appear as a thickened horizontal line. Figure 4.2 shows the depth distribution off the south coast of Long Island. The near-shore parts of Figurew 4.5 and 4.2 are generally similar and show bottom slopes of the order of 10^{-2}. The flatness of the bottom of a typical coastal ocean basin would lead one to think that constant depth models should be reasonably realistic, except very close to shore where the relative changes in depth are rapid (change in depth relative to local depth). In a way this is true and the phenomena revealed by an analysis of constant depth models have their observable counterparts in real basins. However, there are also other modes of motion which only occur in basins of variable depth. Furthermore, some phenomena, present over whatever depth distribution, have their character qualitatively modified by depth variations, sometimes in quite subtle and unexpected ways. Particularly interesting effects arise from an interaction of depth variations and the Earth's rotation.

Considerable insight into the dynamical role of topography may be gained from suitably simple variable depth models, at first supposing the fluid to be homogeneous again. This approach may be expected to yield realistic results except in boundary layers where strong isopycnal surface tilts are likely to develop. In later sections of this chapter the effects of stratification are briefly reconsidered in variable depth models.

4.1. WIND SETUP OVER VARIABLE DEPTH

One class of topographic effects is related to the simple fact that the force of the wind distributed over a shallow layer of fluid produces larger average accelerations than the same force acting over deep water. The wind setup solution, Equation (2.1), shows the free surface slope to be inversely proportional to the depth of a constant depth basin.

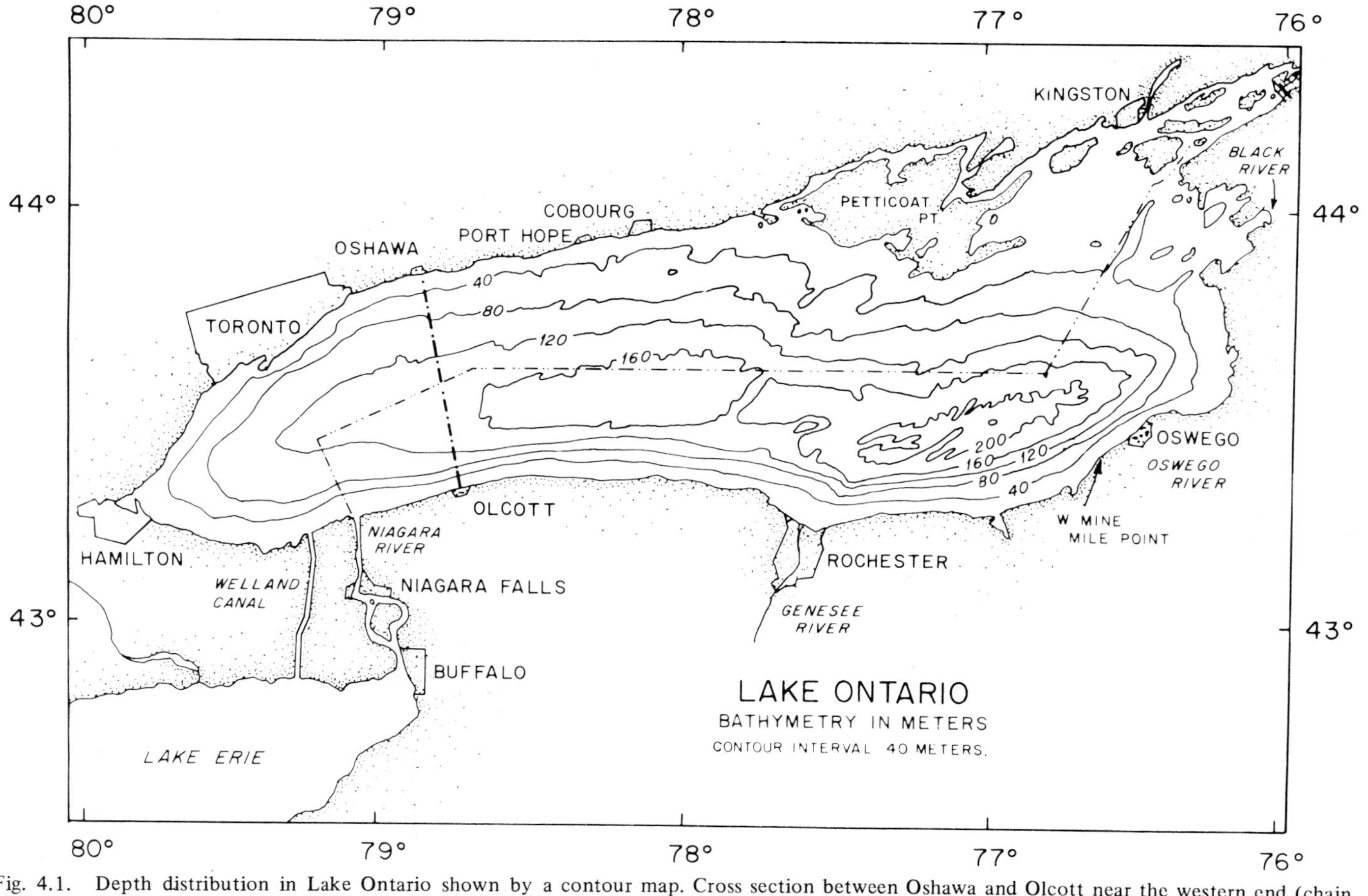

Fig. 4.1. Depth distribution in Lake Ontario shown by a contour map. Cross section between Oshawa and Olcott near the western end (chain dotted line) is also shown later in Figure 4.5.

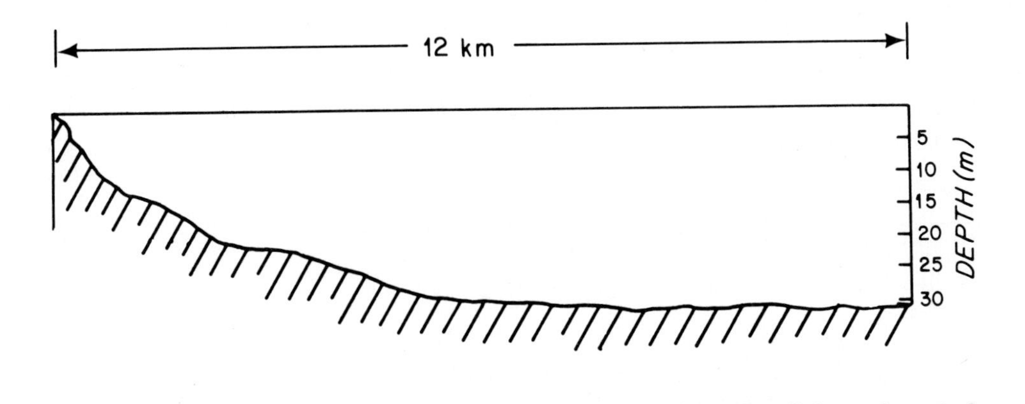

Fig. 4.2. Typical depth distribution close to an open coast: bottom profile off the south coast of Long Island.

One of the simplest questions to ask is, whether an analogous solution exists over variable depth, and if so, what depth one should use in calculating setup. The hyperbolic nature of the governing equations suggests that the response of a realistic (variable depth) basin to suddenly imposed wind stress will still consist of a static solution (i.e., setup), plus seiches, but both presumably of a more complex structure than found for constant depth models. At least in suitably small basins, or for suitably short times, it is also likely that earth rotation effects remain unimportant. Thus one may begin the investigation of topographic influences by seeking a variable depth analog of setup, at first neglecting the Coriolis force, as well as bottom friction, as in the elementary discussion of setup and seiche in Chapter 2.

Consider therefore a closed basin of arbitrary horizontal shape and depth distribution, over which a wind stress constant in time (but arbitrarily distributed in the horizontal plane) is imposed suddenly at $t = 0$. The linearized transport equations with Coriolis force and bottom stress neglected are

$$\begin{aligned} \frac{\partial U}{\partial t} &= -gH\frac{\partial \zeta}{\partial x} + F_x, \\ \frac{\partial V}{\partial t} &= -gH\frac{\partial \zeta}{\partial y} + F_y, \\ \frac{\partial U}{\partial x} + \frac{\partial V}{\partial y} &= -\frac{\partial \zeta}{\partial t}, \end{aligned} \tag{4.1}$$

where now $H = H(x, y)$. These equations may be shown to be hyperbolic, so that the response of the basin to a suddenly imposed force may be expressed as the sum of a particular or forced solution and a series of oscillations. For the particular solution one may try the nearest analog of the wind setup which is some suitable *time-independent* distribution of surface level perturbation $\zeta(x, y)$. Since after $t = 0$ also the forcing is time-independent, U and V can be at most linear in time according to the first and second of

(4.1). If the initial state is rest, one may set therefore

$$U = At, \qquad V = Bt, \tag{4.2}$$

where $A(x, y)$ and $B(x, y)$ are depth-integrated accelerations, constant in time (in a model without bottom friction). Equations (4.1) now reduce to

$$A = -gH\frac{\partial \zeta}{\partial x} + F_x,$$

$$B = -gH\frac{\partial \zeta}{\partial y} + F_y, \tag{4.3}$$

$$\frac{\partial A}{\partial x} + \frac{\partial B}{\partial y} = 0.$$

The boundary conditions are that the normal components of the transport vanish at the shores at all times t, so that the same condition applies to the accelerations A, B. If, subject to these conditions, Equations (4.3) may be solved, the solution yields a time-independent surface level distribution accompanied by a nontrivial transport distribution, increasing linearly with time.

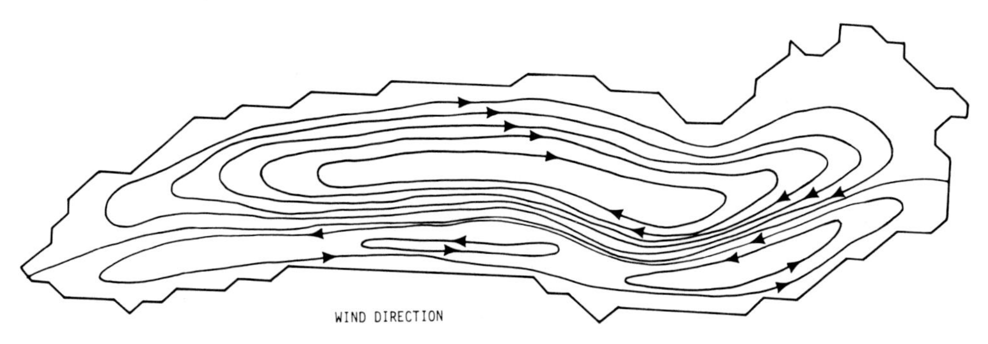

Fig. 4.3. Typical result of numerical model calculations of wind driven, frictionally controlled flow in closed basins, showing transport streamlines. From Rao and Murthy (1970). Basin shown is Lake Ontario, driving is by eastward wind stress. A slightly smoothed version of the depth distribution shown in Figure 4.1 was used in the model calculations.

It is not difficult to demonstrate that such solutions exist. Equations similar in form to (4.3) arise in modelling frictionally controlled flow in a basin with complex depth distribution. A number of numerical studies have been published illustrating the solutions of these equations for different types of forcing. An example relating to Lake Ontario due to Rao and Murthy is shown in Figure 4.3. Although the Coriolis force was taken into account in the model calculations, its influence in a long and narrow basin such as Lake Ontario is relatively minor, so that the flow pattern shown is closely similar to a solution of Equations (4.3) under uniform eastward wind. The transport streamlines shown in the figure are contours of constant ψ, where ψ is related to the transport components as

$$U = \frac{\partial \psi}{\partial y}, \qquad V = -\frac{\partial \psi}{\partial x}. \tag{4.3a}$$

Taken as representing a solution of Equation (4.3), the streamlines represent transport at a fixed (short) time after the imposition of wind stress. The appearance of closed gyres in similar numerically calculated flow fields is typical and deserves further discussion.

4.1.1. Topographic Gyres

In order to explore the fundamental physical properties of the solutions of Equations (4.3) it is convenient to restrict the discussion to basins of particularly simple geometry, to those which are long and narrow, with parallel depth contours over a substantial 'trunk' section. The wind stress is supposed acting *along* these same depth contours, the main question of interest being how the transport is distributed in a cross-section *perpendicular* to the wind.

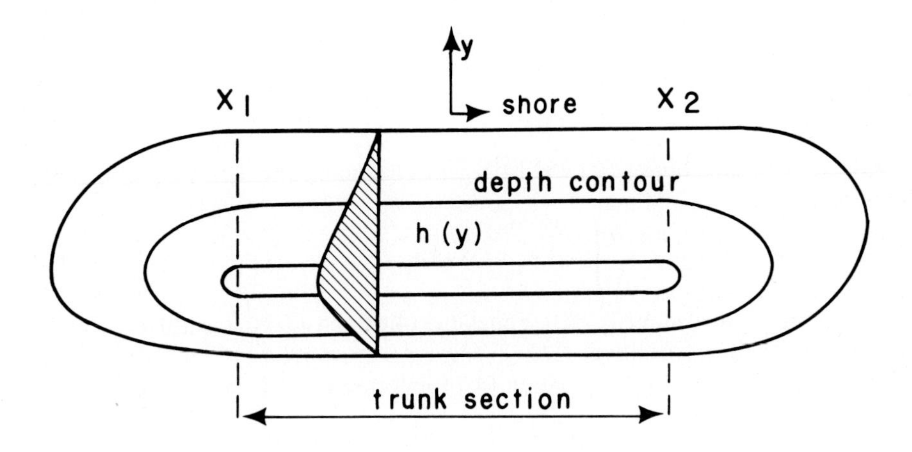

Fig. 4.4. Schematic diagram of long and narrow shallow sea with regular 'trunk' section, in which depth contours are parallel. From Csanady (1973).

An idealized model of a long and narrow basin is sketched in Figure 4.4, which is seen to represent the main features of the previous example, Lake Ontario, reasonably faithfully. The x axis is placed along the length of the basin and the forced motion is considered produced by a constant wind stress acting along this same axis $F_x = F$, $F_y = 0$, suddenly imposed at time zero. To be determined is the transport distribution in a cross-section (x = constant plane) within the trunk region.

At any cross-section (whether in the trunk region or not) the continuity equation yields with time-independent surface level distribution:

$$\int_{y_1}^{y_2} A \, dy = 0, \tag{4.4}$$

where y_1 and y_2 are the coordinates of the shores. In the trunk region, $x_1 < x < x_2$, it is reasonable to seek a solution wherein transport is parallel to the boundaries and to the depth contours, i.e., $B = 0$. Because also $F_y = 0$, the second of Equations (4.3) shows $\partial\zeta/\partial y = 0.$, i.e., the surface elevation is constant in a cross-section. The first of the same equations yields

$$A = -gH\frac{d\zeta}{dx} + F, \qquad x_1 < x < x_2. \tag{4.5}$$

Integrating this equation over the cross-section, and observing Equation (4.4), it is possible to calculate the elevation gradient:

$$\frac{d\zeta}{dx} = \frac{Fb}{gS}, \qquad x_1 < x < x_2, \tag{4.6}$$

where $b = y_2 - y_1$ is the width of the basin and S is its cross-sectional area:

$$S = \int_{y_1}^{y_2} H\,dy. \tag{4.6a}$$

Returning to Equations (4.2) and (4.5), one may now write the transport in the trunk region as

$$U = Ft\left[1 - \frac{Hb}{S}\right], \qquad x_1 < x < x_2. \tag{4.7}$$

It may be verified directly that Equations (4.6) and (4.7) constitute a solution of Equations (4.1) for the trunk region of a long and narrow basin.

The simple relationship of Equation (4.7) is illustrated in Figure 4.5 using the depth

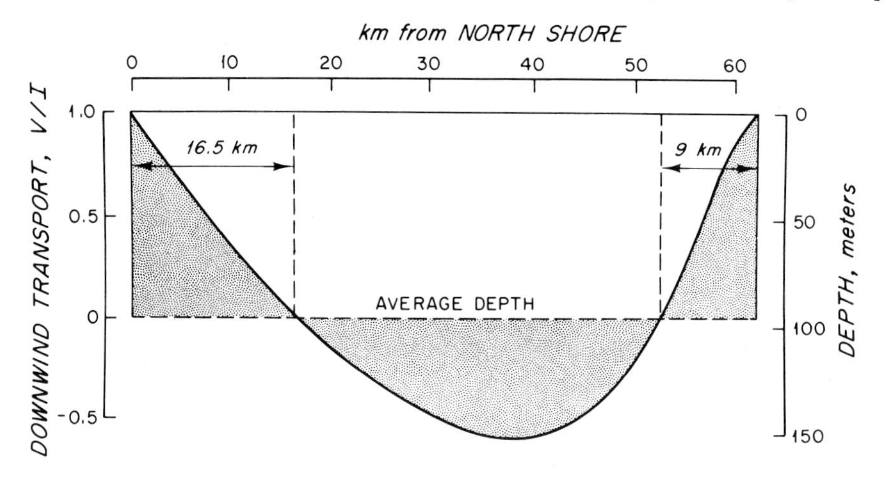

Fig. 4.5. Distribution of wind-driven transport in cross-section of a long and narrow sea as given by frictionless, inertial model (left-hand scale, the unit being the impulse $I = Ft$). Transport distribution is simply a rescaled depth distribution.

distribution of Lake Ontario at the Oshawa-Olcott section (Figure 4.1) as an example. The transport distribution may be simply described as a rescaled and displaced depth distribution. The calculated transport is zero where $H = H_a = S/b$, which is the *average* depth of the section. The elevation gradient $\mathrm{d}\zeta/\mathrm{d}x$ is the same as would be produced by the wind in a basin of constant depth, equal to the cross-sectional average depth. Along the locus of the cross-sectional average depth the wind stress and pressure gradient are in exact balance and no downwind or upwind flow is produced. In shallower water the wind-stress is greater than the total gravity force $gH\ \mathrm{d}\zeta/\mathrm{d}x$ due to surface slope, and the water accelerates downwind. In deep water the pressure gradient dominates and a return flow develops.

The transport distribution in the remainder of the basin may now also be elucidated qualitatively. Transport streamlines (defined by Equation (4.3a) in the trunk region may be determined to correspond to the calculated transport pattern. This yields relatively densely spaced lines both at the center and at the shores, pointing in opposite directions, however. In the end regions the streamlines must close: the details depend on the depth distribution, but the appearance of a 'double-gyre' pattern follows regardless of these details. Figure 4.6 illustrates this qualitative inference, which is also in accord with numerical calculations for more realistic examples, see Figure 4.3. The location of the closed gyres is determined by the precise depth distribution.

Equation (4.7) also shows that the downwind transport is proportional to Ft, which is the impulse of the wind-stress. Near shore, where the depth is much less than the average depth, the transport is nearly equal to the windstress impulse, in the frictionless model. The physical reason for this is that the adverse pressure gradient is determined by the average depth. The total pressure-gradient force is proportional to the local depth and in the shallow shore zone is thus negligible in comparison with the wind stress. Therefore virtually all of the latter's impulse is absorbed by the water column, increasing its total momentum almost exactly by the amount of the wind stress impulse.

The calculated velocity distribution is characterized by strong coastal currents and relatively weak return flow. Realistic values of wind stress impulse are a few times 1 $\mathrm{m}^2\ \mathrm{s}^{-1}$. A wind stress of 0.1 Pa acting for 10 hr produces an impulse of 3.6 $\mathrm{m}^2\ \mathrm{s}^{-1}$. In a basin of an average depth of about 100 m, waters shallower than 10 m hardly 'feel' the pressure

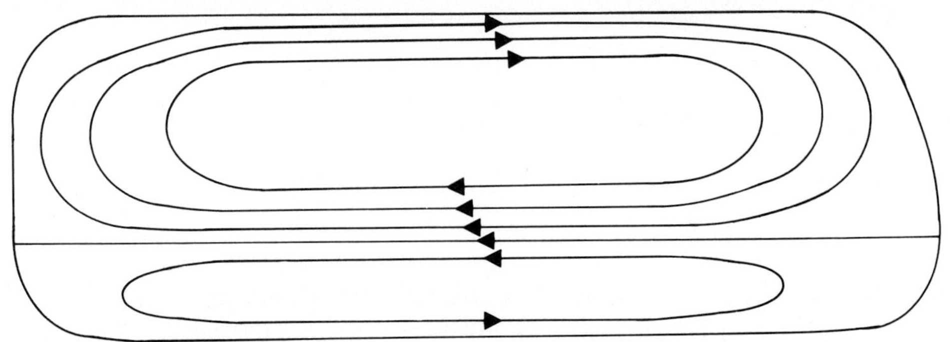

Fig. 4.6. General appearance of wind-driven transport streamlines in long and narrow basins. From Csanady (1973).

gradient force of wind setup, as already pointed out. Distributing the corresponding transport of 3.6 $m^2\ s^{-1}$ over a depth of 10 m results in an average velocity of 36 cm s^{-1}. This indeed is of the same order of magnitude as the speed of typical observed coastal currents. By contrast, average return flow velocities (along the upwind legs of the topographic gyres) are an order of magnitude less, because a similar transport is there distributed over a depth an order of magnitude greater. Now it so happens that many different physical factors produce velocities of the order of 3 cm s^{-1} in the coastal ocean. Thus the return flow is likely to be submerged into a very noisy background. On the other hand, the directly wind-driven coastal currents stand out as easily identifiable, distinct phenomena.

4.2. WINDWARD AND LEEWARD SHORES

The focus of the discussion in the previous section was the transport streamline pattern: the pressure field given by (4.6) for the trunk section is simply a longshore gradient of magnitude F/H_a, with H_a section-average depth. Where H_a is substantial, this represents a logical and simple extension of the notion of wind setup. At the ends of a closed basin, however, the depth reduces to zero and the linearized equations (4.1) predict a singularity in the surface elevation distribution. This has repeatedly been noted in the literature without further discussion. Although from a physical point of view the singularity does not turn out to be particularly disturbing, it deserves a brief discussion here.

The essence of the problem is retained in a simple windward or leeward shore model consisting of a semi-infinite inclined-plane beach (Figure 4.7b). The shoreline is the y-axis and the depth is a linear function of distance from shore:

$$H = sx. \tag{4.8}$$

The wind stress acts normal to shore, $F_x = F$, $F_y = 0$ along a windward, $F_x = -F$ along a leeward shore. With Earth rotation effects still neglected Equations (4.1) simplify to, for a windward shore and a time-independent pressure field:

$$0 = -\ gsx\frac{\partial \zeta}{\partial x} + F. \tag{4.9}$$

This simplification arises because there is no reason why longshore transports or pressure gradients should appear. The continuity equation with $\partial\zeta/\partial t = 0$, and the boundary condition $U = 0$ at the shore than imply the above result. Equation (4.9) at once integrates to

$$\zeta = \frac{F}{gs}\ln\frac{x}{x_r}, \tag{4.10}$$

where x_r is an integration constant, some convenient reference distance.

As $x \to 0$, Equation (4.10) has a logarithmic singularity, suggesting that the physics of the problem is inadequately represented by Equation (4.9). One approximation in the derivation of this equation was that the depth $H = sx$ was supposed large compared to the

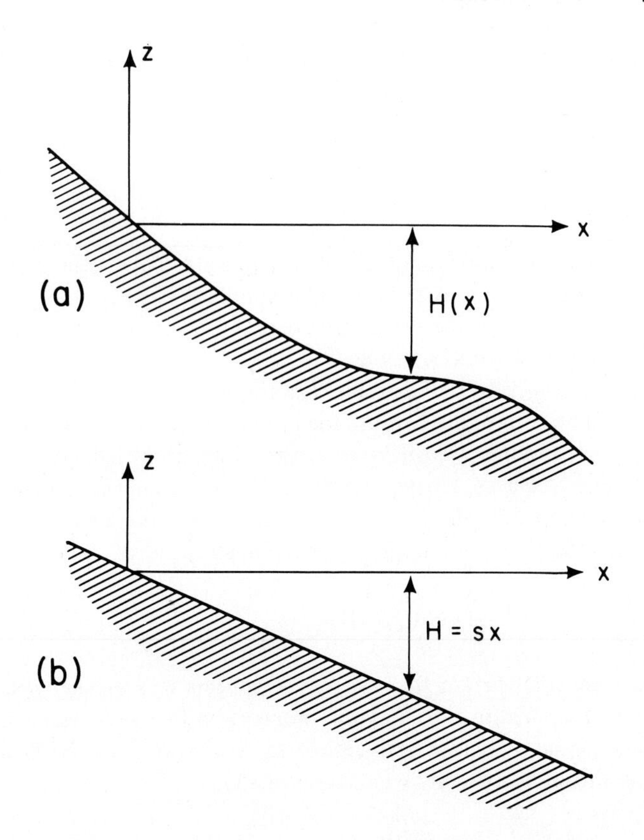

Fig. 4.7. Models of coastal zone used frequently in theoretical models: (a) general two dimensional model, depth a monotonic function of offshore distance only, constant along lines parallel to the coast; (b) inclined plane beach model.

surface displacement ζ. As $H \to 0$ this is clearly no longer true, an amended momentum balance being

$$g(\zeta+sx)\,\frac{\partial\zeta}{\partial x} = F. \tag{4.9a}$$

In a suitably small neighborhood of the point $x = 0$ an approximate solution of (4.9a) is

$$\zeta^2 = \frac{2Fx}{g} + \text{constant}. \tag{4.10a}$$

Equation (4.9a) may be integrated step by step, beginning with (4.10a) near the shore, choosing the integration constant to be zero. Very soon the solution becomes indistinguishable from that of Equation (4.10): the nonlinear correction is significant to $x =$ order (F/gs^2) only, or at most a few meters from shore even for a bottom slope s as low as 10^{-3}.

Over a leeward beach ($F_x = -F$) Equation (4.9a) has an asymptotic solution:

$$\left.\begin{aligned} \zeta &= \frac{F}{gs}, \\ \frac{\partial \zeta}{\partial x} &= -s, \end{aligned}\right\} \quad (x \to 0). \tag{4.10b}$$

This is a thin film of fluid supposedly held along the slope against gravity by wind stress. The range of distance x over which the nonlinear correction is significant is the same over the leeward as over the windward coast, i.e., it is in practice millimeters to a few meters depending on the bottom slope.

In terms of its relevance to the coastal ocean this entire discussion of Equation (4.9a) is more or less academic, not only because the nonlinear corrections are insignificant, but also because a more important correction to the momentum balance is the horizontal momentum flux due to waves. Writing $-\rho\overline{u'^2}$ for the wave-induced Reynolds stress normal to the coast, (Longuet–Higgins and Stewart, 1964) the first of Equations (4.1) amended for wave effects, but still neglecting bottom stress, is:

$$H\frac{\partial \overline{u'^2}}{\partial x} = -gH\frac{\partial \zeta}{\partial x} + F_x. \tag{4.11}$$

Most of the wave momentum flux-divergence appears in a breaker zone of a typical width of $L_b = 100$ m. Within this zone the momentum flux reduces from $\rho\overline{u'^2}|_d$, associated with the incoming deepwater waves, to zero. The wave momentum flux induces a surface slope in the absence of any wind, from (4.11):

$$\frac{\partial \zeta}{\partial x} = -\frac{1}{g}\frac{\partial \overline{u'^2}}{\partial x} \cong -\frac{\overline{u'^2}|_d}{gL_b}. \tag{4.11a}$$

Given the typical magnitudes $u' = 1$ m s^{-1}, $L_b = 100$ m this 'wave runup' slope is 10^{-3}, the total runup over the breaker zone being $\overline{u'^2}/g$ or typically 0.1 m.

In very shallow water the frictional forces associated with the breaking of the waves also become important and usually dwarf the wind stress. As everyday observation shows, wave orbital velocities directly at the shore are 0.1 m s^{-1} and more even in very small waves, present in all but the lightest of breezes. A moderate onshore wind of speed 7 m s^{-1} (which evokes a kinematic stress of $F \cong 1$ cm^2 s^{-2}) usually generates waves of orbital velocities of 1 m s^{-1}, and corresponding bottom stresses some four orders of magnitude greater than F. Under these circumstances there is of course no thin film creeping up a beach (as Equation (4.10b) would suggest) because the force of the wind on the water is transferred by the waves directly to the bottom, along with a much greater drag force drawn from the wave momentum flux*.

Disregarding other complications very close to shore, the dynamics of the breaker zone is reasonably supposed dominated by wave momentum flux. This zone has been the

---

* One may also observe on a leeward beach that water tossed out over the beach by wind and waves percolates back through the sand, adding further to the complexity of beach processes.

subject of detailed investigation by coastal engineers. In the large-scale coastal circulation problem the breaker zone plays no significant role: the pressure gradients associated with wave-driven longshore currents, as well as wave runup, decay within a distance of order 100 m from the coast. In what follows, attention will be directed exclusively on the coastal ocean outside such breaker zones, i.e., in practical terms on water more than, say, 3 m deep. In this depth Equation (4.9a) already gives a typical offshore gradient of only 3×10^{-6}, so that there is no difficulty with too close an approach to the logarithmic singularity. For simplicity, in many of the models to be discussed below the boundary conditions will be imposed at $H = 0$, which sometimes will lead to the appearance of a logarithmic singularity. All such results should be interpreted as valid only from some finite depth on, of the order of a few meters.

4.3. SEICHES IN VARIABLE DEPTH BASINS

In Chapter 2, the initial establishment of wind setup in a constant depth basin was discussed in some detail. The question arises, is that discussion qualitatively valid when the basin has variable depth? If Equations (4.1) are taken as the basis (i.e., if one neglects the Coriolis force) the effects of depth variations turn out to be minor. This may be shown as follows, restricting the discussion for simplicity to the case of long and narrow basins.

Let the x-axis coincide with the locus of maximum depth in each cross section, a curve known as 'Talweg'. The curvature of this axis is supposed sufficiently gentle not to affect the dynamical argument. As Equation (4.6) gives the pressure field in terms of section-average properties, width b and area S (Equation (4.6a)), a similar approach is adopted for the flow and an along-axis total transport Q is introduced:

$$Q = \int_{y_1}^{y_2} U \, dy. \tag{4.12}$$

Integrating Equations (4.1) with respect to the cross-basin coordinate y, and supposing as before $V = 0$ everywhere, one arrives at

$$\begin{aligned} \frac{\partial Q}{\partial t} + gS \frac{\partial \zeta}{\partial x} &= Fb, \\ \frac{\partial Q}{\partial x} + b \frac{\partial \zeta}{\partial t} &= 0. \end{aligned} \tag{4.13}$$

Equation (4.6) is a particular or forced solution of these equations. A series of free solutions may also be found for the homogeneous equations. These have to satisfy the boundary condition:

$$Q = 0, \quad (x = 0, L) \tag{4.14}$$

where L is the length of the basin. Free solutions having the character of seiches (standing waves) result from the Ansatz:

$$Q = q(x) \sin \sigma t, \qquad \zeta = Z(x) \cos \sigma t, \tag{4.15}$$

where σ is an appropriate free oscillation frequency and $q(x)$, $Z(x)$ are amplitude distributions. Mathematically the problem is identical with that of the vibrating string have variable mass per unit length (Courant and Hilbert, 1953). With the substitution (4.15), the homogeneous Equations (4.13) reduce to:

$$-\frac{\sigma}{gS} q = \frac{dZ}{dx}, \qquad b\sigma Z = \frac{dq}{dx}. \tag{4.16}$$

These equations are easily solved by numerical means, replacing the derivatives by central differences. Starting from one end of the basin, $x = 0$, solutions are calculated for trial values of σ until the boundary condition $q = 0$ is satisfied at the other end, $x = L$. In practical cases the solution takes very little computer time. Fundamental as well as higher modes of oscillation may be found in this manner. An excellent example of such calculations is a study of the free oscillations of Lake Erie, by Platzman and Rao (1964). Figure (4.8) illustrates the elevation and transport distribution obtained in this manner for Lake Superior by Rockwell (1966). Also shown in the figure are setup and topographic gyres after Murthy and Rao (1970), as well as the properties of the fundamental seiche as modified by earth rotation, after Platzman (1972).

The extensive mathematical background available on eigenvalue problems of this kind allows one to assert now that the complete solution of the initial value problem for Equations (4.1) may be represented by the particular solution Equation (4.6) and a series of oscillations, much as in the constant depth case. Mathematically, the static setup can be expanded in a series of eigenfunctions, and the particular solution cancelled at time $t = 0$. From a physical point of view, the net result is that the more complex setup distribution given by Equation (4.6) is established in a period of order σ_1^{-1}, where σ_1 is the lowest natural frequency of oscillations.

It is, furthermore, easy to demonstrate that σ_1 calculated for a variable depth basin is the same as σ_1 for a constant depth basin, of a depth equal to some intermediate depth of the variable depth basin. Thus even for long and narrow basins, such as Lake Erie, calculated fundamental seiche periods $2\pi\sigma_1^{-1}$ are generally of the order of ten hours. From the point of view of the circulation problem, wind setup is thus established rapidly, on a time scale of f^{-1} or less, whatever the depth distribution. This à posteriori justifies the neglect of the Coriolis force in Equations (4.1) in calculating wind setup, or rather its initial development.

4.4. VARIABLE DEPTH AND EARTH ROTATION

The more subtle effects of topography arise from an interplay of earth rotation and depth variations on time scales of f^{-1} and longer. The general nature of these effects is governed

a)

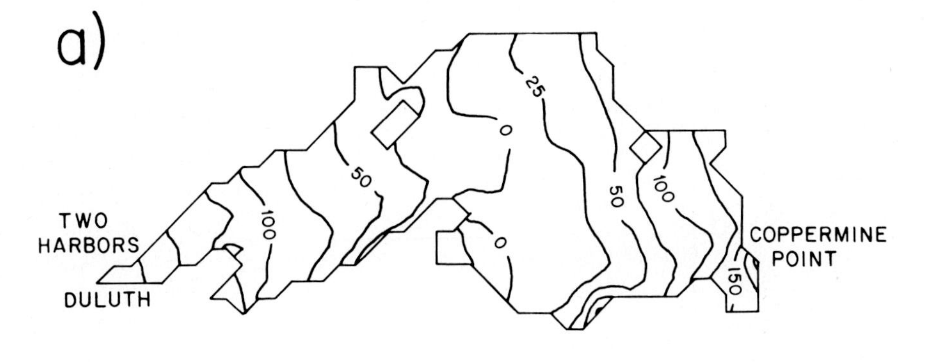

b)

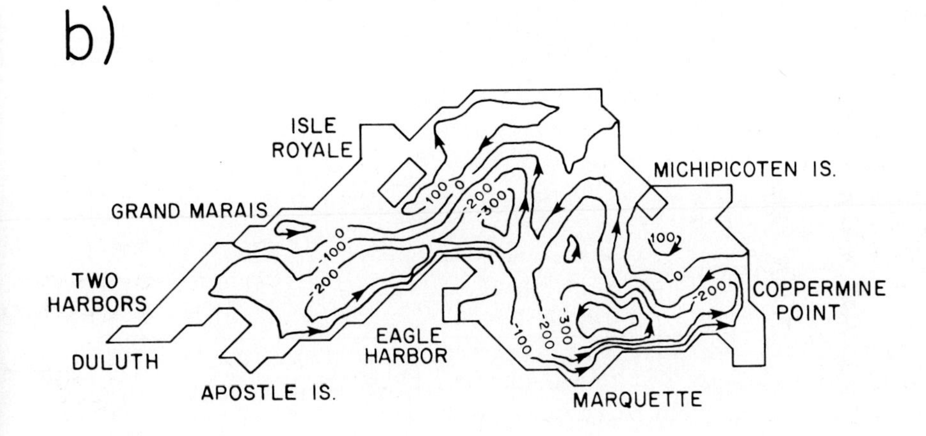

c)

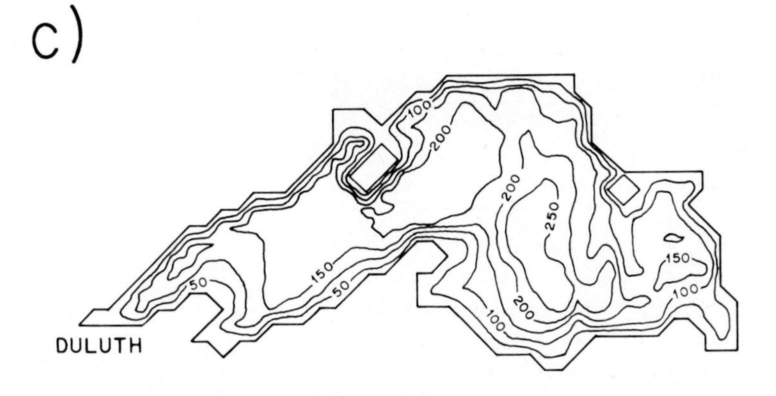

Fig. 4.8. Contour lines of numerically calculated wind setup (a), and transport streamlines of topographic gyres (b) in Lake Superior, for smoothed depth distribution (c). After Murthy and Rao (1970).

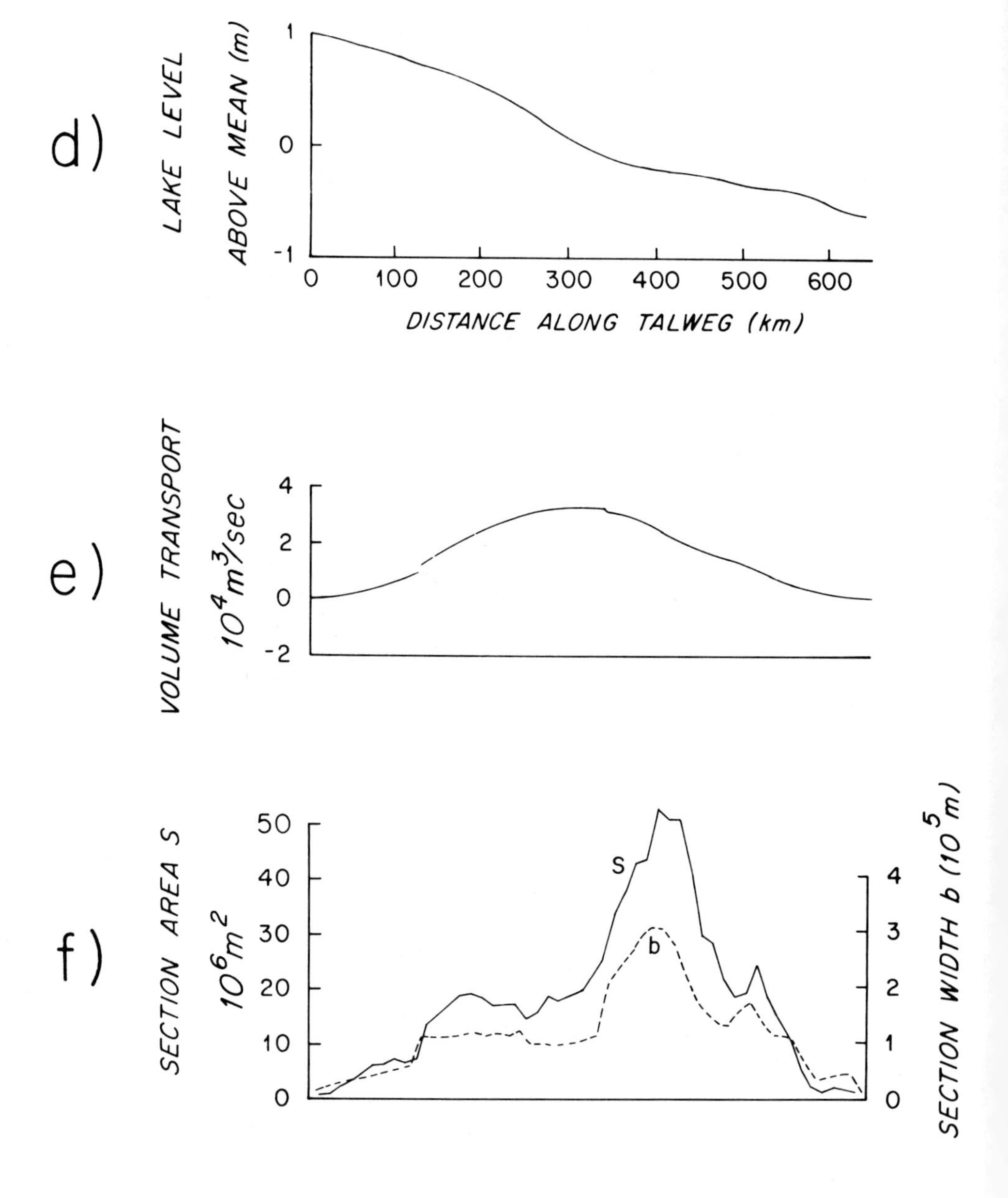

Fig. 4.8 (continued). Calculated distribution of fundamental seiche amplitude along Talweg of Lake Superior (d), and volume transport across perpendicular sections (e). Section areas and widths are shown at bottom (f). After Rockwell (1966).

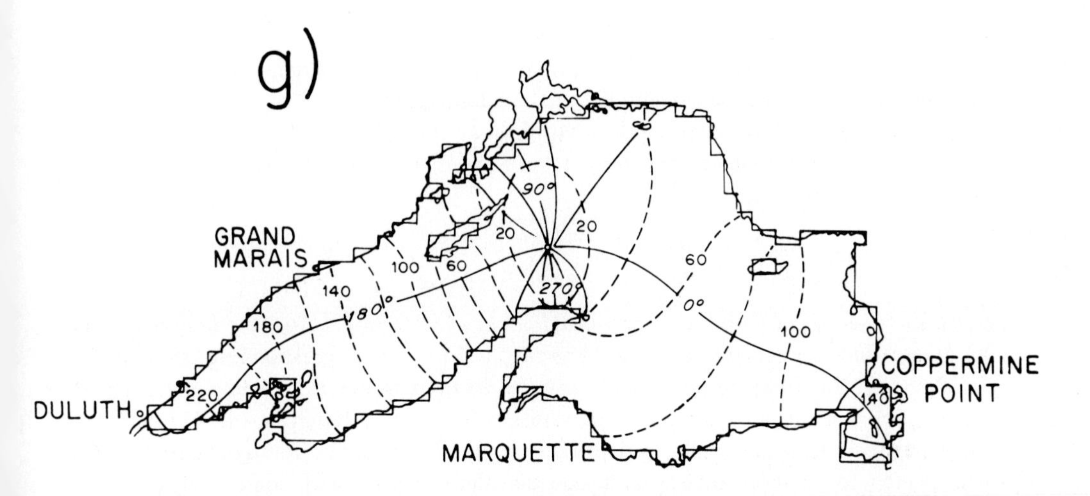

Fig. 4.8 (continued, (g)). Fundamental seiche amplitude and phase distribution, calculated with earth rotation effect taken into account. From Platzman (1972).

by a depth integrated vorticity tendency balance derived from the transport equations for variable depth. The equations are, with bottom friction neglected:

$$\begin{aligned}\frac{\partial U}{\partial t} - fV &= -gH\frac{\partial \zeta}{\partial x} + F_x,\\ \frac{\partial V}{\partial t} + fU &= -gH\frac{\partial \zeta}{\partial y} + F_y,\\ \frac{\partial U}{\partial x} + \frac{\partial V}{\partial y} &= -\frac{\partial \zeta}{\partial t}.\end{aligned} \tag{4.17}$$

Consider an arbitrary depth distribution $H(x, y)$ and take curl on Equations (4.17):

$$\frac{\partial}{\partial t}\left(\frac{\partial V}{\partial x} - \frac{\partial U}{\partial y}\right) - f\frac{\partial \zeta}{\partial t} = -g\left(\frac{\partial H}{\partial x}\frac{\partial \zeta}{\partial y} - \frac{\partial H}{\partial y}\frac{\partial \zeta}{\partial x}\right) + \frac{\partial F_y}{\partial x} - \frac{\partial F_x}{\partial y}. \tag{4.18}$$

The right-hand side is zero if the curl of the wind stress vanishes and

$$\zeta = \zeta(H, t) \tag{4.19}$$

i.e., if surface elevation is constant over isobaths (at any given t). In such a case, on a time integration of (4.18) the result found in a constant depth basin is recovered (Equation

(2.10)):

$$\frac{\partial V}{\partial x} - \frac{\partial U}{\partial y} = f\zeta \tag{4.20}$$

having supposed that the motion starts from rest. One physical interpretation of this equation is that the vorticity of the depth-averaged motion is generated through vortex stretching, due solely to the rise and fall of the sea surface.

Equation (4.18) shows that depth variations may similarly lead to vorticity generation. The vorticity tendency dependent on bottom slope is given by the first term on the right, which may be written as

$$f\left(u_g \frac{\partial H}{\partial x} + v_g \frac{\partial H}{\partial y}\right),$$

where $u_g = -(g/f)(\partial\zeta/\partial y)$, $v_g = (g/f)(\partial\zeta/\partial x)$ are geostrophic velocity components. To the extent that the geostrophic flow 'runs into a slope', the corresponding stretching or squashing of the fluid columns gives rise to vorticity generation. When the vorticity tendency associated with bottom slope vanishes (i.e., where Equation (4.14) holds, and the geostrophic flow is parallel to depth contours), the dynamics of coastal currents may be expected to be more or less as in constant depth basins, with some straightforward modifications on account of different geometry. In the contrary case some entirely new phenomena may be expected to arise, as will indeed be seen in later sections of this chapter.

4.4.1. Sudden Cross-Shore Wind

Consider at first the forcing of motions over a semi-infinite inclined plane beach by impulsive cross-shore wind, uniform in space:

$$\begin{aligned} F_x &= F, \qquad t \geqslant 0 \\ F_y &= 0. \end{aligned} \tag{4.21}$$

The depth distribution is as given by Equation (4.8), the domain of interest being $x \geqslant 0$. There is no reason why an alongshore sea-level gradient $\partial\zeta/\partial y$ should arise in this problem, so that the right-hand side of (4.18) vanishes, and no special complications due to variable depth need be anticipated. Equation (4.20) becomes for this simple model

$$\frac{\partial V}{\partial x} = f\zeta. \tag{4.22}$$

A non-oscillatory solution analogous to wind setup in a constant depth basin is sought. The results in Section 2.4 suggest that the oscillatory components of the motion should be eliminated ('filtered out') if the condition is imposed that integrated cross-shore accelerations vanish:

$$\frac{\partial U}{\partial t} = 0. \tag{4.23}$$

The first of Equations (4.17) together with (4.22) now provide two equations for V, ζ from which V may be eliminated to leave:

$$\frac{\partial}{\partial x}\left(gH\frac{\partial \zeta}{\partial x}\right) - f^2\zeta = 0. \tag{4.24}$$

For a beach of constant slope, substituting Equation (4.8), one finds

$$gsx\frac{\partial^2 \zeta}{\partial x^2} + gs\frac{\partial \zeta}{\partial x} - f^2\zeta = 0. \tag{4.25}$$

This equation may be brought to a particularly simple form by introducing a slope length scale:

$$R_s = \frac{gs}{f^2}. \tag{4.26}$$

Writing $\xi = x/R_s$, Equation (4.25) becomes

$$\xi\frac{\partial^2 \zeta}{\partial \xi^2} + \frac{\partial \zeta}{\partial \xi} - \zeta = 0. \tag{4.25a}$$

The general solution of this homogeneous equation is

$$\zeta = AK_0(2\xi^{1/2}) + BI_0(2\xi^{1/2}), \tag{4.27}$$

where A, B are integration constants, which may, however, be functions of time, and K_0, I_0 are modified Bessel functions. In order to keep the solution physically realistic at $\xi \to \infty$ one has to set $B = 0$. The boundary condition at the coast is zero normal transport, $U = 0$, which by the second of Equations (4.17) also implies in this case of no longshore stress or pressure gradient:

$$V = 0, \quad (x = 0). \tag{4.28}$$

The first of Equations (4.17), with the Ansatz (4.23), yields now

$$\lim_{x\to 0}\left(gsx\frac{\partial \zeta}{\partial x}\right) = F. \tag{4.29}$$

This enables the determination of the integration constant A in (4.27), the final result being

$$\zeta = -\frac{F}{gs}K_0(2\xi^{1/2}). \tag{4.30}$$

This has a logarithmic singularity at $x \to 0$, the physical significance of which has been discussed in detail in Section 4.2. The solution is realistic from some small finite distance x on. For $x >> R_s$ the elevation ζ becomes exponentially small, so that R_s plays a role similar to the radius of deformation R in a constant depth model.

From the first of Equations (4.17) the longshore transport is determined to be:

$$V = -\frac{F}{f}\left[1 - \xi^{1/2} K_1(2\xi^{1/2})\right]. \tag{4.31}$$

Because V and ζ are constant in time, $U = 0$ everywhere on account of the shore boundary condition and the continuity equation, (3rd of (4.17)). This of course applies to the non-oscillatory solution only. As discussed in Chapter 2, the longshore current V builds up in a period of order f^{-1}, when an appropriate distribution of cross-shore transport $U(x, t)$ first establishes the elevation distribution $\zeta(x)$.

As anticipated on the basis of the vorticity balance, the general character of the wind setup solution over variable depth differs little from what was found over constant depth. The differences are due to the constriction of onshore flow by the reducing depth. The long-shore transport is zero at the coast and it asymptotically becomes the Ekman transport at large distances. The scale of the distances in this model is R_S, a very large quantity in practice, typically 1000 km or more, even larger than the radius of deformation R over oceanic depths. A constant slope model is unrealistic over such distances. It is easy to find solutions for piecewise constant slope models patched to a constant depth abyss, representing continental shelves of differing character (see e.g., Csanady, 1974a). The results differ generally little from those of constant depth models.

4.4.2. Longshore Wind

The analogous non-oscillatory solution for longshore wind stress suddenly applied over a semi-infinite sea of constant bottom slope is now also easily written down. The forcing is described by:

$$\begin{aligned} F_x &= 0, \\ F_y &= F, \qquad (t \geqslant 0). \end{aligned} \tag{4.32}$$

In the absence of any y-variability in forcing, no longshore pressure gradient appears. Thus again there is no vorticity tendency associated with either bottom slope or wind-stress curl and (4.19), (4.20), and (4.22) apply. The Ansatz (4.23) is again used to filter out oscillatory motions and leads to the differential equation (4.24) over arbitrary $H(x)$, or (4.25) over a beach of constant slope, with solution given by (4.27). The boundary conditions which apply are, however, now different: $U = 0$ at $x = 0$, while the second of (4.17) yields:

$$V = Ft, \qquad (x = 0). \tag{4.33}$$

Substituting into the first of (4.17) yields the shore boundary condition on the pressure field:

$$\lim_{x \to 0} gsx \frac{\partial \zeta}{\partial x} = fFT. \tag{4.34}$$

This is almost the same as (4.29) the difference being that the right-hand side is multiplied by ft. Therefore

$$\zeta = -\frac{Fft}{gs} K_0(2\xi^{1/2}). \tag{4.35}$$

Equation (4.17) now also yield the corresponding transports

$$V = Ft\xi^{1/2}K_1(2\xi^{1/2}),$$
$$U = \frac{F}{f}\left[1 - \xi^{1/2}K_1(2\xi^{1/2})\right]. \tag{4.36}$$

These results are again closely analogous to those found for a constant depth model. At $x \ll R_s$, for example, $V = Ft$, $U = 0$, all of the wind-impulse appears as longshore momentum. However, over vanishing depth the longshore velocity has to become infinite as H^{-1}, and this leads to a logarithmic singularity in the surface elevation because the cross-shore pressure gradient has to balance a Coriolis force tending to infinity. The physically most unrealistic aspect of this result is that bottom friction does not in fact allow the velocity to become very large and terminates the linear-in-time growth phase of the longshore transport after a frictional adjustment period which is quite short in shallow water. Effects of bottom friction are discussed further in chapter 6.

4.4.3. Interior Velocities

The distribution of the velocity within the water column which accompanies the above calculated pressure fields is also very much as found earlier for the constant depth case. As in Section 2.7, the velocity may be resolved into a pressure field induced distrbution (u_1, v_1) and a frictionally induced one (u_2, v_2), the latter only being a function of depth (in a homogeneous fluid). As long as bottom friction is negligible, so supposed in the above solutions of the global problem, the frictionally induced field consists solely of a surface Ekman layer, in depths much greater than D. In shallower water bottom friction cannot be neglected even for short periods so that the transport distributions given by Equation (4.36) are unrealistic at $x \to 0$. The modifications due to bottom friction are, however, straightforward as their later discussion will show (Chapter 6).

In deeper water, the pressure field-induced velocities are, for the case of sudden longshore wind:

$$u_1 = -\frac{F}{fH}\xi^{1/2}K_1(2\xi^{1/2}), \qquad v_1 = \frac{Ft}{H}\xi^{1/2}K_1(2\xi^{1/2}) \tag{4.37}$$

which follows from (4.36) on subtracting Ekman transport from U and dividing the transports by depth. At distances x small compared to R_s there is thus significant cross-shore velocity u_1, (adjustment drift) meaning that the water column below the Ekman layer is stretched or squashed. The vorticity should change correspondingly. This may be checked by writing down the vorticity tendency equation, not for the transport components, but for the pressure field-induced velocities u_1, v_1. Deleting the forcing terms from Equation (4.17), dividing by depth H and *then* taking curl one finds indeed:

$$\frac{\partial}{\partial t}\left(\frac{\partial v_1}{\partial x} - \frac{\partial u_1}{\partial y}\right) = \frac{f}{H}\frac{\partial \zeta}{\partial t} + \frac{f}{H}\left(u_1 \frac{\partial H}{\partial x} + v_1 \frac{\partial H}{\partial y}\right). \tag{4.38}$$

The second term on the right expresses the vortex stretching effect due to bottom slope, the first term that due to surface level change. Substituting from (4.35) and the first of (4.37) one finds

$$\frac{\partial}{\partial t}\left(\frac{\partial v_1}{\partial x} - \frac{\partial u_1}{\partial y}\right) = -\frac{Fs}{H^2}\left[\xi K_0(2\xi^{1/2}) + \xi^{1/2} K_1(2\xi^{1/2})\right]. \tag{4.39}$$

Close to shore, $\xi \ll 1$, the square bracketed expression on the right is unity, so that the vorticity tendency is $-Fs/H^2 = -F/(sx^2)$, i.e., strongly concentrated near the shore. Equation (4.39) is also of course in accord with the second of (4.37), two differentiations of which yield $\partial^2 v_1/\partial t \partial x$, i.e., the left-hand side of (4.39). The first term in the square bracketed expression in (4.39), proportional to $\partial\zeta/\partial t$, is small compared to the second term for $\xi < 1$. At realistic distances from the shore the vortex stretching term due to bottom slope therefore dominates. It is easy to appreciate the physical content of this result: the upward or downward surface velocity, $\partial\zeta/\partial t$, is typically of order 10^{-4} cm s^{-1}. This is the case when a 10 cm s^{-1} coastal current, 10 km wide, adjusts to geostrophic equilibrium in a period f^{-1}: the sea level at the shore has to rise or fall by 1 cm. If the current is generated by wind stress of 0.1 Pa, acting over a 100 m deep water column, the cross-shore velocity u_1, which is the adjustment drift, is of order 1 cm s^{-1}. Even over a bottom slope as low as 10^{-3} the corresponding $u_1\ \partial H/\partial x$ term is 10^{-3} cm s^{-1}, or an order of magnitude larger than $\partial\zeta/\partial t$.

4.5. VORTICITY WAVES

It is now appropriate to turn to cases in which the depth variations also affect the *depth-integrated* vorticity balance. Consider a topographic gyre described in Section 4.1.1, supposing it to be established by a sharp wind impulse, in a period T, short enough to neglect the Coriolis force, so that the results derived in that section apply. A transport distribution is established, which for a long and narrow basin is given by Equation (4.7). From this one calculates the depth-integrated vorticity distribution:

$$\frac{\partial V}{\partial x} - \frac{\partial U}{\partial y} = \frac{FT}{H_a}\frac{\mathrm{d}H}{\mathrm{d}y}, \tag{4.40}$$

where $H_a = S/b$ is the section-average depth. In this model the x-axis was chosen to coincide with the 'Talweg' (locus of maximum depth) of the basin, with the wind stress applied along positive x. To the right of the wind $\mathrm{d}H/\mathrm{d}y$ is positive and relatively large near the coast, to the left it is negative. Figure 4.9 illustrates schematically the resulting distribution of depth-integrated vorticity.

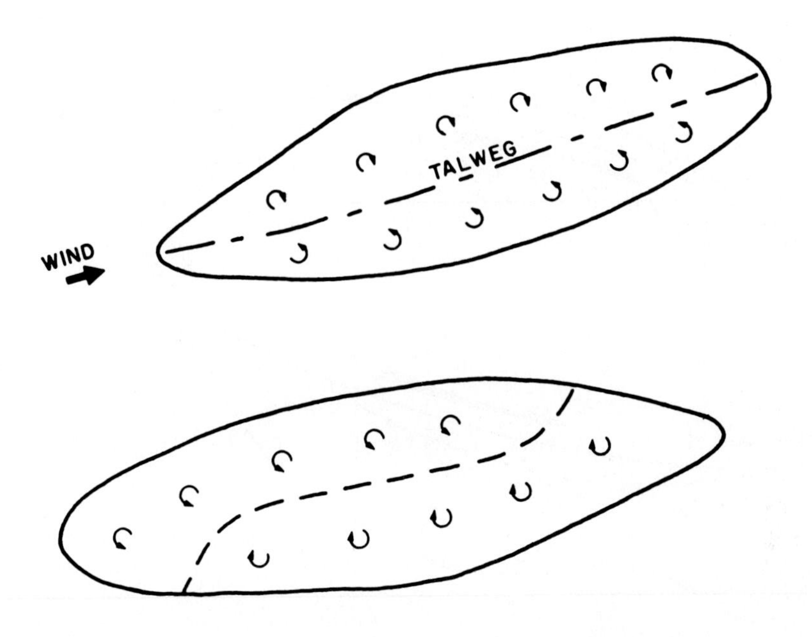

Fig. 4.9. Vorticity distribution associated with freshly generated topographic gyre in closed basin (top). Talweg ('valley way') is locus of maximum depth in each cross-section. Bottom: vorticity distribution at a later time. From Csanady (1974d).

After the wind impulse the fluid is left to itself, with a transport distribution sketched in Figure 4.6 and a vorticity distribution much as shown in Figure 4.9. Under the influence of the Coriolis force, the pressure distribution adjusts to geostrophic equilibrium, so that sea level rises along the coast to the right of the wind, sinks to the left (Figure 4.10). A longshore pressure gradient arises in this manner across the end sections of the basin, driving from the right-hand shore to the left-hand one. At the coast, where the normal transport vanishes, and in the absence of surface or bottom stress, the equations of motion show that the longshore pressure gradient generates longshore transport. This clearly modifies the initial pattern sketched in Figure 4.6: at the downwind end a counterclockwise coastal current is generated, at the upwind end a clockwise one. In effect the coastal current in the right-hand counterclockwise cell of the topographic gyre is extended further counterclockwise, that in the left-hand clockwise cell similarly, but 'backwards', i.e., also in a counterclockwise sense.

The depth-integrated vorticity balance is governed by Equation (4.18), with the wind-stress curl set zero. The vortex stretching term on the right is now certainly not zero. At the downwind end of the basin the fluid crosses depth contours, see Figure 4.6, in both cells of the topographic gyre, moving from shallow coastal regions toward deeper parts of the basin. The corresponding stretching of the vortex lines appears as a positive right-hand side in Equation (4.18) and leads to the generation of positive depth-integrated

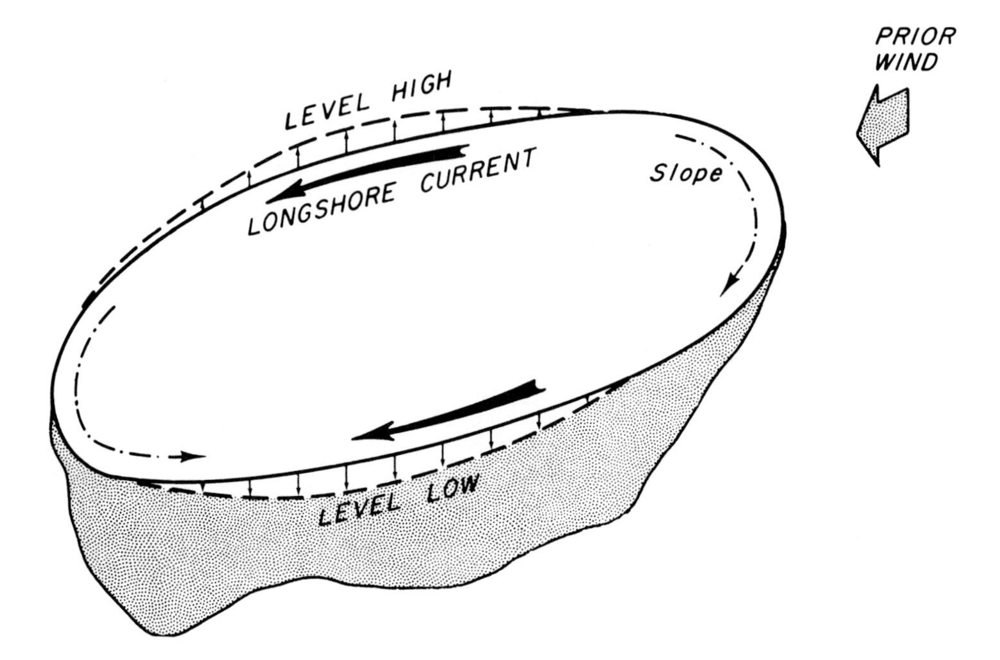

Fig. 4.10. Coastal sea levels accompanying freshly generated topographic gyre. Chain dotted arrows ('SLOPE') show direction of pressure gradient force acting on near-shore water columns.

vorticity. At the upwind end of the basin negative vorticity is generated. After a short period the vorticity distribution is modified in the manner also sketched in Figure 4.9.

These qualitative inferences may be supported by some crude quantitative estimates as follows. Suppose that the coastal current initially generated by the wind-impulse has typical velocity v_0 and that this current occupies a coastal strip of width l. The coastal elevation/depression ζ_0 necessary for geostrophic balance is then

$$\zeta_0 = \frac{f}{g} v_0 \, l. \tag{4.41}$$

In a basin of perimeter length P the longshore variation of the elevation pattern accompanying the topographic gyre has a lowest wavenumber $k = 2\pi/P$, so that the longshore elevation gradient at the ends of the basin is:

$$\frac{\partial \zeta}{\partial y} = \lambda k \, \zeta_0 = \frac{\lambda k f v_0 \, l}{g}, \tag{4.42}$$

where λ is a constant of order unity. The corresponding acceleration $g \ \partial\zeta/\mathrm{d}y$ generates a longshore velocity equal to the initial velocity v_0 in a period $v_0/(g \ \partial\zeta/\partial y)$. The reciprocal of this is a typical frequency σ of the process:

$$\sigma = \lambda k f l. \tag{4.43}$$

The linear propagation velocity of the pattern along the coast should be:

$$c = \frac{\sigma}{k} = \lambda f l. \tag{4.44}$$

For the typical values $f = 10^{-4}\ \mathrm{s}^{-1}$, $l = 10$ km one finds the propagation velocity c to be of order 1 m s^{-1}, σ of order $10^{-5}\ \mathrm{s}^{-1}$, for a typical k of $10^{-5}\ \mathrm{m}^{-1}$.

The mode of motion inferred in the above discussion:

(1) is a free mode, taking place in the absence of forcing, once an initial vorticity distribution is generated;

(2) depends for its existence on depth variations and earth rotation, being related to the vortex stretching phenomenon;

(3) has a characteristic frequency typically low enough to be of interest in the circulation problem.

Such a mode of motion is best described as a 'vorticity wave'. A simple analytical model of a vorticity wave may be constructed by supposing homogeneous fluid, a straight infinite coast $x = 0$, isobaths parallel to the coast, $H = H(x)$, and postulating that longshore transport variations of wavelength $2\pi/k$ have somehow been generated. In a sense the latter step corresponds to unwrapping the perimeter of a closed basin containing a topographic gyre into a straight line, on the reasonable hypothesis that the curvature of the coastline has no first order dynamical effect, centrifugal forces being negligible given ordinary velocities and smooth coastlines. In such a model the vortex stretching may be supposed primarily due to depth variations. The vorticity wave is then known as a 'topographic' wave. A simple topographic wave model is discussed in the next few sections.

4.5.1. Topographic Wave in a Coastal Strip

Consider a strip-like domain of the coastal ocean with parallel depth contours as illustrated schematically in Figure 4.11. All natural basins have a central portion of more or less constant depth. This is presented in the model by supposing $H = H_0$ = constant for $x \geqslant l$. The general discussion of the previous section has shown that vorticity waves have frequencies low compared to the inertial frequency f, so that $\partial U/\partial t$ may be dropped from the first transport equation. This step may also be thought of as a device for filtering out oscillations of inertial frequency and higher. Furthermore, the earlier discussion of longshore current generation showed that vortex stretching due to bottom slope, over realistic beach slopes of 10^{-2}–10^{-3}, overwhelms vortex stretching due to surface elevation changes. Thus a second reasonable approximation in a topographic wave model is to drop the $f\ \partial\zeta/\partial t$ term in the vorticity balance. The model should therefore reveal the novel aspects of the flow field due to variable depth more or less in isolation.

With the two postulates introduced the vorticity tendency equation (4.18) becomes:

$$\frac{\partial^2 V}{\partial t \partial x} = -g \frac{\mathrm{d}H}{\mathrm{d}x} \frac{\partial \zeta}{\partial y} \tag{4.45}$$

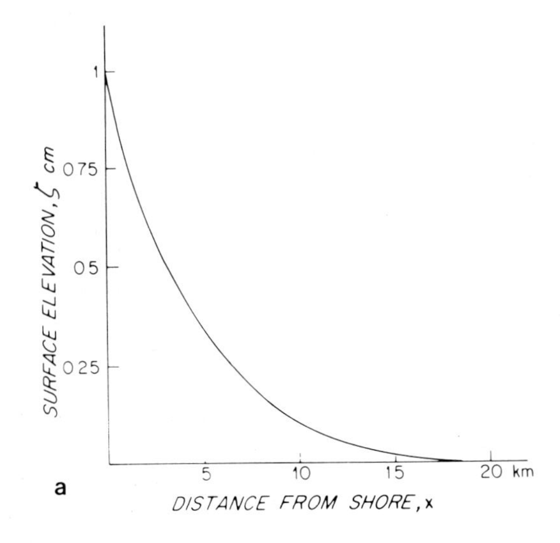

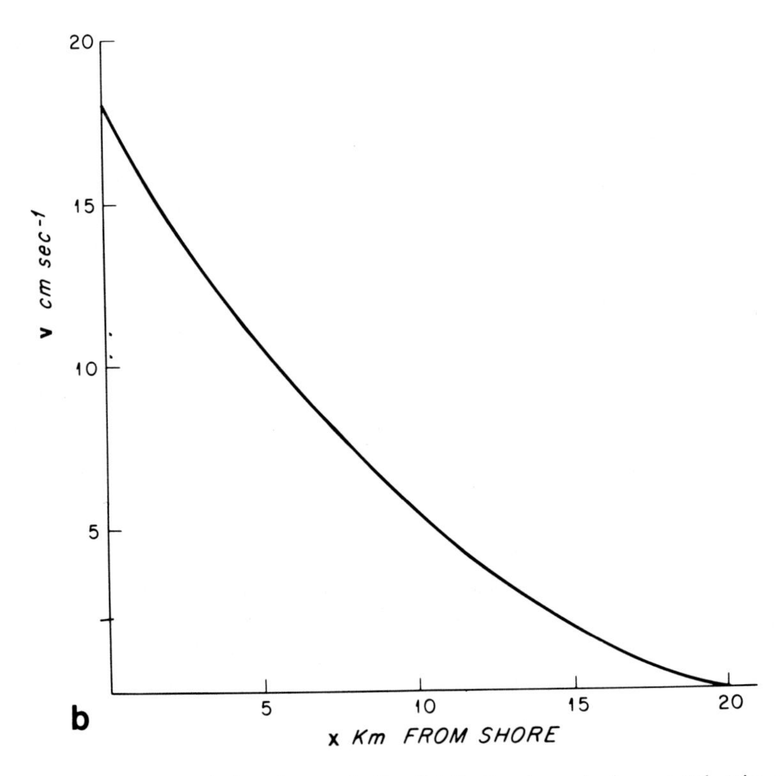

Fig. 4.11a–b. Elevation and alongshore velocity distribution in a sloping coastal strip of 20 km width, for the lowest mode topographic wave. Coastal elevation amplitude was arbitrarily assumed to be 1 cm.

having dropped the $\partial U/\partial t$ and the $f\ \partial\zeta/\partial t$ term and supposed $H = H(x)$. The first of Equations (4.17) gives, with the postulate (4.23) and no cross-shore wind stress:

$$fV = gH\frac{\partial\zeta}{\partial x}. \tag{4.46}$$

From these two equations the longshore transport is easily eliminated to yield:

$$\frac{\partial^2}{\partial t\partial x}\left(H\frac{\partial\zeta}{\partial x}\right) + f\frac{\mathrm{d}H}{\mathrm{d}x}\frac{\partial\zeta}{\partial y} = 0. \tag{4.47}$$

The boundary condition at the shore, $U = 0$, gives with the second of (4.17), after the longshore wind stops acting:

$$\frac{\partial V}{\partial t} = -gH\frac{\partial\zeta}{\partial y}, \quad (x = 0). \tag{4.48}$$

Substituting (4.46) this may be reduced to a condition on surface level:

$$\frac{\partial^2\zeta}{\partial t\partial x} = -f\frac{\partial\zeta}{\partial y}, \quad (x = 0). \tag{4.49}$$

Over the constant depth portion of the model, $x > l$, Equation (4.47) becomes degenerate, so that the boundary condition $\zeta \to 0$ at infinity is equivalent to

$$H\frac{\partial\zeta}{\partial x} = 0, \quad (x = l). \tag{4.50}$$

The longshore velocity thus reduces to zero at the outer edge of the coastal strip. It may be recalled that, in the general discussion of the previous section the vorticity wave was also supposed confined to a strip of finite width l. The arguments of that section depend on the existence of such a length, characterizing the 'trapping width' of a coastal current.

The solution of Equation (4.47) was discussed by Gill and Schumann (1974) under quite general conditions. The elevation is supposed separable into a cross-shore distribution $Z(x)$ (length) and a nondimensional amplitude function of y and t, $\phi(y, t)$:

$$\zeta = \phi(y, t)\ Z(x). \tag{4.51}$$

The condition of separability is found to be:

$$\frac{\partial\phi/\partial t}{\partial\phi/\partial y} = c = -\frac{fZ(x)\ \mathrm{d}H/\mathrm{d}x}{\mathrm{d}(H\ \mathrm{d}Z/\mathrm{d}x)/\mathrm{d}x} \tag{4.52}$$

with c = a positive constant of the dimension of velocity. The equation for ϕ is, from this:

$$c\frac{\partial\phi}{\partial y} - \frac{\partial\phi}{\partial t} = 0 \tag{4.53}$$

which has the general solution:

$$\phi = \phi(y + ct) \tag{4.54}$$

representing a wave of arbitrary wave-form propagating toward negative y. Recall here that the domain of interest chosen was $x \geqslant 0$, so that a wave propagating to negative y leaves the coast to the right (in the northern hemisphere) exactly as a Kelvin wave. In a closed basin such a wave propagates around the perimeter in a cyclonic direction.

The equation governing the $Z(x)$ distribution is

$$c \frac{\mathrm{d}}{\mathrm{d}x}\left(H \frac{\mathrm{d}Z}{\mathrm{d}x}\right) + f \frac{\mathrm{d}H}{\mathrm{d}x} Z = 0. \tag{4.55}$$

The boundary condition at the coast reduces to

$$\frac{c}{f}\frac{\mathrm{d}Z}{\mathrm{d}x} + Z = 0, \qquad (x = 0) \tag{4.56}$$

or equivalently, using (4.55):

$$\lim_{x \to 0}\left(H \frac{\mathrm{d}^2 Z}{\mathrm{d}x^2}\right) = 0, \qquad (x = 0). \tag{4.56a}$$

The boundary condition at the outer edge of the coastal strip is

$$H_0 \frac{\mathrm{d}Z}{\mathrm{d}x} = 0, \qquad (x = l). \tag{4.57}$$

Equations (4.55) to (4.57) pose an eigenvalue problem, possessing solutions for a discrete set of positive values of c only, c_1, c_2, c_3, ... which depend on the topography of the coastal strip. The equations also show at once that, for similar depth distributions, c_k are proportional to fl, as inferred in the earlier general discussion of vorticity waves.

4.5.2. Inclined Plane Beach – Coastal Strip Model

In order to illustrate further the character of the solutions, consider the simple case of an inclined plane beach occupying a limited coastal strip:

$$\begin{aligned} H &= sx, \qquad (0 \leqslant x \leqslant l) \\ H &= sl, \qquad (x > l). \end{aligned} \tag{4.58}$$

Equation (4.55) now simplifies to

$$\xi \frac{\mathrm{d}^2 Z}{\mathrm{d}\xi^2} + \frac{\mathrm{d}Z}{\mathrm{d}\xi} + Z = 0, \tag{4.59}$$

where

$$\xi = \frac{fx}{c}. \tag{4.60}$$

The solution, satisfying the boundary condition at the coast, (4.56), is

$$Z = AJ_0(2\xi^{1/2}), \tag{4.61}$$

where J_0 is a Bessel function and A is an arbitrary amplitude. The second boundary condition at the edge of the coastal strip, Equation (4.57), can be satisfied only for certain values of c, such that

$$J_1\left(2\sqrt{\frac{fl}{c}}\right) = 0. \tag{4.62}$$

From tables of the Bessel function $J_1(\ )$ it may be determined that

$$\frac{c_k}{fl} = \frac{1}{3.67}, \quad \frac{1}{12.3}, \quad \frac{1}{25.87}, \quad \text{etc.} \tag{4.63}$$

for $k = 1, 2, 3, \ldots$, these corresponding to the successive zeros of $J_1(\ )$. The longshore velocity is, from (4.46):

$$\frac{V}{H} = \frac{g}{f}\frac{\partial \zeta}{\partial x} = -\frac{gA\phi}{(cfx)^{1/2}} J_1\left(2\sqrt{\frac{fx}{c}}\right). \tag{4.64}$$

The value of this at the coast is

$$\frac{V}{H} = -\frac{gA\phi}{c}, \qquad (x = 0). \tag{4.64a}$$

The lowest wave mode, $k = 1$, has the simplest structure with monotonic variation of V/H, and has the highest celerity:

$$c_1 = \frac{fl}{3.67},$$

where the shore amplitude of the elevation is $A\phi = 1$ cm, the corresponding longshore velocity amplitude is $-3.67\ g/fl$, or about 18 cm s^{-1} for $l = 20$ km.

With the same width coastal strip $c_1 = 0.54$ m s^{-1}. Over a much wider strip the celerity becomes proportionately larger. The coastal elevation amplitude increases with l for the same longshore velocity amplitude, geostrophic balance of a broader current requiring this. Figure 4.11 illustrates the distribution of elevation and longshore velocity in the lowest mode topographic wave, $k = 1$ over the inclined plane beach – coastal strip model, with $l = 20$ km.

Returning now to the topographic gyre pattern of Figure 4.6, which in Section 4.5 was inferred to propagate cyclonically around a basin, one observes that the inclined plane beach – coastal strip model used above is able to account for the propagation of an arbitrary initial distribution of elevation only over the *sloping* part of the model, i.e., at $x \leqslant l$. The topographic gyre pattern may be thought expanded in terms of the eigenfunctions, each representing a topographic wave mode. However, the return flow leg of the topographic gyre within the flat bottomed central basin cannot be so treated and one concludes that the coastal strip model is overidealized. Physically, the central basin return flow leg of the topographic gyre may also be expected to initiate oscillatory motions (perhaps of low frequency), which can only be analyzed using a more realistic complete basin model. To be sure, the dominant observable motions in the topographic gyre occur near the coasts, and the propagation of these can be adequately represented by the coastal strip model. The picture is nevertheless incomplete without some insight into basin-wide topographic mode oscillations.

4.5.3. Basin-wide Vorticity Wave Model

The possible occurrence in realistic basins of vorticity waves has been recognized in classical hydrodynamics, where they were sometimes referred to as 'second class motions' (Ball, 1965). In discussing the oscillations in a circular paraboloid, Lamb (1932) identifies a 'comparative slow oscillation' which tends to steady rotational motion as $f \to 0$. This remains the simplest model of a basin-wide vorticity wave and will be briefly discussed here.

Let the depth of a circular basin vary as

$$H = H_0 \left(1 - \frac{r^2}{r_0^2}\right), \tag{4.65}$$

where r_0 is the radius of the shores, and H_0 is the central, maximum depth. Polar coordinates are conveniently introduced, so that the transport equations applying to free oscillations are as written down in Equation (2.73), but for variable depth, and with the forcing terms deleted. On taking curl, the following vorticity tendency equation emerges

$$\frac{\partial}{\partial t}\left[\frac{\partial (rV_\phi)}{\partial r} - \frac{\partial V_r}{\partial \phi}\right] = -g \frac{dH}{dr}\frac{\partial \zeta}{\partial \phi} \tag{4.66}$$

having supposed $\partial\zeta/\partial t \cong 0$ for the slow oscillations in question. The transport components may be eliminated after a divergence operation on (2.73) and yield

$$\frac{\partial}{\partial t}\left[\frac{\partial}{\partial r}\left(rH\frac{\partial \zeta}{\partial r}\right) + \frac{H}{r}\frac{\partial^2 \zeta}{\partial r^2}\right] = f\frac{dH}{dr}\frac{\partial \zeta}{\partial \phi}. \tag{4.67}$$

From the general discussion of Articles 193 and 212 of Lamb (1932) the following solution is singled out:

$$\zeta = A\left(\frac{r}{r_0}\right)\left(1 - \frac{3}{2}\frac{r^2}{r_0^2}\right)\cos(\phi + \sigma t), \tag{4.68}$$

where A is an arbitrary amplitude (half of which is the shore elevation amplitude) and σ is the frequency, given to a high degree of approximation by, neglecting a term of order σ^2/f^2:

$$\frac{f}{\sigma} = -7 - \frac{f^2 r_0^2}{gH_0}. \tag{4.69}$$

For typical basins, $r_0 << f^{-1}(gH_0)^{1/2}$, so that $\sigma \cong -0.14f$, or indeed small compared to the inertial frequency. The negative value of σ means that the wave propagates in the positive ϕ or cyclonic direction, much as a Kelvin wave.

A transport stream function may be introduced in view of the approximation $\partial\zeta/\partial t$, by:

$$V_r = -\frac{1}{r}\frac{\partial\psi}{\partial\phi}, \qquad V_\phi = \frac{\partial\psi}{\partial r}. \tag{4.70}$$

To the same approximation as before this may be calculated to be

$$\begin{aligned}\psi &= \frac{gH}{f}\left(\zeta - \frac{\sigma}{f}r\frac{\partial\zeta}{\partial r}\right) = \\ &= A\frac{gH}{f}\left(1 - \frac{r^2}{r_0^2}\right)\frac{r}{r_0}\left[1 - \frac{\sigma}{f} - \frac{3r^2}{2r_0^2}\left(1 - 3\frac{\sigma}{f}\right)\right]\cos(\phi + \sigma t).\end{aligned} \tag{4.71}$$

The elevation contours and streamlines of this vorticity wave model are illustrated in Figures 4.12a and b, for $t = 0$. The pattern propagates around the perimeter unchanged, with angular speed σ.

Noteworthy is the presence of coastal currents, directed toward $\phi = \pi/2$ along the diameter $\phi = 0$, and return flow in the center. The entire pattern is similar to that of a topographic gyre and is clearly likely to be excited by a wind impulse toward $\pi/2$. A further interesting characteristic is that velocities at fixed points rotate clockwise near the perimeter as the pattern propagates, anticlockwise near the center.

Lamb's results have been extended to an elliptical basin by Ball (1965). The basin-wide vorticity wave corresponding to the one just discussed is illustrated by Saylor *et al.* (1980) for such an elliptical model. Figure 4.13 shows an illustration of such a wave taken from the latter paper.

Higher mode vorticity waves (of higher azimuthal or radial wave numbers) have characteristics that approach those of coastally trapped vorticity waves more and more as the mode number increases.

4.6. STRATIFIED FLUID OVER REALISTIC TOPOGRAPHY

It remains to discuss the interaction of stratification and depth variations, a subject that combines the difficulties encountered in the present chapter with those of the previous one. Early in Chapter 3 it was pointed out that in a stratified fluid internal material surfaces move vertically through distances of the order of tens of meters both in wind-forced and free (wave-like) motions. The attendant stretching and squashing of fluid columns is clearly an effective vorticity generating mechanism. This mechanism was seen to act with maximum intensity within a coastal boundary layer scaled by the internal radius of deformation, a distance of typically 5–10 km. Within the same distance range vorticity generation associated with bottom slope is also effective, so that a satisfactory theoretical model has to take into account vorticity tendencies due to both isopycnal movements and bottom slope.

By and large, satisfactory theoretical models of stratified flow over variable depth have yet to be developed. Existing models are all based on linearized theory discussed in detail in Chapter 3, with depth variations included either as boundary conditions, for a

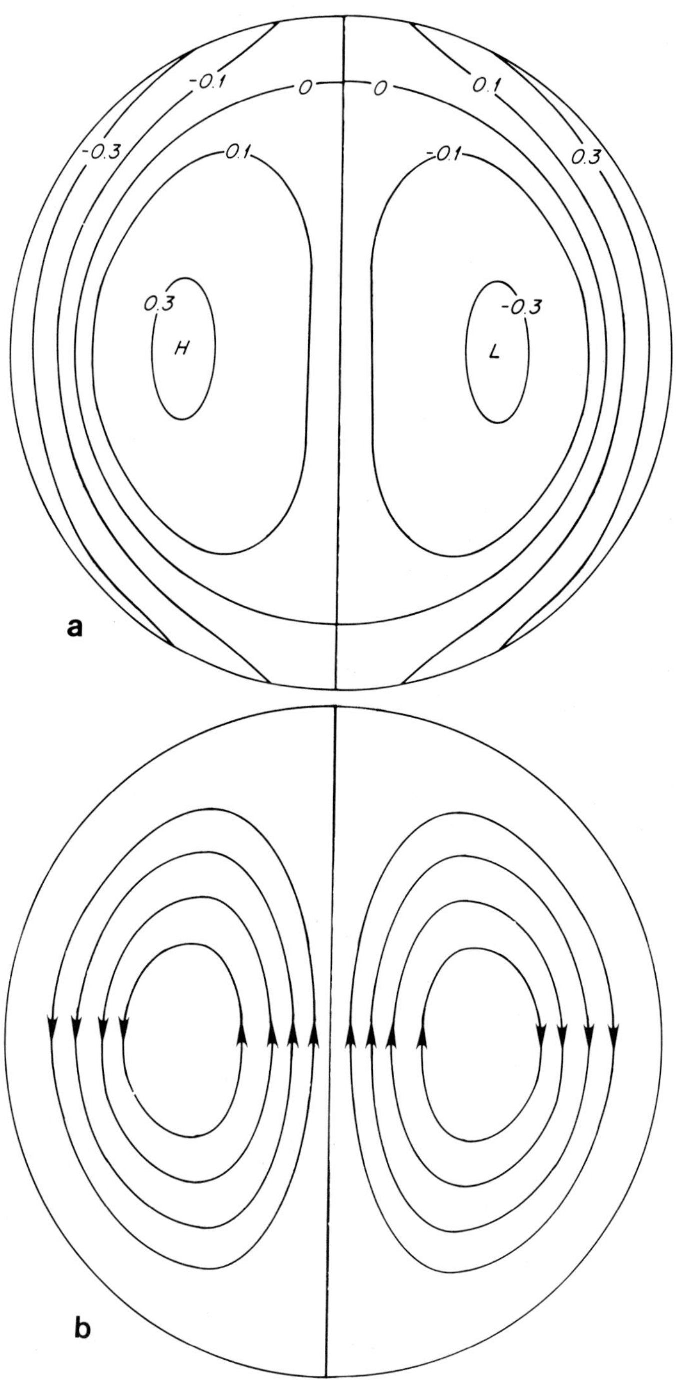

Fig. 4.12a–b. Elevation distribution, (a), and transport streamline pattern, (b), in basin-wide vorticity wave model in circular paraboloid. Note general similarity to topographic gyres discussed earlier.

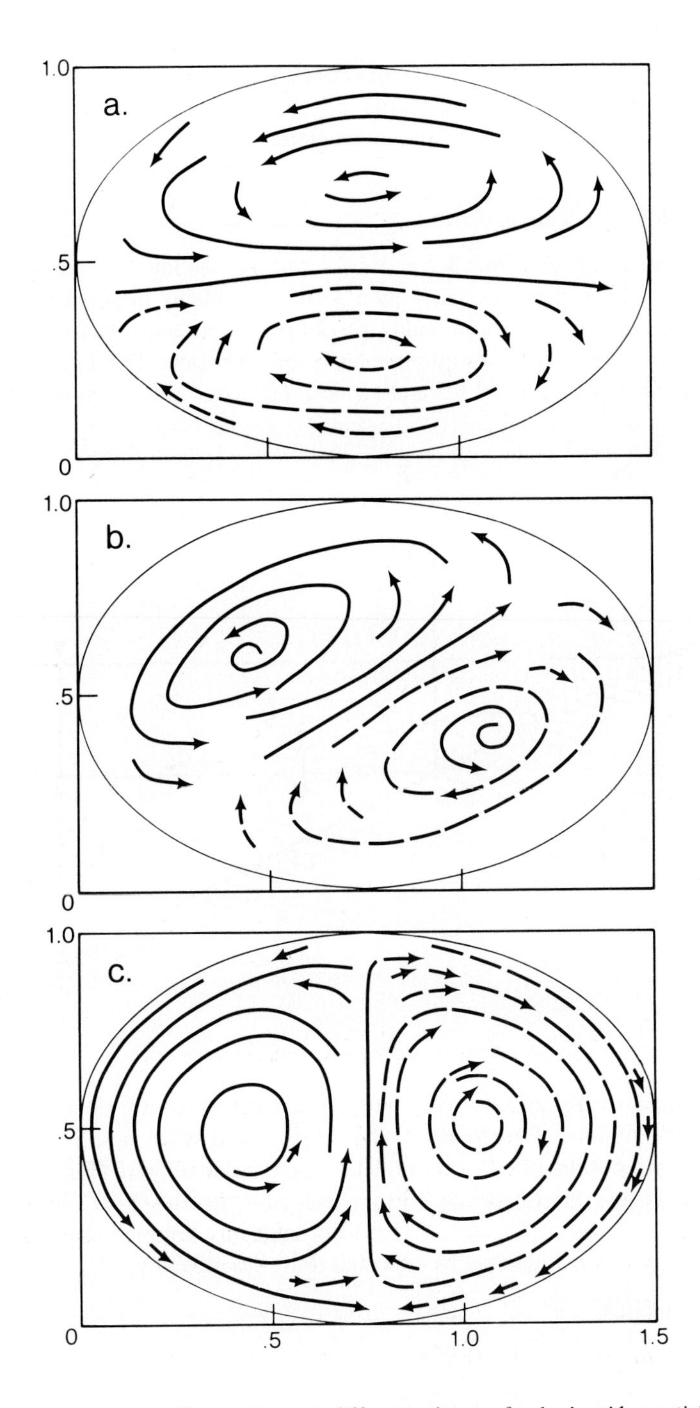

Fig. 4.13. Transport streamline pattern at different phases of a basin-wide vorticity wave in an elliptical paraboloid. From Saylor *et al.* (1980).

continuously stratified fluid column, or directly in the transport equations for a two-layer fluid. Even within the severe limitations of linearization the theoretical difficulties have not been fully resolved yet, and key simple models have not yet crystallized. The discussion below is therefore brief, cursory and incomplete.

4.6.1. Coastal Jet Over Sloping Beach

The linearized models of Sections 3.7 and 4.4.2 can be combined into a two-layer fluid model over an inclined plane beach, Figure 4.14. A suddenly imposed longshore wind may be expected to generate a developing longshore current, with the longshore momentum absorption of the top layer presumably predominating. The calculations follow closely those in Sections 3.7 and 4.4.2 and can be summarized quickly (following Csanady, 1977a).

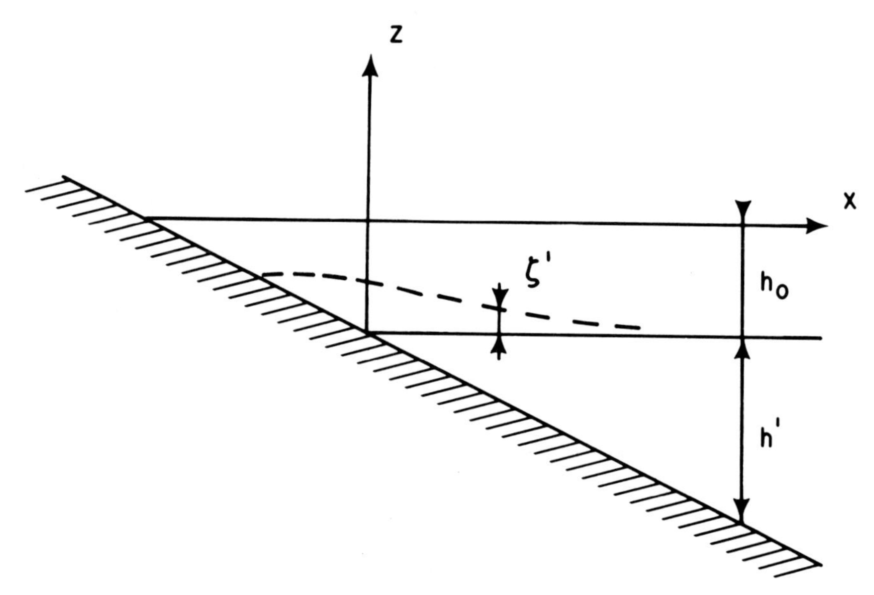

Fig. 4.14. Two layer fluid over inclined plane beach.

To the accuracy of linearized theory and of the hydrostatic approximations the two layer transport equations (Equations (3.48)) remain valid when h and h' are functions of the horizontal coordinates. In the model to be considered here (Figure 4.14) $x = 0$ where the pycnocline intersects the bottom and only the two-layer portion $x \geqslant 0$ is considered; there $h = h_0$ = constant, $h' = sx$. For suddenly imposed longshore wind the response of the two-layer fluid may be expected to be described by:

$$\begin{aligned} U &= U(x), & U' &= U'(x), \\ V &= B(x)t, & V' &= B'(x)t, \\ \zeta &= W(x)t, & \zeta' &= W'(x)t. \end{aligned} \tag{4.27}$$

Here B and B' are depth-integrated longshore accelerations in top and bottom layers, and W and W' are vertical velocities of surface and pycnocline, all functions of x only.

On substitution into Equations (3.48) the longshore accelerations and vertical velocities may be eliminated and the problem reduced to two equations for the cross-shore transport components U and U':

$$-\frac{gh}{f}\frac{d^2}{dx^2}(U + U') + fU = F,$$
$$-\frac{gh'}{f}\frac{d^2}{dx^2}(U + U') + \frac{g\epsilon h'}{f}\frac{d^2U}{dx^2} + fU' = 0, \tag{4.73}$$

where F is the kinematic longshore wind stress. In analogy with the barotropic and baroclinic modes of the two-layer, constant depth model, solutions are sought to Equations (4.73) consisting of two additive components, U_1 and U_2, U'_1 and U'_2. The first component is to have (nearly) constant velocity, the second (nearly) zero total transport:

$$U = U_1 + U_2, \qquad U' = U'_1 + U'_2,$$
$$\frac{U'_1}{h'} - \frac{U_1}{h} = 0\left(\epsilon\frac{U_1}{h_0}\right), \tag{4.74}$$
$$U_2 + U'_2 = 0\,(\epsilon U_2).$$

After some manipulations Equations (4.73) may now be reduced to two uncoupled differential equations:

$$\frac{d^2}{dx^2}(U_1 + U'_1) - \frac{f^2}{g(h+h')}(U_1 + U'_1) = -\frac{fF}{g(h+h')},$$
$$\frac{d^2U_2}{dx^2} - \frac{f^2}{g\epsilon}\left(\frac{1}{h} + \frac{1}{h'}\right)U_2 = -\frac{fF}{g\epsilon h}. \tag{4.75}$$

The first of these is the same equation that can be obtained from (4.17) for total cross-shore transport in a homogeneous fluid of (variable) depth $H = h + h'$. Its solution is as discussed in Section 4.4.2 and need not be considered further. The second equation (4.75), for $h' = sx$ can be reduced to

$$\frac{d^2U}{dx^2} - \left(\frac{1}{4} + \frac{k}{x}\right)U = -\frac{1}{4} \tag{4.76}$$

which is a non-homogeneous Whittaker equation (Babister, 1967). The variables in (4.76) are nondimensional, given by

$$U_* = \frac{fU_2}{F}, \qquad x_* = \frac{2x}{R_2},$$
$$k = \frac{h_0}{2sR_2}, \qquad R_2 = \frac{(\epsilon g h_0)^{1/2}}{f}, \tag{4.77}$$

with the stars dropped in (4.76). A relatively simple explicit solution for (4.76) may be found for the specific value of the parameter $k = 0.5$, which results, for example if

$$\begin{aligned} s &= 3 \times 10^{-3}, \\ \epsilon &= 2 \times 10^{-3}, \\ h_0 &= 18 \text{ m}, \\ f &= 10^{-4} \text{ s}^{-1}, \end{aligned}$$

all of which are reasonable 'typical' quantities. The corresponding internal radius of deformation R_2 is 6 km.

Once U_2 has been calculated, pycnocline vertical velocity W'_2 and the top-layer longshore acceleration B_2 follow from:

$$\begin{aligned} W'_2 &= \frac{\mathrm{d}U_2}{\mathrm{d}x}, \\ B_2 &= -\frac{g\epsilon h h'}{f(h+h')}\frac{\mathrm{d}^2 U_2}{\mathrm{d}x^2}. \end{aligned} \tag{4.78}$$

The solution for B_2 is illustrated in Figure 4.15 using the above typical parameters. Physically, B_2 represents the difference between the homogeneous fluid response (uniform acceleration throughout the water column) and the stratified response. In other words, B_2/F is the fraction by which the top layer acceleration exceeds its share of the uniform response. For instance, in Figure 4.15, at 6 km from the pycnocline-bottom intersection the depths of top and bottom layers are equal. Uniform response would be 0.5 F for each layer's (depth-integrated) acceleration. The actual split is different by $0.09F$, i.e., $V = 0.59Ft$, $V' = 0.41Ft$, which is a relatively mild difference, considerably less than one would find in a constant depth model with $h' = h_0$. One concludes that bottom slope *reduces* the concentration of momentum in the top layer of a coastal jet.

The physical reason for this behavior emerges on considering the motion of the bottom layer. The pycnocline velocity W'_2 at $x = 0$ is found to be in this model.

$$W'_2 = \left(\frac{\pi}{2} - 1\right)\frac{F}{fR_2}, \quad (x = 0). \tag{4.79}$$

A rising pycnocline, given the kinematics of the model, implies a shoreward velocity $u' = -W'/s$, which is with $k = 0.5$, $R_2 = h_0/s$:

$$u' = -\left(\frac{\pi}{2} - 1\right)\frac{F}{fh_0}, \quad (x = 0). \tag{4.80}$$

This is an adjustment drift velocity generating longshore acceleration, hence

$$v' = \left(\frac{\pi}{2} - 1\right)\frac{Ft}{h_0}, \quad (x = 0). \tag{4.81}$$

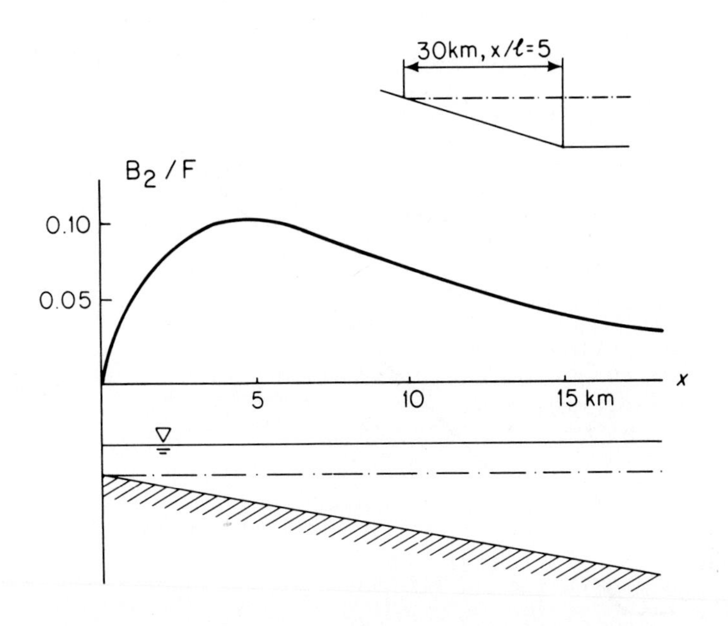

Fig. 4.15. Baroclinic mode contribution to top layer transport in inclined plane beach model, forced by impulsive longshore wind. Ordinate unit is longshore wind stress, equal also to depth integrated total acceleration. From Csanady (1971).

The depth of the bottom layer at $x = 0$ is zero, so that $v = Ft/h_0$ in the top layer. Bottom layer velocity is more than half this large, and one notes that the contrast between top and bottom layer velocity is considerably reduced, already at the pycnocline-bottom intersection, as compared to the constant depth case. The physical reason is that the pycnocline-bottom intersection moves shoreward (or offshore, for oppositely directed wind) as the pycnocline rises (or falls). The consequent adjustment drift generates longshore flow in the same direction as the wind, partly compensating for the effect of stratification on momentum distribution within the water column.

4.6.2. Vorticity Waves With Stratification and Bottom Slope

In the discussion of the internal Kelvin wave model in Section 3.9.1 it was pointed out that the large vertical movements of the pycnocline expected in these waves, of an amplitude as excited by even modest wind impulses, lead to rapid generation of vorticity. The typical thermocline slope associated with a velocity contrast of 0.5 m s^{-1} between top and bottom layers is of the order of 3×10^{-3}, or much the same as the typical bottom slope. Thus the kinematic effect of either thermocline or bottom slope on fluid columns moving in a cross-shore direction is about the same, if a large amplitude Kelvin-like wave occurs over sloping bottom. One surmises that effects of stratification and of bottom slope in models of coastally trapped vorticity waves are coupled.

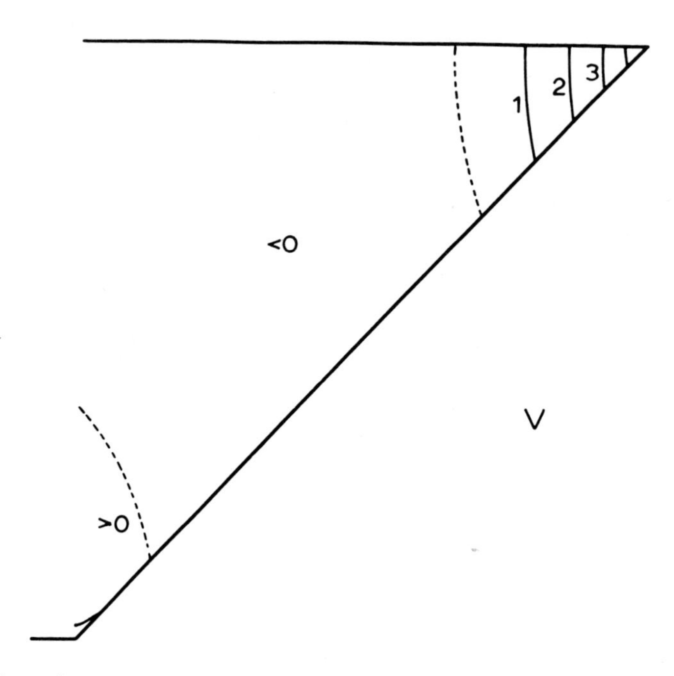

Fig. 4.16. Lines of constant longshore velocity amplitude in hybrid topographic-Kelvin wave at low stratification. Pattern is nearly the same as in a homogeneous fluid. From Huthnance (1978).

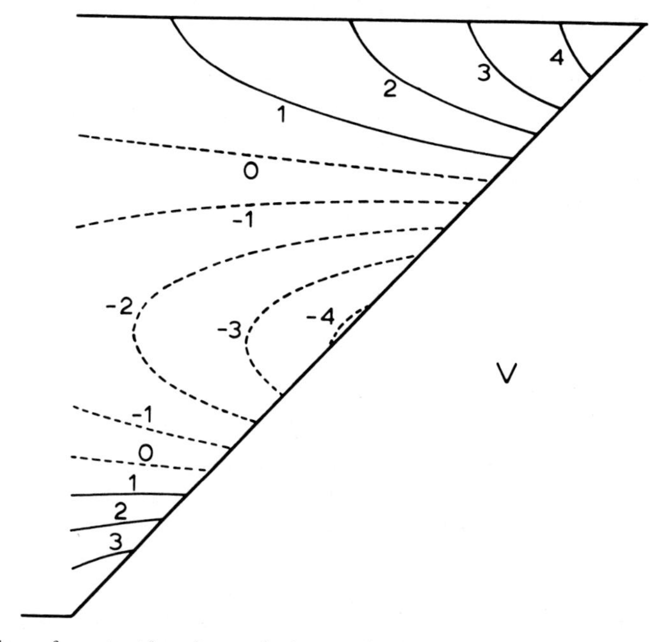

Fig. 4.17. Lines of constant longshore velocity amplitude in hybrid-topographic-Kelvin wave at high stratification. Pattern is much as in a constant depth model internal Kelvin wave.

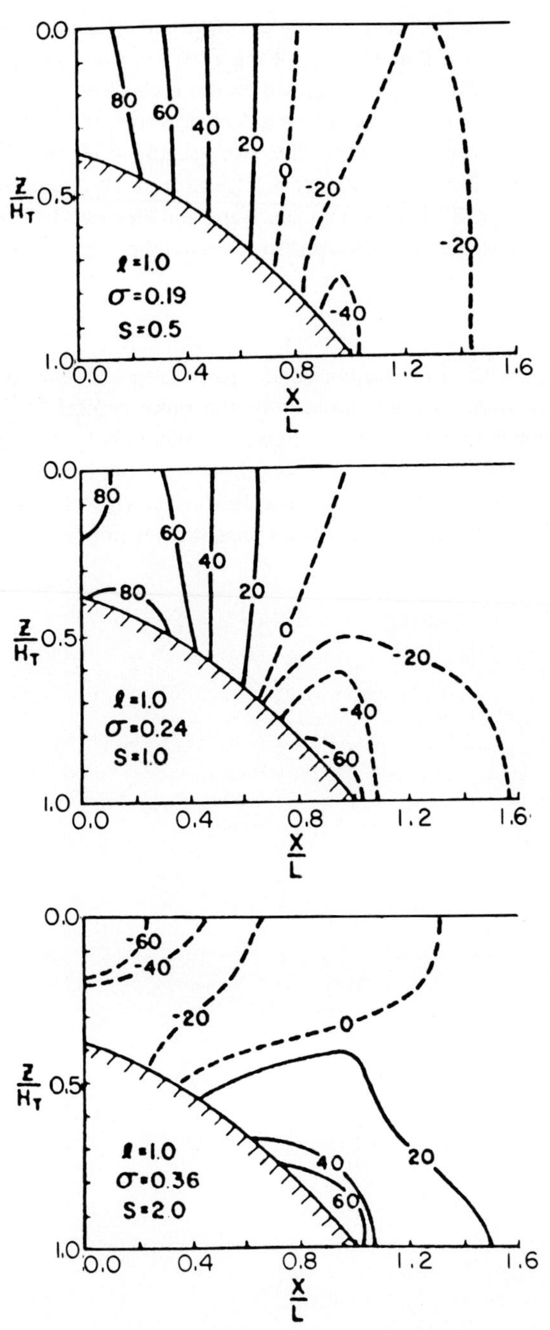

Fig. 4.18. Hybrid topographic-Kelvin waves (longshore velocity distribution as in previous two figures) in a model of Wang and Mooers (1976), for three cases of different stratification, increasing from top to bottom.

The complexity of the problem was indeed revealed in a number of recent investigations, notably those of Allen (1975), Wang (1975), Wang and Mooers (1976) and Huthnance (1975, 1978). While some trapped modes are nearly the same as homogeneous fluid topographic waves, and others as internal Kelvin waves, some of the practically most important modes are hybrids between the two. Figure 4.16 from Huthnance (1978) illustrates a topographic-like mode, Figure 4.17, also from Huthnance (1978), an internal Kelvin-wave like mode, and Figure 4.18 from Wang and Mooers (1976) also a hybrid. The criterion separating these classes is the stratification number:

$$S = \frac{\epsilon g H}{f^2 l^2}, \tag{4.82}$$

where l is the scale width of the sloping shelf region. Large values of S apply only to very narrow shelves (l of order 5 km), while with the more typical $l = 20$ km or more the stratification number is small. There are, however, many complications, the precise geometry of the coastal region as well as the longshore wavenumber playing important roles.

The theory of coastally trapped waves is still under vigorous development (see e.g., Mysak, 1980) and it is reasonable to expect that further important insights will emerge in the near future.

CHAPTER 5

Transient Coastal Currents

5.0. INTRODUCTION

In the context of the coastal circulation problem, 'transient' currents will be taken to mean wind-induced motions of a time scale much longer than f^{-1}, but shorter than the time in which bottom friction comes to control the pattern of flow. The foregoing three chapters have developed simple theoretical models for gaining physical insight into the structure and dynamics of such transient currents. In the present chapter an attempt will be made to connect the theoretical results quantitatively to observational evidence.

The theoretical investigations have revealed many subtle and complex phenomena, which one cannot expect to verify without comprehensive, suitably well documented observations. As short a time as a decade ago such observational detail was not available. Attempts have been made to deduce the structure of currents along a given piece of coastline, for example, from records collected at a single mooring, but these have not been convincing or conclusive. Not until larger cooperative experiments had been carried out and their results analyzed did a suitably broad base of empirical knowledge accumulate for a conclusive theoretical interpretation.

Two of the large-scale experiments which have yielded detailed and well documented evidence on transient currents were the International Field Year on the Great Lakes (IFYGL) on Lake Ontario, 1972 and 1973, and the Coastal Upwelling Experiment (CUE) off the Oregon coast (early seventies).

Coastal waters off the North American Pacific coast and in the Great Lakes are subject to frequent wind impulses and respond primarily in the inertial rather than the frictional mode. The reason is partly that tidal motions are absent in the Great Lakes, weak enough over the Pacific shelf not to enhance bottom friction significantly. By contrast, the east coast continental shelf of North America north of Cape Hatteras is subject to strong tides, which generate relatively large bottom stress and lead to a dominance of the response to weather-cycle forcing by bottom friction rather than inertia, as will be discussed in greater detail in later chapters. At the same time, the Oregon shelf is not subject to remote influences strong enough to submerge local response to forcing, unlike, say, the South Atlantic Bight Shelf, the outer portion of which is under the direct influence of fluctuating events in the Gulf Stream system. It is fortunate that detailed observations have been carried out in two shallow seas with relatively simple dynamical response characteristics. The discussion in this chapter will relate exclusively to these two environments, which may be regarded as relatively pure prototypes.

Even in such a favorable environment, well documented examples of clear-cut isolated wind-stress episodes are rare. A few of these have been documented, however, and will be discussed below, along with other statistical (climatological) evidence on currents, sea level, isopycnal surface behavior, etc. The objective here is to connect these observations to the simple theoretical models of Chapters 2, 3, and 4, rather than to give a full account of the observational evidence. Further detail may be found in the papers cited.

5.1. LONGSHORE VELOCITY AND TRANSPORT

Water movement observations during the International Field Year on the Great Lakes in Lake Ontario included a detailed survey of five coastal transects around the lake, with 1 km spatial resolution and with a frequency of once per day, weather permitting (Csanady, 1973, 1974, 1976; Csanady and Scott 1974, 1980). Various aspects of the IFYGL results have been discussed by Bennett (1977), 1978); Birchfield and Hickie (1977); Blanton (1974, 1975); Boyce (1974, 1977); Marmorino (1978, 1979); Simons (1973, 1974, 1975, 1976) and others, see references in the quoted papers. Figure 5.1 shows the location of the coastal transect observations around Lake Ontario.

A few clear-cut wind impulses were observed during IFYGL following relatively quiescent periods. The coastal transect observations on the days immediately after the storm provided observational evidence on inertial response with unique spatial detail. Two wind impulses occurred on August 6 and August 9, 1972. Preceeding these storms several

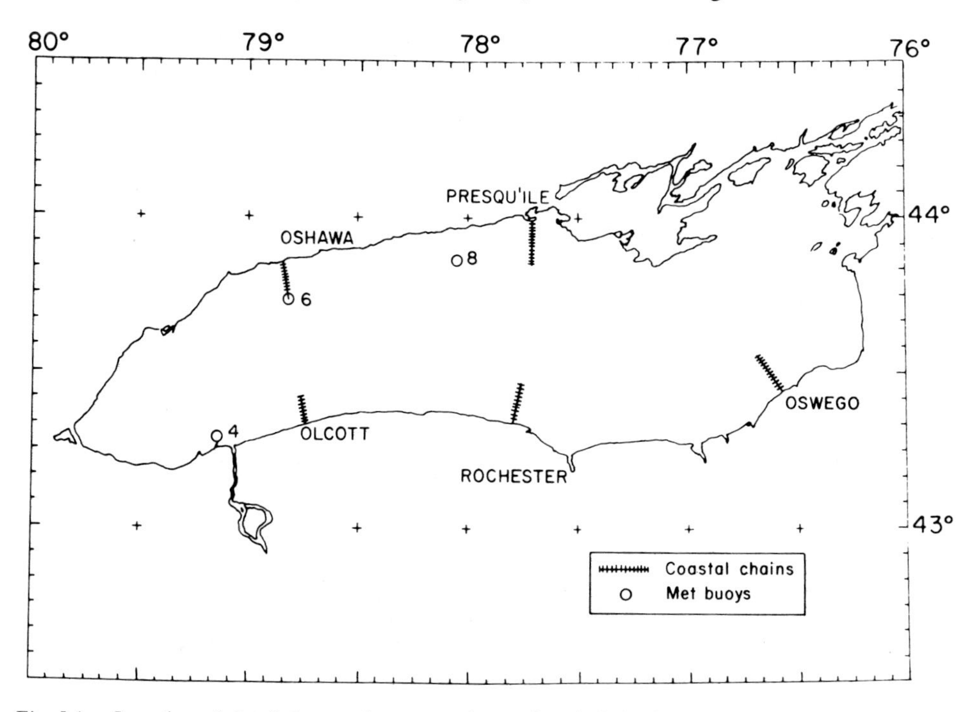

Fig. 5.1. Location of detailed coastal transect observations in Lake Ontario during the International Field Year on the Great Lakes (IFYGL).

quiescent days allowed coastal currents to decay to negligible velocities. During both storms the wind stress was directed along the EW or long axis of the lake; during the first episode on August 6, in the eastward, during the second, on August 9, in the westward direction. The wind-stress impulse of the first storm was estimated at $I = 2.5 \text{m}^2 \text{ s}^{-1}$, the second at $I = -9 \text{ m}^2 \text{ s}^{-1}$, the easterly direction being taken as positive.

Current surveys in the Oshawa-Olcott or western transect were taken on August 8 and 10, half a day or so after the storms had died down. Figure 5.2 shows the longshore transport V determined by these observations as a function of distance from shore.

Along the south shore at about 7 km from shore, and along the north shore approximately at 9 km, the longshore transport V peaked fairly close to the value predicted by the inertial response theory ($V \cong I$). It also dropped to zero at about the locus of section-average depth, 9 and 16 km respectively from the shores, which, according to theory, is a result of wind setup. The low values of transport near shore may be attributed to bottom friction which presumably limited longshore velocities to moderate values and also destroyed longshore motion in shallow water between the storm and the beginning of the survey. Nearshore frictional effects are discussed further in Chapter 6. By and large, these and other similar observations in Lake Ontario show beyond doubt the first-order correctness of the inertial response theory, especially as expounded in Chapter 4.

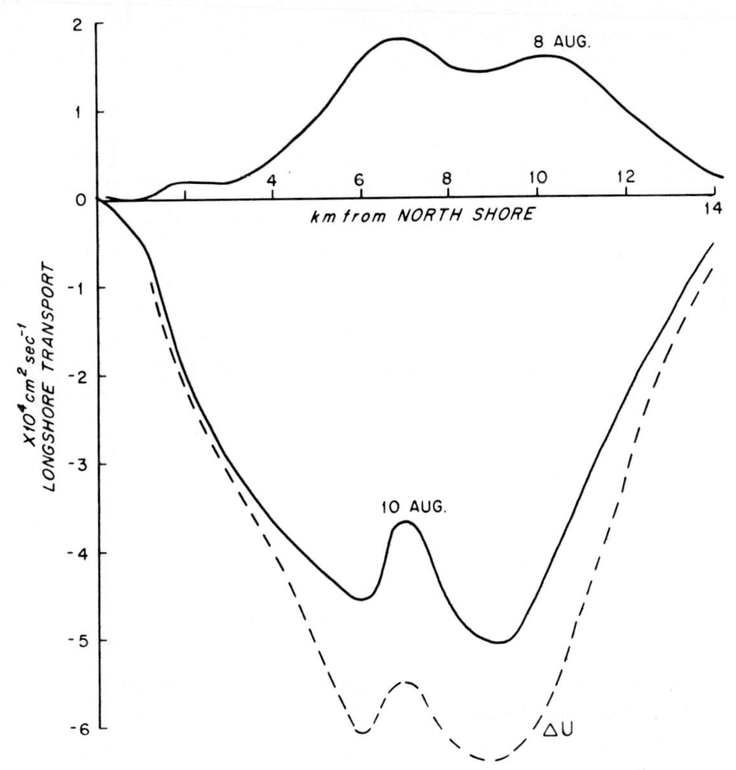

Fig. 5.2a. Observed longshore transport versus distance from shore at Oshawa, following storms of August 6 and August 9. From Csanady (1973).

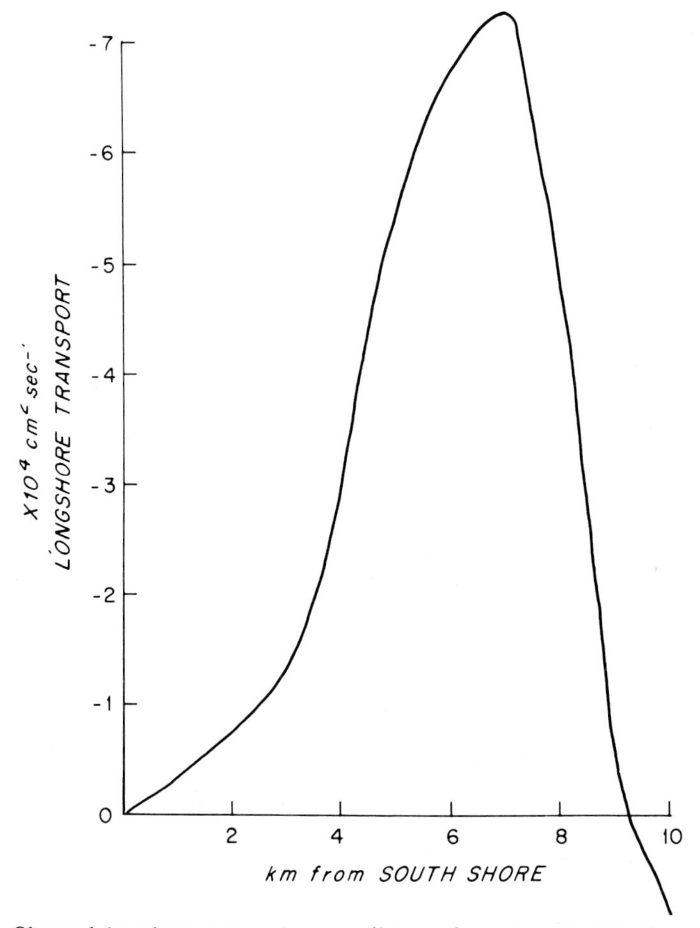

Fig. 5.2b. Observed longshore transport versus distance from shore at Olcott coastal transect, following storm of August 9. From Csanady (1973).

The Oregon shelf has also been subject to intensive observational studies, now for almost two decades (Collins *et al.*, 1968; Collins and Patullo, 1970; Huyer and Patullo, 1972; Halpern, 1974; Cutchin and Smith, 1973; Huyer *et al.*, 1974; Smith, 1974; Huyer *et al.*, 1974; Halpern, 1976; Mooers *et al.*, 1976a; Kundu and Allen, 1976; Huyer *et al.*, 1978, 1979; with further references given in these papers). Much of the Oregon work has been oriented toward the understanding of the seasonal upwelling cycle and its biological implications, but a considerable amount of evidence was also accumulated on the dynamics of wind driven transient currents.

Several studies on the Oregon shelf, notably Smith (1974) showed conclusively that longshore wind impulses were associated with longshore current fluctuations distributed more or less evenly over the water column near shore, where the pycnocline comes to the surface, as well as outside the range of isopycnal distortion, e.g., along the 100 m isobath. Figure 5.3 from Smith (1974) demonstrates the latter point. In such deeper

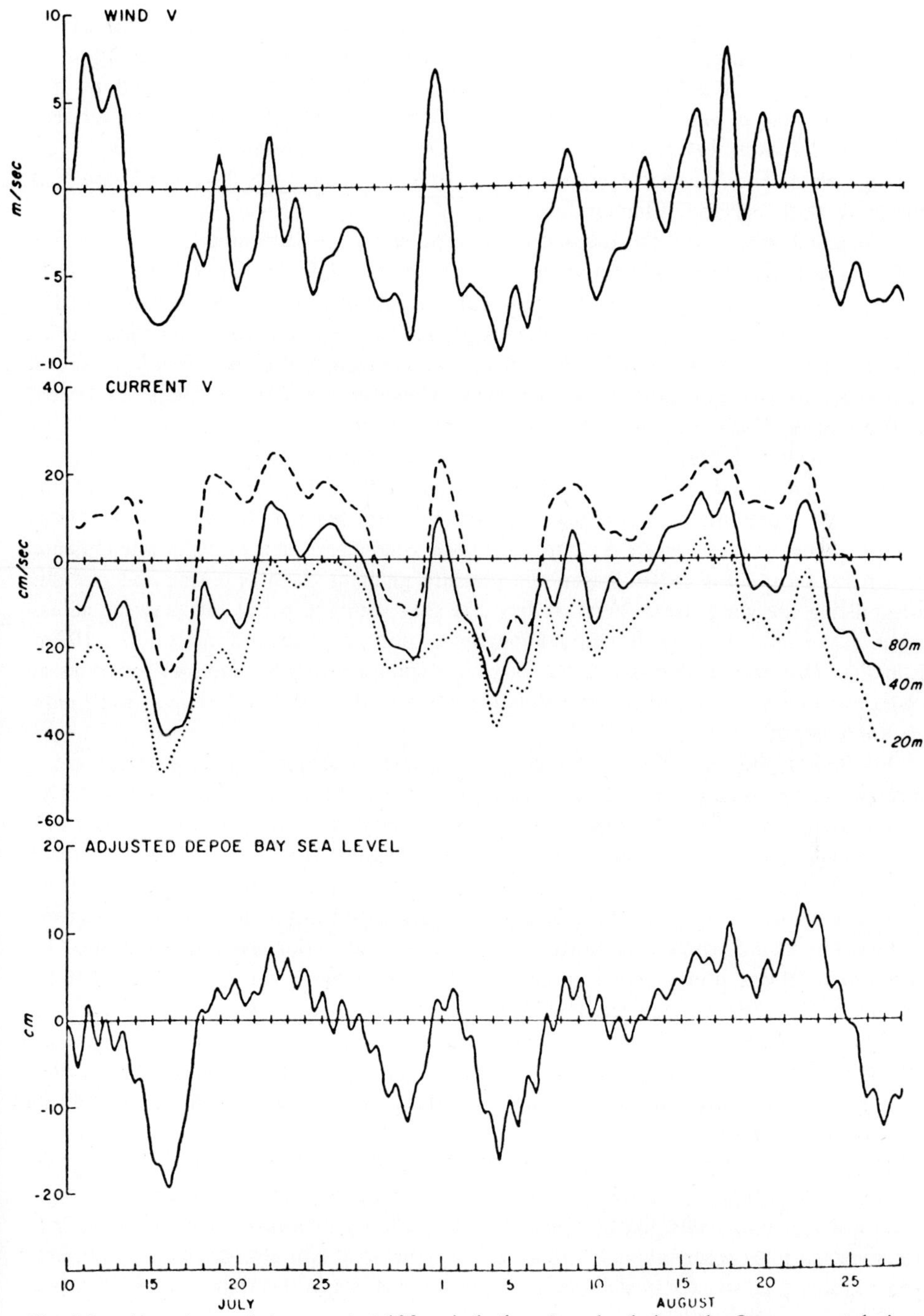

Fig. 5.3. Alongshore wind, current at 100 m isobath and sea level along the Oregon coast during a 1972 observation period. From Smith (1974). Copyright American Geophysical Union.

water opposing pressure gradients are important and one does not expect the simple equivalence of transport V and wind impulse I to hold. However, the fluctuations in both V and I are seen to be correlated and to be of the same order of magnitude. It is remarkable that in spite of pronounced stratification, and the presence of a surfaced pycnocline, coastal waters should respond in certain respects essentially as if they were homogeneous. The observed facts are thus in striking accord with the general results of the theoretical analysis in Chapter 3.

A detailed analysis of the longshore and cross-shore momentum balances of Oregon shelf currents has been carried out by Allen and Kundu (1978). Figure 5.4 taken from this analysis of currents along the 100 m isobath i.e., about 12 km from the coast, at 80 m depth, shows a direct comparison of the longshore Coriolis force due to adjustment drift, fu, with the observed acceleration $v_t = dv/dt$. The adjustment drift was clearly important and responsible for a portion of the observed v_t. However, it is also true that if adjustment drift alone had been responsible for longshore acceleration, one would have $Y_t = v_t + fu = 0$. In fact this quantity was found to be different from zero, see the third diagram in Figure 5.4. Some of the flucuations of Y_t may be noise, but a smoothed version presumably represents a fluctuating longshore pressure gradient force. There is a weak inverse relationship between Y_t and longshore wind-stress, suggesting a phenomenon similar to setup in a closed basin, the pressure gradient force opposing and partially neutralizing the wind stress. Very crudely, the split seems to be 50–50 between adjustment drift-related Coriolis force, and longshore pressure gradient force (at the 100 m isobath). This means that a 0.2 Pa longshore wind stress (for example) generates an opposing sea level gradient of approximately $\partial\zeta/\partial y = 10^{-7}$, as it would in a closed basin of 200 m depth.

Frictionless, inertial adjustment theory is not able to account for the development of a setup-like opposing pressure gradient along a long open coast. Longshore gradients arise, according to this theory, because of the nonuniformity of forcing, initially in phase with the forcing, and then propagating slowly in the cyclonic direction. Along the Oregon coast, this would not be expected to produce results such as shown in Figure 5.4. However, the theory of *frictional* equilibrium flow (developed in the next chapter, see especially Section 6.5.2) shows that bottom friction *does* lead to the development of opposing pressure gradients pretty much as observed. A reasonable conjecture is then that the observed Y_t is due to such frictional effects, which must be present at least very close to shore (as already pointed out in connection with Figure 5.2). To verify this conjecture it would be necessary to investigate theoretically inertial adjustment with friction taken into account, and observe the nearshore flow structure in greater detail than has so far been done off Oregon.

Huyer *et al.* (1978) and Hickey and Hamilton (1980) further analyzed the dynamics of transient currents on the Oregon shelf. They confirmed Smith's (1974) results and specifically demonstrated that a model forced *locally* by longshore wind stress accounts for most current observations at mid-shelf. On the long and straight west coast shelf, where forcing is by large-scale weather systems, this is exactly what one would expect from inertial response theory.

In a particularly revealing study of the onset of the summer upwelling regime off

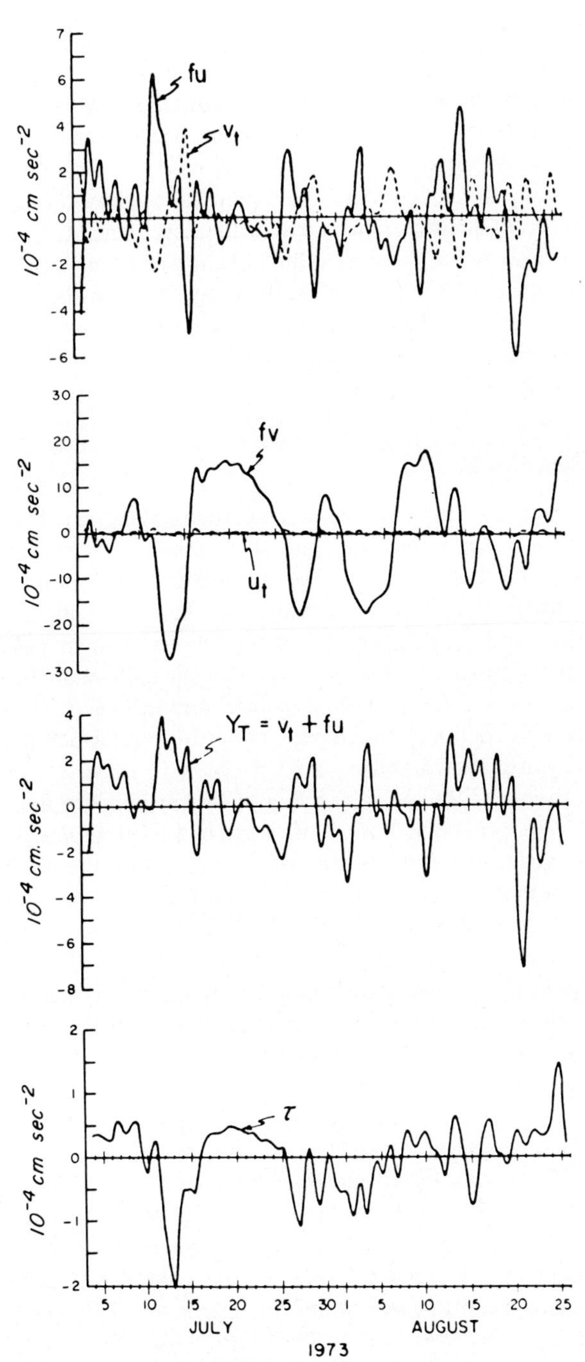

Fig. 5.4. Terms in longshore and cross-shore momentum balance, observed along Oregon coast in 1973. Bottom graph is wind stress, others as defined in text. From Allen and Kundu (1978).

Oregon Huyer *et al.* (1979) analyze the details of the water column's response to a major southward wind-stress event. They find that the coastal sea level responds to the wind with a time lag of about 2 hr, while the longshore current along the 100 m isobath develops another 7 hr later. These values are again very much as one would expect on the basis of the geostrophic adjustment models.

On the whole, both the Oregon and the Great Lakes observations demonstrate the overriding importance of inertial adjustment in the development of longshore currents. Moreover, the response is homogeneous fluid-like ('barotropic' in a sense reasonably generalized from a constant depth model) sufficiently far from the coast, where neither frictional influences nor the inclination of isopycnals interfere. One point of disagreement, however, is the appearance of setup along an open coast, which is presumably a global effect of bottom friction.

5.2. COASTAL SEA LEVEL

The cross-shore momentum balance of wind-driven transient currents should be dominated by the Coriolis force of longshore flow and the cross-shore pressure gradient, i.e., by approximate geostrophic balance. This has repeatedly been reported in observational studies of coastal sea level in various locations. An important quantitative question is, what is the effective width l of the coastal current? For the Great Lakes, simple inertial models suggest l of the order of 10 km, for the Oregon shelf some 30 km, although, as pointed out in Section 4.4.2 the actual coastal level response cannot be determined without taking into account frictional effects very close to shore, which limit the magnitude of longshore velocity in shallow water.

Following the storms of August 6 and August 9 over Lake Ontario, coastal currents of 0.2–0.3 m s^{-1} were generated along both north and south shores. The eastward wind stress should have produced a seal level rise of order 2 cm on the north shore, a similar depression on the south shore. Following the westward storm this should have changed to a 3 cm or so depression on the north shore, a similar rise on the south shore. Figure 5.5 from Simons (1975) shows observed sea level at one north shore location (Cobourg) and two south shore ones (Rochester and Oswego). Burlington is at the western end of the lake. The results are what one expects from theory, except that the level at Cobourg responds only little following the first (weaker) storm.

Along the Oregon shelf the sea level signal is larger, and it tracks longshore velocity to a striking extent, see again Figure 5.3. Many other studies have corroborated this result of Smith (1974). A somewhat puzzling point is that the effective width l of the current has to be taken to be about 60 km to explain observed sea level fluctuations, rather wider than what may be considered shelf width: at 60 km distance from the coast the water depth over this shelf is some 500 m. However, since inertial response theory is unrealistic in very shallow water in any case ($\zeta \to \infty$ at $x = 0$ according to Equation (4.35)) these observations cannot be quantitatively related to inertial models, except in an order of magnitude sense.

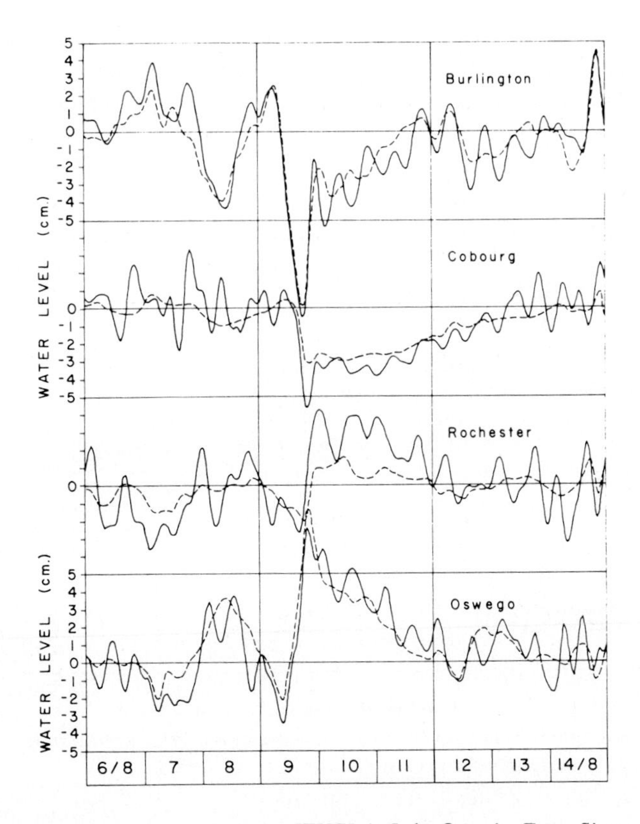

Fig. 5.5. Observed water levels during IFYGL in Lake Ontario. From Simons (1975).

5.3. UPWELLING, DOWNWELLING AND COASTAL JETS

Intense upwelling events are known to occur in a number of coastal locations, notably in the Great Lakes and along the Oregon coast. Early reports described the hydrography of upwelling (Church, 1945; Ayers *et al.*, 1958; Smith *et al.*, 1966), while later systematic studies in the course of large-scale cooperative experiments provided detailed information also on longshore and cross-shore currents. In the course of these investigations some clear-cut upwelling events have been documented, produced by a local alongshore wind impulse, which may be compared with the inertial response theory.

A spectacular example of impulsively produced upwelling was observed at Oshawa, on the north shore of Lake Ontario on October 10, 1972. A massive eastward wind-stress impulse acted on the lake between October 7 and October 9, of a total strength of I = 27 m^2 s^{-1}. A realistic two-layer idealization of the density structure during this episode is

$$\epsilon = 0.87 \times 10^{-3},$$

$$h_t = 25 \text{ m},$$

$$h_b = 75 \text{ m}.$$

This gives an internal radius of deformation of R_i = 4 km. The minimum impulse required to raise the thermocline to the surface (according to theory) is about 13 $m^2 s^{-1}$. The larger impulse of 27 $m^2 s^{-1}$ should cause the thermocline-surface intersection to move about 3 km away from the coast.

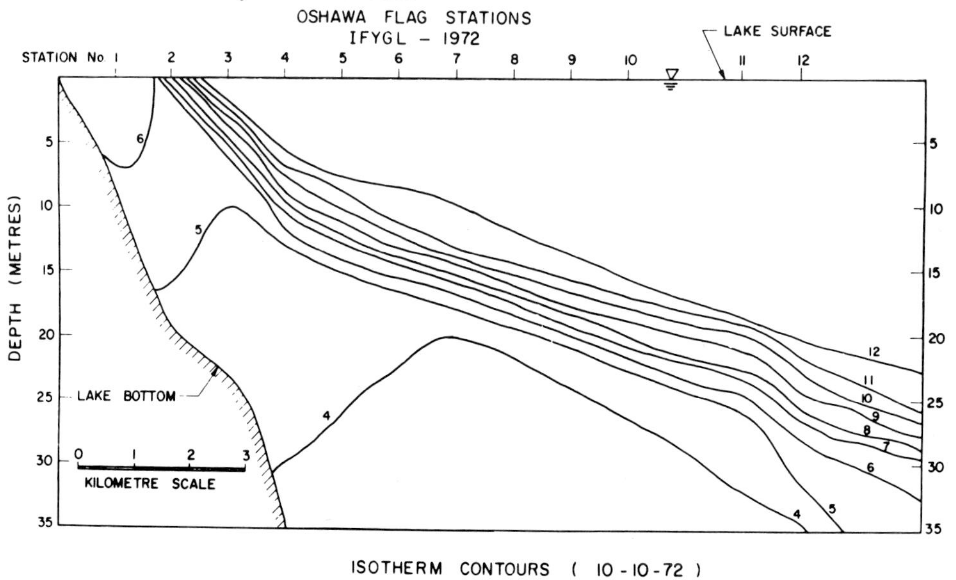

Fig. 5.6a. Observed isotherms in Lake Ontario at Oshawa during IFYGL, following a strong eastward wind impulse in early October. From Csanady (1976).

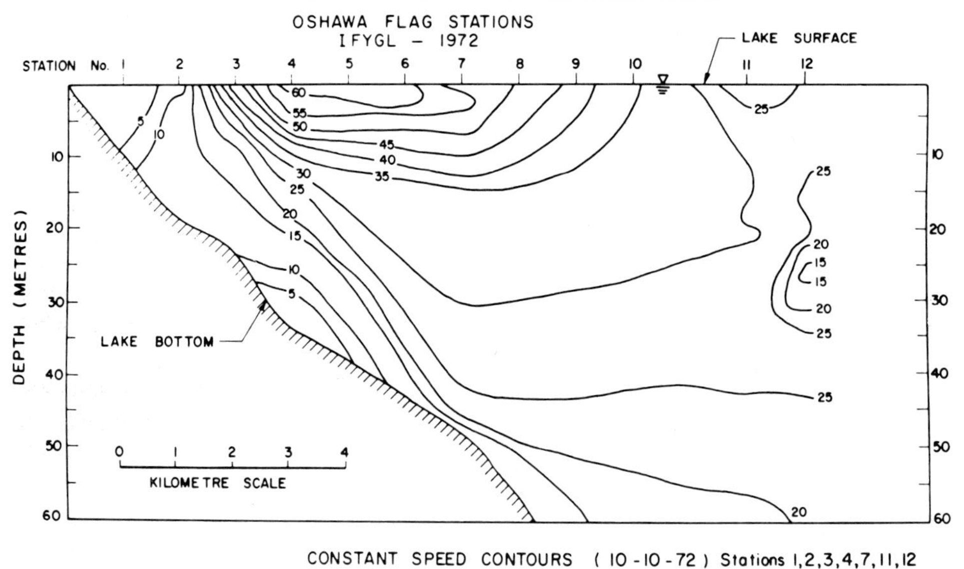

Fig. 5.6b. Observed longshore velocities in Lake Ontario at Oshawa during IFYGL, following a strong eastward wind impulse in early October. From Csanady (1976).

Figure 5.6 shows the distribution of isotherms and constant longshore velocity contours in a cross-isobath transect observed some 12 hr after the cessation of strong eastward winds. The *e*-folding scale of the thermocline structure is clearly close to the theoretical value of about 4 km. The offshore displacement of the front is about as calculated, a little over 2 km. The maximum difference in longshore velocity between warm and cold layers is also close to the theoretical value of 53 cm s^{-1}, although a two-layer constant depth model is difficult to relate accurately to the nearshore portion of the observed flow structure where bottom influences appear to be significant.

Prior to this episode the thermocline was not flat, having adjusted to *westward* flow on a previous occasion, so that isotherms dipped downward near shore by several meters. From the large vertical displacements of the isotherms one may infer that the cold layers below the thermocline were displaced shoreward by several kilometers. This (northward) adjustment drift is legitimately held responsible for the generation of strong (30 cm s^{-1}), (eastward) longshore flow in the cold layers extending to some 14 km from shore.

The results of a somewhat weaker eastward wind impulse (I = 3 cm^2 s^{-1}) are illustrated in Figure 5.7 at all five coastal transects operated during IFYGL. This occurred earlier in the season, 22–23 July acting upon a shallow thermocline, some 6–10 m in depth only. An upwelling developed on the north shore, downwelling along the south shore. The coastal jet associated with the downwelling was more distinct and carried a depth-integrated transport about equal to the impulse, i.e., 3 m^2 s^{-1}, while its *e*-folding scale was about 4 km, equal to the internal radius of deformation. With a thermocline as shallow as found along the north shore, upwelling apparently cannot be associated with a strong coastal jet (possibly on account of internal friction). The main characteristics of inertial response are nevertheless abundantly evident along both the north and south shores.

Along the Oregon coast, the surface outcropping of isopycnals during the summer is a well known event. Early theoretical ideas (Yoshida, 1967) and the interpretation of early observations in terms of these ideas (Smith *et al.*, 1971; Mooers *et al.*, 1976) tended to focus on a search for a steady flow pattern accompanying upwelling. One important result established by the extensive Oregon coast studies of the early seventies was that upwelling episodes are essentially transient events, contrary to the early ideas and impressions, even if a succession of such events occurs each summer. In between the southward wind impulses (which generate the upwelling events) the pycnocline relaxes somewhat, but does not usually reach a horizontal equilibrium position before the next impulse. Thus there is something like a quasi-steady summer mean state characterized by a weakly inclined pycnocline, from which the upwelling events start. Halpern (1974) shows such a partially relaxed thermocline observed on August 26, 1972 and an upwelling event which developed two days later (Figure 5.8). Reference back to Figure 5.3 shows that strong southward winds started blowing on the 24th and by the 28th built up an impulse of about 40 m^2 s^{-1}. Distributing the wind impulse over a 30 m top layer depth (as suggested by the nearshore portion of the survey on August 26) one deduces that an impulse of about 30 m^2 s^{-1} would be required to bring the pycnocline to the surface. The actually observed 40 m^2 s^{-1} impulse should have displaced it seaward by several kilometers, as was indeed observed.

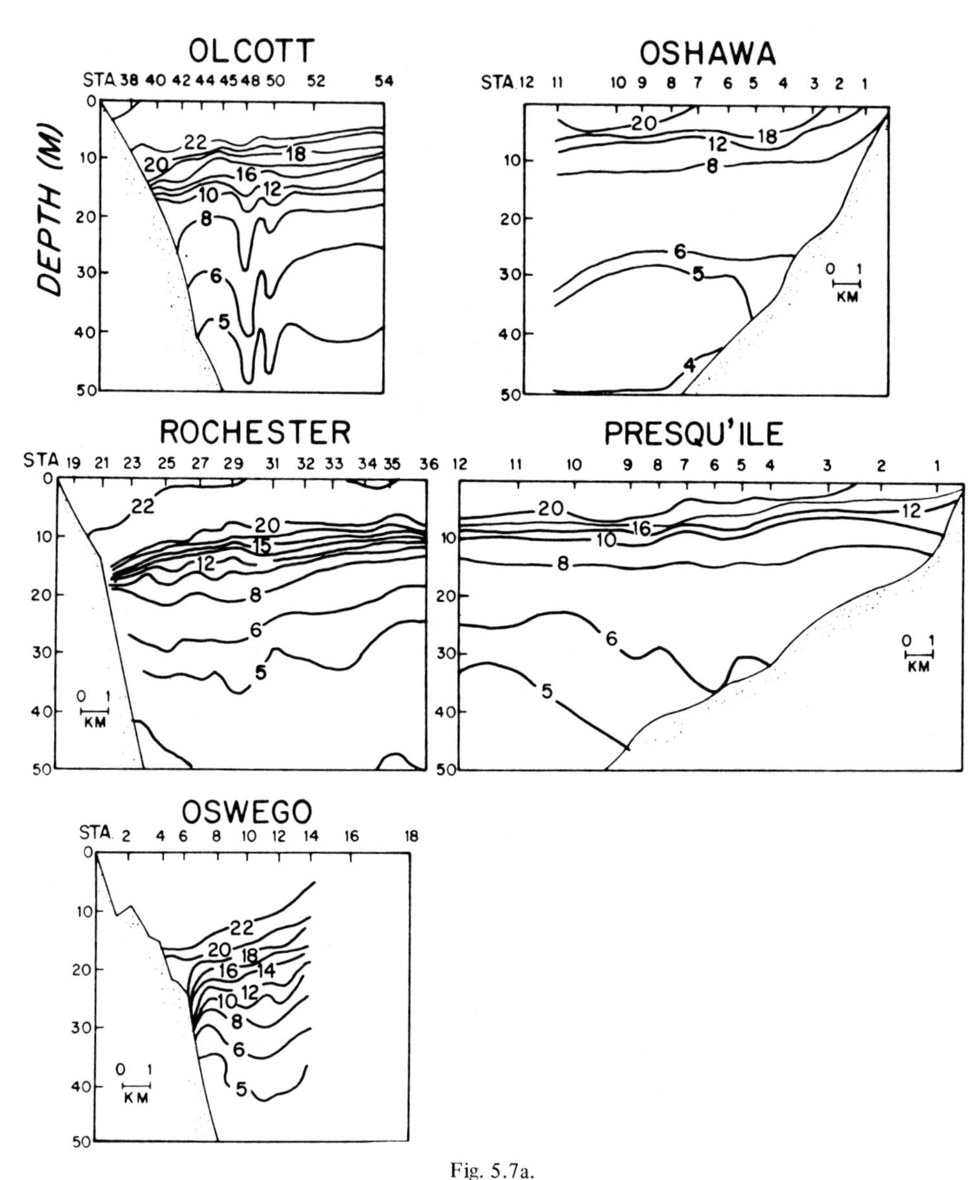

Fig. 5.7a.

Fig. 5.7a–b. Isotherms and constant longshore velocity contours in five coastal transects of Lake Ontario following a moderate eastward wind impulse in July 1972. From Csanady and Scott (1974).

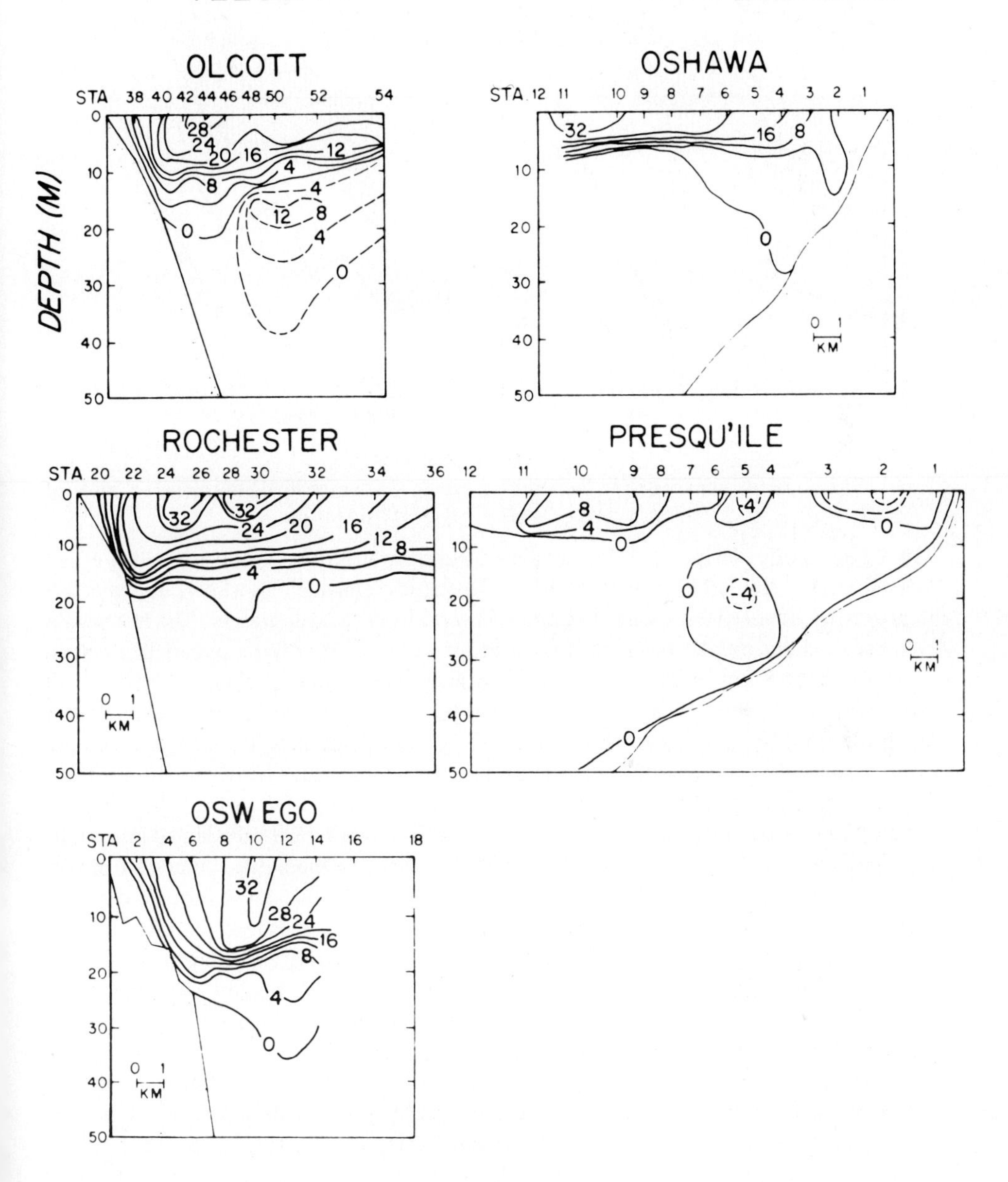
CROSS SECTIONS OF LONGSHORE COMPONENT OF VELOCITY
DATE: JULY 23, 1972
OLCOTT
OSHAWA
ROCHESTER
PRESQU'ILE
OSWEGO
STA
DEPTH (M)
KM

Fig. 5.7b.

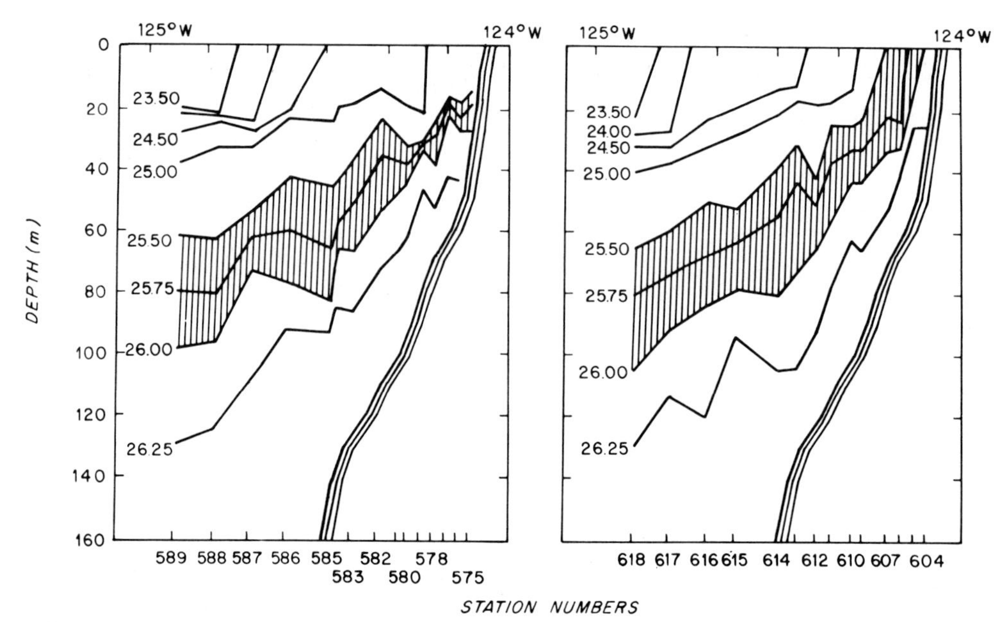

Fig. 5.8. Response of isopycnals off the Oregon coast to southward wind stress impulse, showing isopycnals before and after. From Halpern (1974).

In a later study, Halpern (1976) analyzed the structure of a coastal upwelling event observed during July 1973 in greater detail. This study convincingly shows all expected characteristics of inertial response to impulsive wind to have been present. The isopycnals developed a characteristic uptilt and the surface outcropping of the pycnocline center moved to about 8 km offshore, following a wind stress impulse of about 60 m^2 s^{-1}. A strong coastal jet with a peak velocity of 0.8 m s^{-1} was observed in the surface layer. Taking the equilibrium pycnocline depth as 80 m, the density defect $\epsilon = 2.5 \times 10^{-3}$, the bottom layer depth as infinite, one calculates a baroclinic radius of deformation of R_2 = 10 km, and an internal wave propagation velocity $c_2 = fR_2$ of 1 m s^{-1}. The observed *e*-folding scale of the uptilt was 16 km or larger than the theoretical value of 10 km, while the maximum coastal jet velocity was somewhat less than the theoretical value of 1 m s^{-1}, as already mentioned. When the upwelling developed, the surface layer moved offshore, while bottom layers compensated and, at greater depth, had onshore velocities of the order of 5 cm s^{-1}. As already pointed out earlier, the longshore Coriolis force of this appeared to be partially balanced by a longshore pressure gradient, but the unbalanced part accelerated bottom currents in deep water, much as shown in Figure 5.4.

The relatively large discrepancy between calculated and observed radii of deformation is presumably a consequence of relaxation effects, as well as of somewhat arbitrary choices of parameters in a two-layer model. In a discussion of the quasi-steady 'mean' state of the pycnocline Mooers *et al.* (1976) quote effective *e*-folding distances of 42 km over the shelf and 68 km over the slope. These large distances appear to be controlled mainly by relaxation phenomena. From the point of view of the present discussion, such results exhibit the limitations of an inertial response model.

Results similar to those of Halpern (1976) were obtained earlier by Huyer *et al.* (1974), who analyzed July 1972 transient upwelling events. In another important contribution Huyer *et al.* (1979) have analyzed in detail the onset of the summer upwelling regime in the 1975 season, which occurred between 25 March and 1 April. A total southward wind-stress impulse of about 50 $m^2 s^{-1}$ was exerted in this period, which is barely sufficient to explain the development of upwelling on the basis of inertial response, including the surface outcropping of isopycnals. Although the quantitative evidence is somewhat inconclusive, one is left with the impression that wind effects exerted at some remote location propagated into the study area and assisted the local wind in the development of upwelling. This again points out one limitation of a simple two-dimensional inertial response model. In many other respects, however, the observations reported by Huyer *et al.* (1979) on the first establishment of the outcropping of the pycnocline conform quite well to that model.

5.4. PROPAGATION OF FLOW EVENTS

Over periods long compared to f^{-1} inertial response involves the cyclonic alongshore propagation of a coastal current system originally set up by the wind. If the pressure field accompanying coastal currents is trapped within a nearshore band of width l, it should propagate according to theory along the coast in the cyclonic direction at a speed $c = \lambda f l$, with $\lambda = 0.3$ to 1.0. The principal signature of such a propagating flow event should be strong longshore flow. As for directly wind-driven longshore currents, geostrophic balance implies a relationship between the coastal sea level signal and the longshore velocity amplitude. At mid-latitudes this means that, given a narrow trapped field, $l = 10$ km, a velocity amplitude of $v_0 = 0.1$ m s^{-1} corresponds to a sea level signal of $\zeta_0 = 1$ cm. A sea level signal of this amplitude is small compared to short-term level fluctuations associated with tides, seiches, etc. and its detection is possible only through statistical time series analysis. By contrast, a velocity signal of 0.1 m s^{-1} is of the same order of magnitude as fluctuations due to other causes and is more immediately apparent in any record. Where the trapped wave is of the internal Kelvin wave type, isopycnal movements accompanying it are also large and conspicuous.

A statistical analysis of sea level records from eastern Australia by Hamon (1962) and their interpretation by Robinson (1964) in terms of continental shelf waves originated current interest in coastally trapped waves. Similar early analyses of sea level records from the west coast of North America (Mooers and Smith, 1968; Cutchin and Smith, 1973; Smith, 1974) yielded some evidence to suggest northward (cyclonic) propagation of certain low-frequency signals. Cutchin and Smith (1973) find a clear spectral peak at 0.22 cycles per day in sea level records, with an equally clear phase difference between two stations consistent with northward propagation at 4.1 m s^{-1}. The inferred longshore wavelength is 1620 km, suggesting that the source of the disturbance is a weather system, which typically has dimensions of this order. Cutchin and Smith also calculated a topographic wave propagation velocity in the lowest mode of 4.5 m s^{-1}, very close to the value inferred from sea level records. Longshore velocity observations at a mooring near some of the sea level gauges were also available and showed high coherence in the 0.22

cycles per day band with sea level fluctuations, at an amplitude very close to what one expects from geostrophic balance, given the longshore velocity distribution in the lowest topographic wave, which had a calculated trapping width of 50 km.

Extensive current meter observations on the Oregon–Washington shelf in the early 70-s provided more direct evidence for the northward propagation of flow episodes unrelated to the wind. Figure 5.9 from Kundu *et al.* (1975) shows comparison of winds and currents observed at the central transect of the 1973 CUE-2 experiment. One flow episode centered at July 31 is clearly not wind-driven. Figure 5.10 from Kundu and Allen (1976) shows the northward propagation of the same event over a distance of about 80 km, at a speed of about 1.5 m s^{-1}. Along the 100 m isobath, the current fluctuations

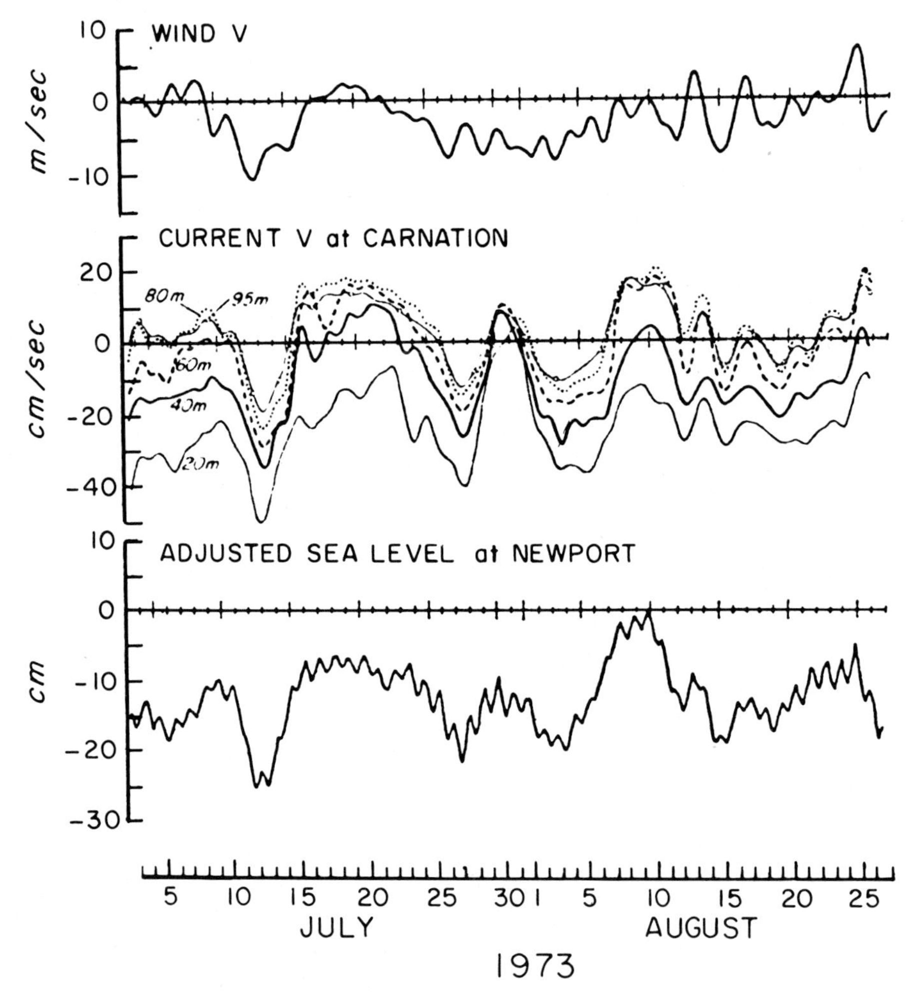

Fig. 5.9. Observations of wind, longshore current along the 100 m isobath, and coastal sea level on the Oregon coast in 1973. Note current event around 31 July unrelated to wind. From Kundu *et al.* (1975).

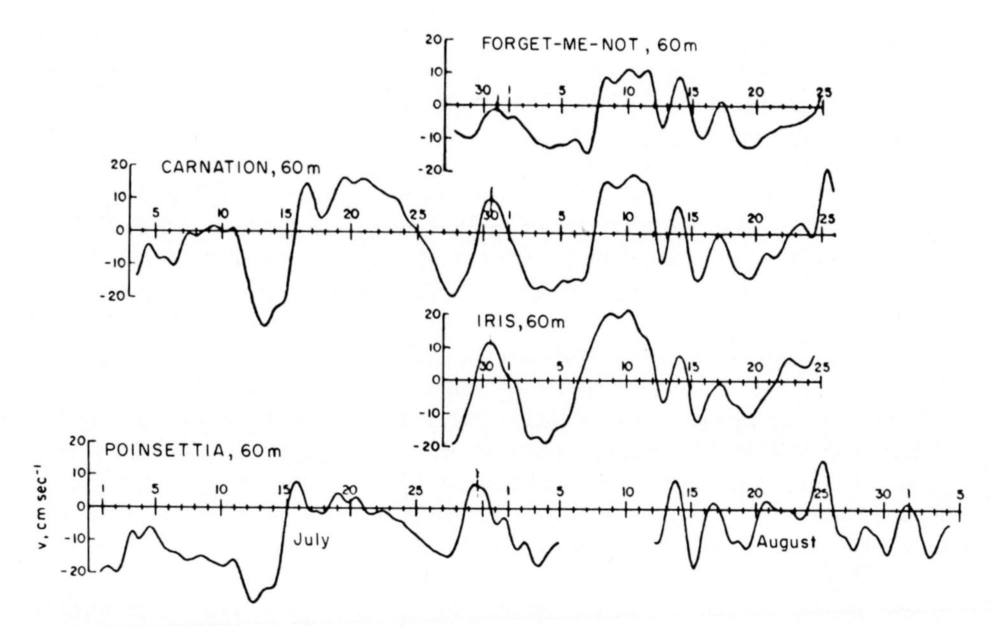

Fig. 5.10. Northward propagation of current event of previous figure. From Kundu and Allen (1976).

involved the entire water column and were strong enough to reverse the direction of the flow for about a two-day period. During the same period the temperature at 40 m depth was higher by about 0.5 °C than before or after, showing that some moderate isopycnal adjustment accompanied the episode. A decomposition into dynamic modes also showed that this event had the modal structure of a hybrid topographic-Kelvin wave with a coastal trapping width of about 30 km.

Statistical analysis of various signals (currents as well as sea level, Huyer *et al.*, 1975; Kundu *et al.*, 1975; Kundu and Allen, 1976) revealed the presence of at least two major wave modes with identifiable northward propagation properties. One mode is apparently close to a 'pure' topographic wave mode, with little influence from stratification, and a propagation velocity of about 5 m s^{-1}, much as found by Cutchin and Smith (1973). The other, slower mode is the hybrid-topographic wave mode just discussed, with a propagation speed of about 1.5 m s^{-1}. At a fixed location the two waves have frequencies respectively of about 0.22 cycles day^{-1} and 0.06 cycles day^{-1}, both consistent with a weather-system size wavelength of about 2000 km. It should be added, however, that longer-period statistical analyses of sea level and current observations taken along the Oregon coast are not always clear-cut, and that there are differences between different observational periods. This is not particularly surprising if one accepts that different wave-like modes are excited at different amplitudes according to random weather events.

A further source of confusion for the statistical analysis is that weather cycles, which excite the flow events, themselves tend to travel southward, in a direction opposite to the travel of coastally trapped waves (Mooers and Smith, 1968; Huyer *et al.*, 1975). There is a directly forced flow field generated by the weather cycles which also travels southward.

It is possible to interpret such a disturbance as a 'forced' wave (see e.g., Clarke, 1977) but a more straightforward point of view is to look upon it as directly (locally) forced wind-driven transient flow, as discussed in earlier sections of this chapter. In the forced wave interpretation the present state of the flow at any location is a result of past wind stress, integrated backward along wave characteristics, i.e., along gradually more distant portions of the coast, in the anticyclonic direction, from where topographic–Kelvin waves come. However, given the large spatial extent of the weather systems involved (order 2000) the distinction between local and non-local forcing becomes more or less academic. As already pointed out, west coast currents and sea level most of the time do behave as if locally forced.

The evidence for cyclonically propagating sea level disturbances, warm and cold fronts, and current reversals is strongest and most detailed in the Great Lakes. The earliest clear demonstration of internal Kelvin wave-like propagation of a warm front around the southern end of Lake Michigan was given by Mortimer (1963). This event followed a southward wind-stress impulse which caused downwelling of warm water on the western shore. Subsequently, the warm front propagated eastward along the southern end of Lake Michigan at a speed of approximately 0.5 m s^{-1}, corresponding to an internal Kelvin wave of a trapping width of $l = 5$ km.

More detailed evidence on the propagation of coastally trapped waves was obtained in Lake Ontario during IFYGL. At the end of July, 1972 a series of eastward wind stress impulses generated a system of coastal jets, associated with appropriate uptilts and down-tilts of the thermocline (Csanady and Scott, 1974, Figure 5.7 above). By the end of the wind stress episode the thermocline tilt and current direction were reversed at some coastal transects. At the end of a further four days of calm weather the reversal propagated virtually around the entire lake (Figure 5.11). The offshore trapping width was clearly seen in the experimental data to be of order 5 km, as suggested by a two layer, internal Kelvin wave model, and the propagation speed was a corresponding 0.5 m s^{-1}. The modal structure of the wave was, however, closer to a hybrid topographic-internal Kelvin wave than to a constant depth model internal Kelvin wave. This is consistent with theoretical estimates of topographic and internal Kelvin wave propagation speeds in Lake Ontario which are very close under summer conditions.

Late in the season (October, 1972) a wave propagation event could be documented on the north shore of Lake Ontario (Csanady, 1976b) which differed from the July event in that it did not involve isopycnal movements. Simultaneously with the development of upwelling along the north shore in the wake of a strong eastward wind impulse on October 10 (Figure 5.6) downwelling and strong eastward i.e., cyclonic coastal jets developed along the south shore. Subsequently, the cyclonic coastal currents propagated to the north shore, where their appearance manifested itself in a spontaneous current reversal and a massive westward flow event in a direction opposite to the wind (Figure 5.12). The thermocline intersected the surface some distance offshore and was apprently unaffected by the longshore pressure gradient which more or less uniformly accelerated the entire water column. The modal structure of this event was thus similar to a 'pure' topographic wave of the linear theory. The propagation speed was close to 0.5 m s^{-1} or about as expected for the observed offshore trapping scale of 10 km.

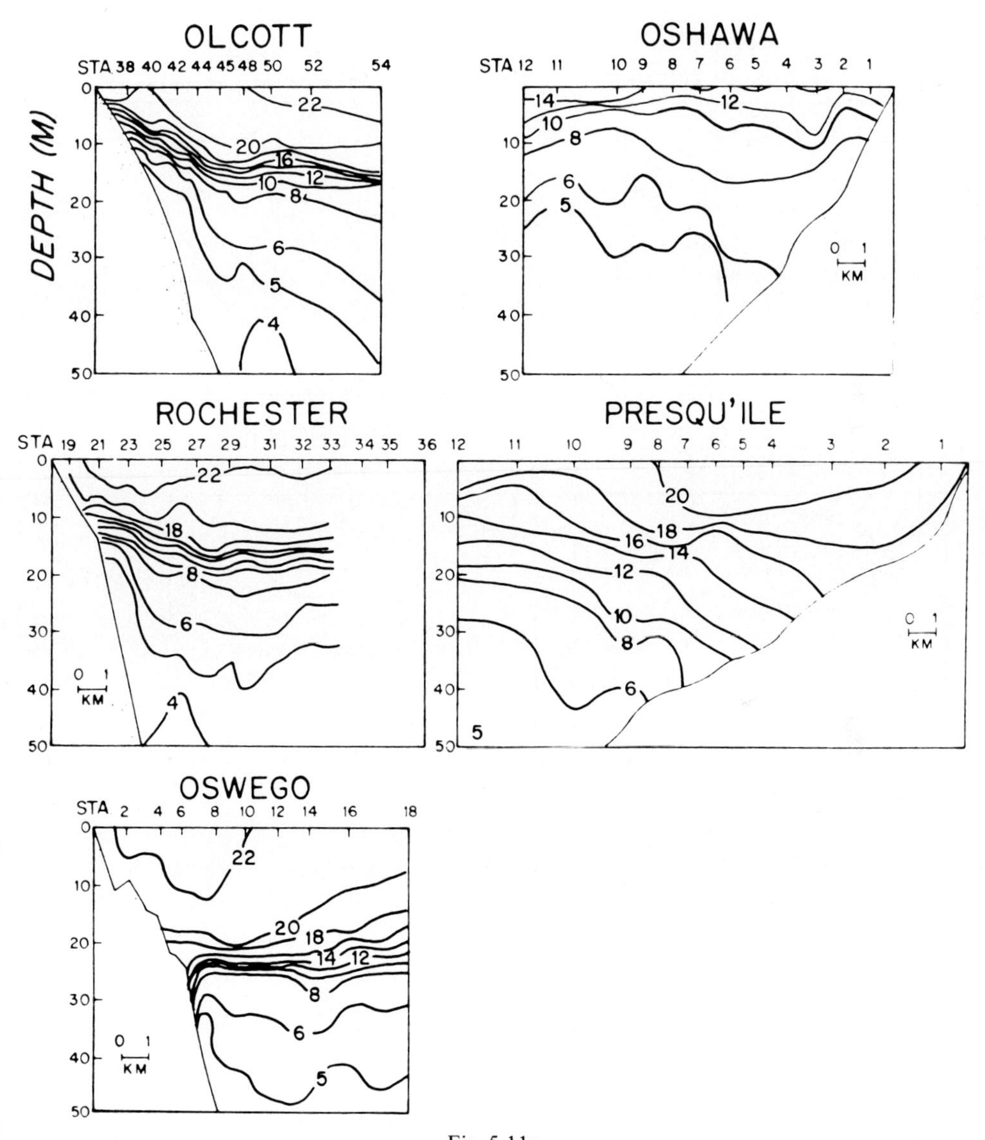

Fig. 5.11a.

Fig. 5.11a–b. Isotherms (a) and constant longshore velocity contours (b) in five coastal transects of Lake Ontario eight days after the observations shown earlier in Figure 5.7. Broken velocity contours indicate reversed flow, i.e., opposite to the exciting wind. From Csanady and Scott (1974).

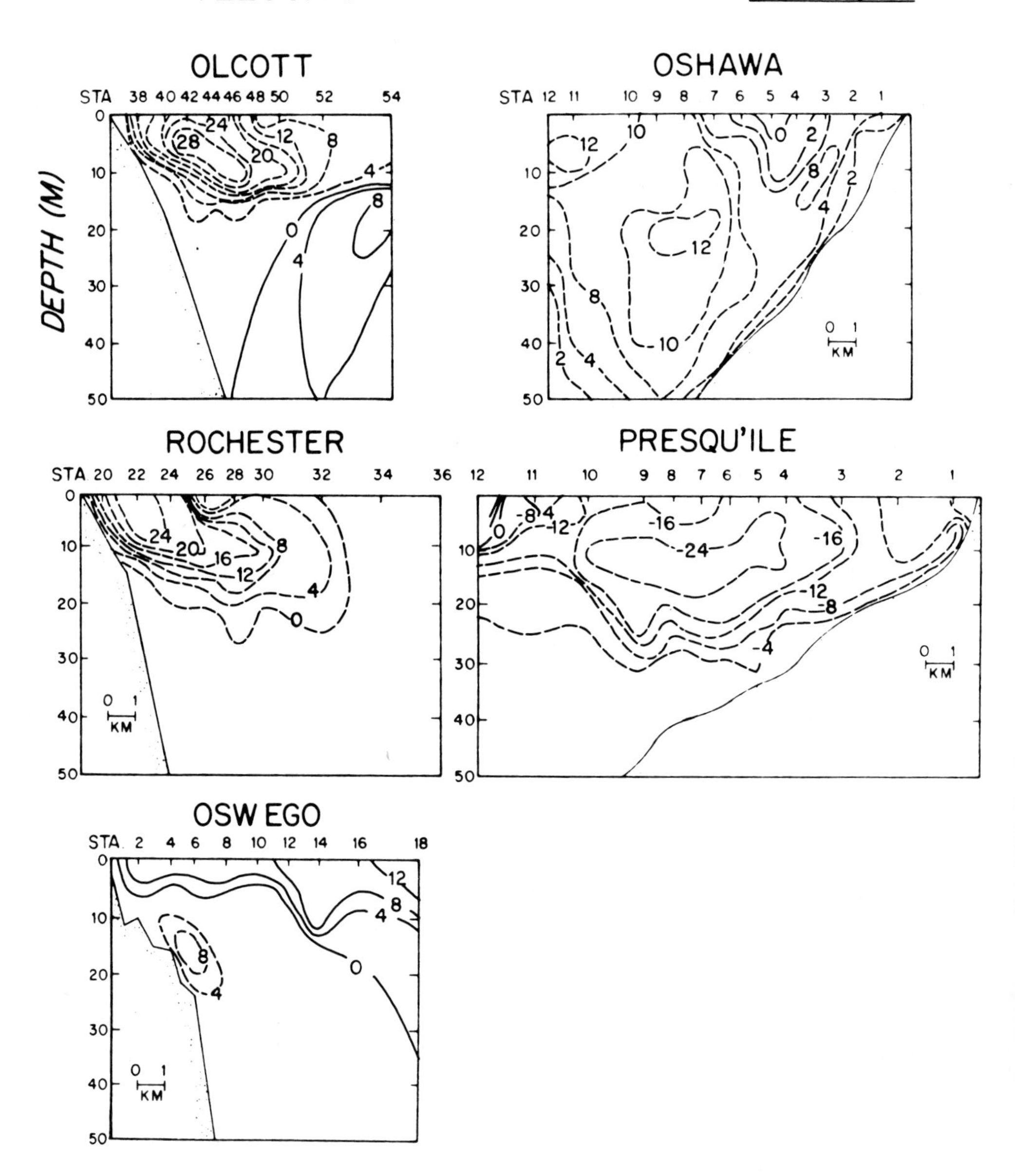

CROSS SECTIONS OF LONGSHORE COMPONENT OF VELOCITY
DATE: JULY 31, 1972
OLCOTT
OSHAWA
ROCHESTER
PRESQU'ILE
OSWEGO
STA
DEPTH (M)
KM

Fig. 5.11b.

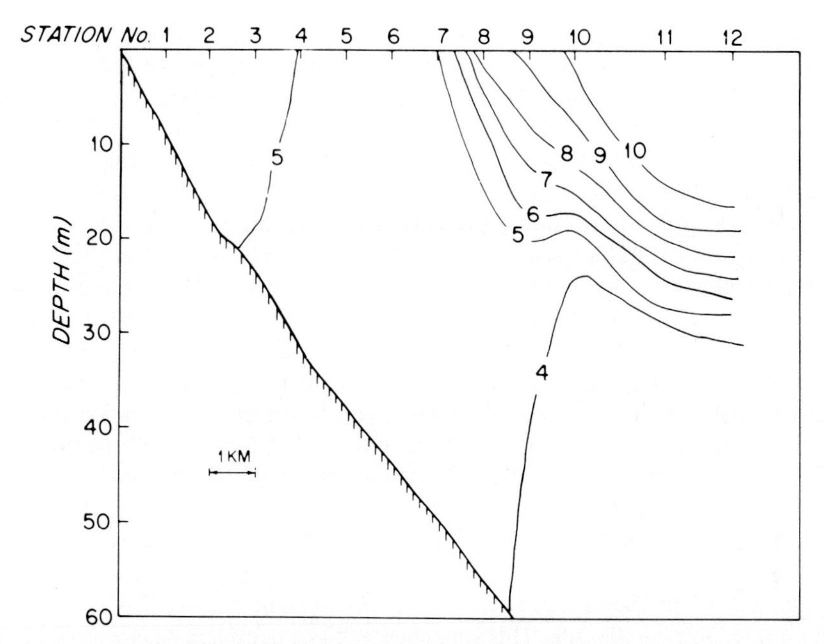

Fig. 5.12a. Isotherms at Oshawa coastal transect in Lake Ontario observed three days after the observations shown earlier in Figure 5.6. From Csanady (1976b).

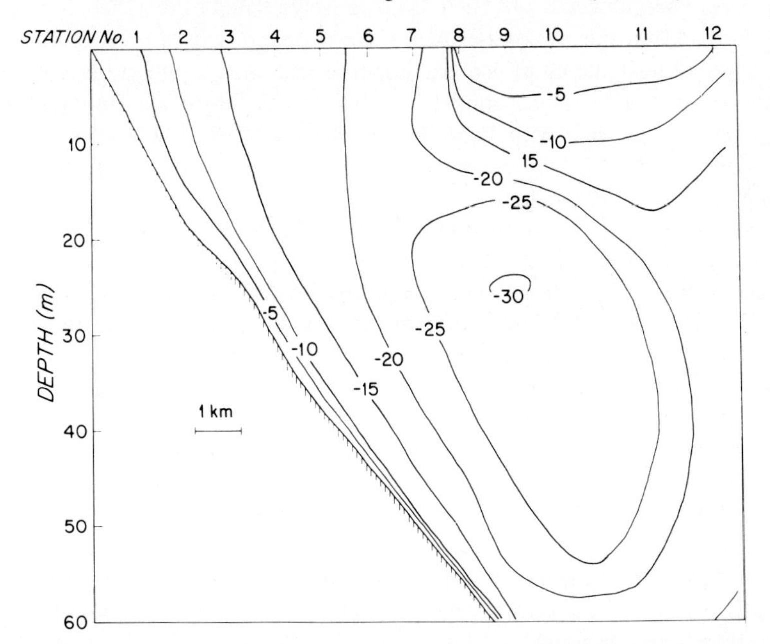

Fig. 5.12b. Constant longshore velocity contours at Oshawa coastal transect in Lake Ontario observed three days after the observations shown earlier in Figure 5.6. Negative longshore velocity values indicate reversed flow, opposite to the wind stress which originally excited the event. From Csanady (1976b).

Later, also the thermocline downtilt propagated to the north shore. The progress of this could be followed on a sequence of surface isotherm maps, obtained by Irbe and Mills (1976) using airbone radiation thermometry. The initial zone of upwelling is shown on a survey taken October 10 (Figure 5.13). Further eastward impulses caused the surface outcropping of the pycnocline to move further offshore and apparently retarded the westward propagation of the warm front along the north shore, see the survey on October 18 (Figure 5.14). During a following quiescent period, however, the front propagated westward (Figure 5.15) at a speed of some 0.2 m s^{-1}, much slower than the preceeding 'pure' topographic wave, but comparable to the internal Kelvin wave propagation velocity for a top layer of 30 m depth, given the small density contrast remaining by this time of the year, about 0.25×10^{-3}.

The surface isotherm maps also highlight some weaknesses of the linear theory and of two-dimensional upwelling models. The eastern end of the cold pool, which the illustrations show to propagate westward, is unlike an internal Kelvin wave, and is more properly described as a propagating front. Recently, Yamagata (1980) has discussed a preliminary theoretical model of such a front. Further development of the theoretical background seems to be necessary before a better understanding of similar phenomena is arrived at.

The illustrations, Figures 5.13–5.15 also show some pronounced wave-like disturbances on the surface front, suggesting hydrodynamic ('baroclinic') instability of an upwelled front in geostrophic equilibrium. This phenomenon has a considerable meteorological and oceanographic literature (see e.g., Pedlosky, 1979). The wave-like wiggles observed on upwelled fronts have, however, not been quantitatively related to this theory so far.

Having pointed out the weaknesses of the linear theory, it is also appropriate to note that the basic characteristics of inertial response and wave propagation are predicted remarkably accurately by linear models. Both internal Kelvin and topographic wave modes are clearly present, with quantitative characteristics closely approximated by simple theoretical models. A hybrid response is observed when the propagation speeds of these two modes are nearly equal.

Although in Lake Ontario the analysis of individual propagation events proved possible and conclusive, statistical analysis of current time-series provides further valuable supporting evidence. A particularly thorough study of this kind by Marmorino (1979), focussing on winter currents, showed a propagation speed of 0.49 m s^{-1} under unstratified conditions. This agrees almost exactly with calculations of 'pure' topographic wave behavior. Marmorino also demonstrated the clockwise rotation of the current vector, which is also a feature of the theoretical model.

During the stratified season, Blanton (1975) has noted periodicities in observed longshore currents of 12–16 days, and pointed out that this greatly exceeds the periods of wind forcing. A period of 15 days corresponds to a wave propagation velocity around the lake of 0.5 m s^{-1}. Blanton further noted that along the north shore of Lake Ontario the currents were predominantly westward, opposite to the long term mean wind stress. He ascribed this result to frequent wavelike propagation of cyclonic flow events, generated originally along the south shore.

In a recent study of currents in southern Lake Michigan Saylor *et al.* (1980) report conclusive evidence for the presence of basin-wide vorticity waves. Pronounced oscillations

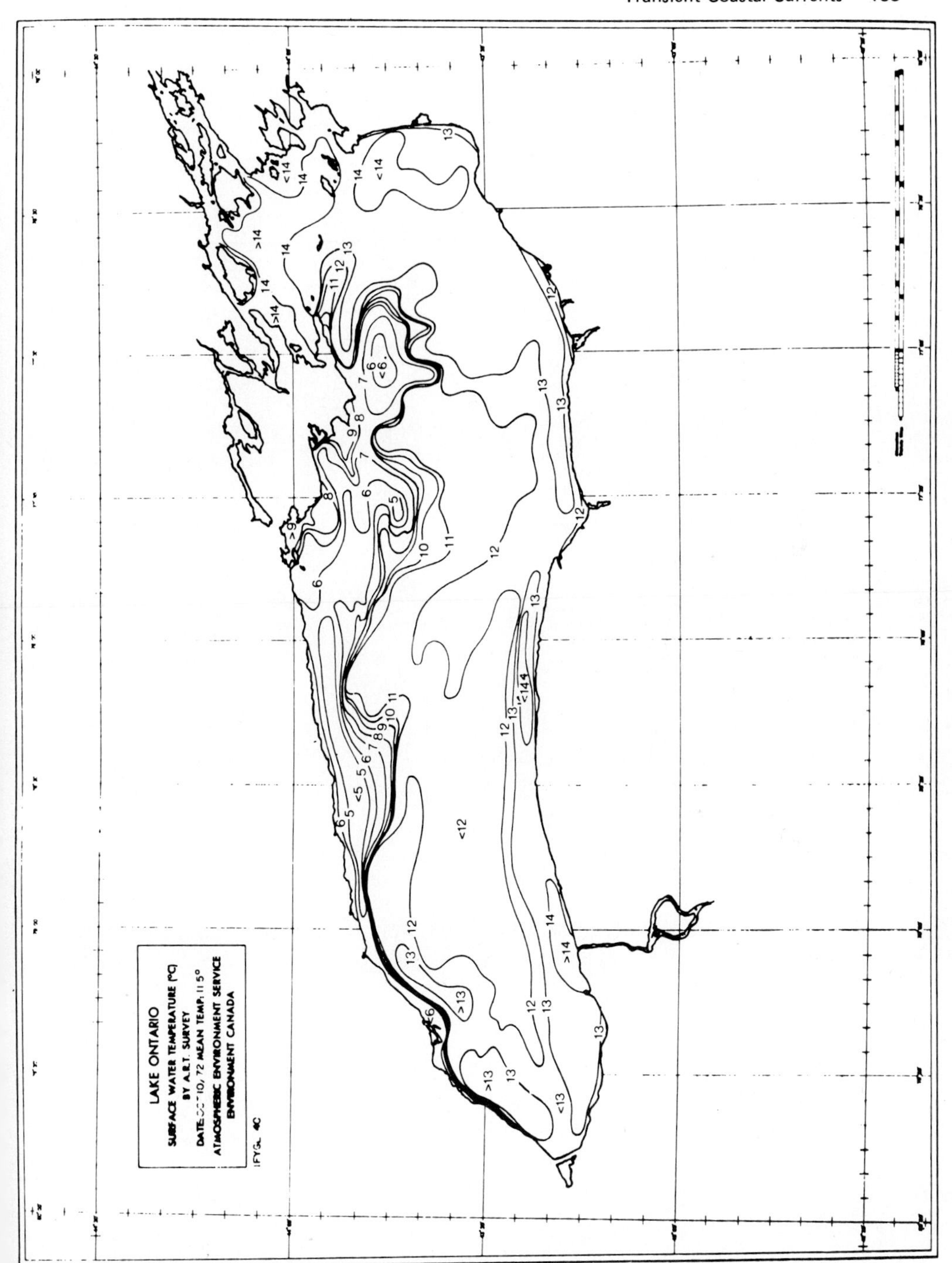

Fig. 5.13. Contours of surface water temperature observed by Airborne Radiation Thermometer (ART) in Lake Ontario during IFYGL, following strong eastward winds. From Irbe and Mills (1976).

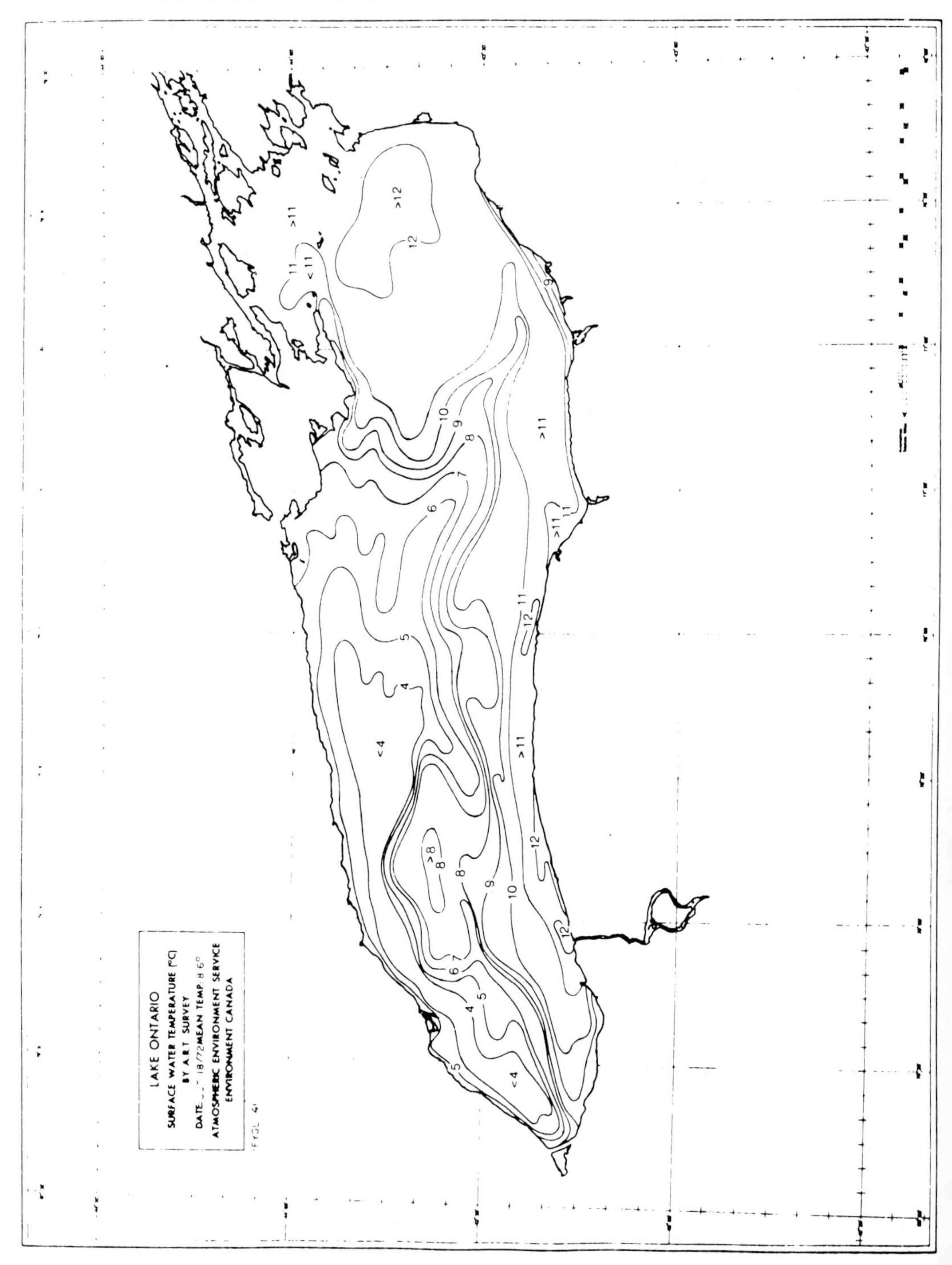

Fig. 5.14. As previous figures, but 8 days later, following further strong eastward winds. From Irbe and Mills (1976).

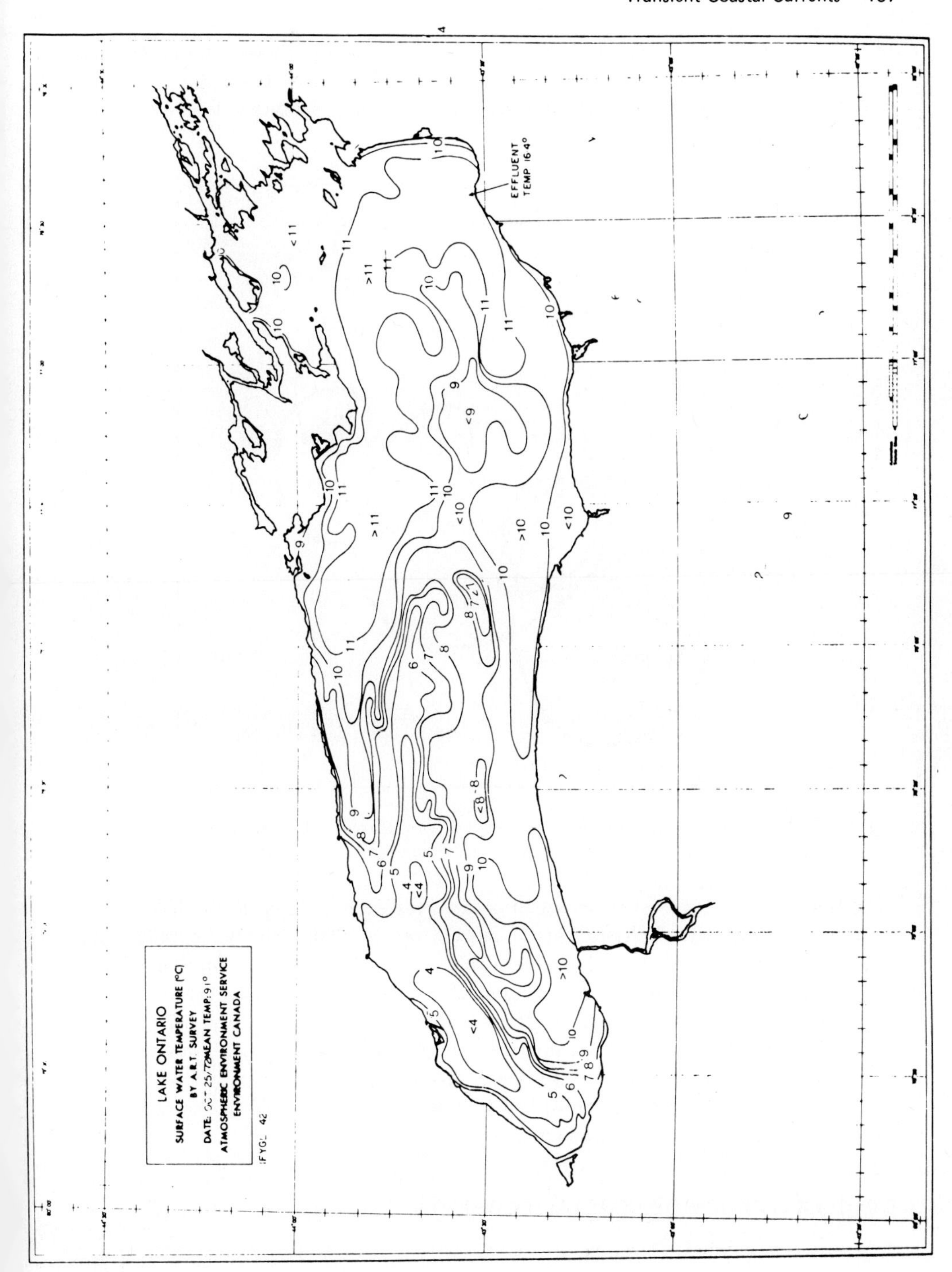

Fig. 5.15. As previous two figures, 7 days later again, following quiescent period. From Irbe and Mills (1976).

of 4 days period were observed at all stations of a central transect of the southern basin, including especially deep stations ($\sim$150 m) in the middle. At these deep stations an almost pure rotational mode was found (see progressive vector diagram, Figure 5.16) with a *cyclonic* rotation.

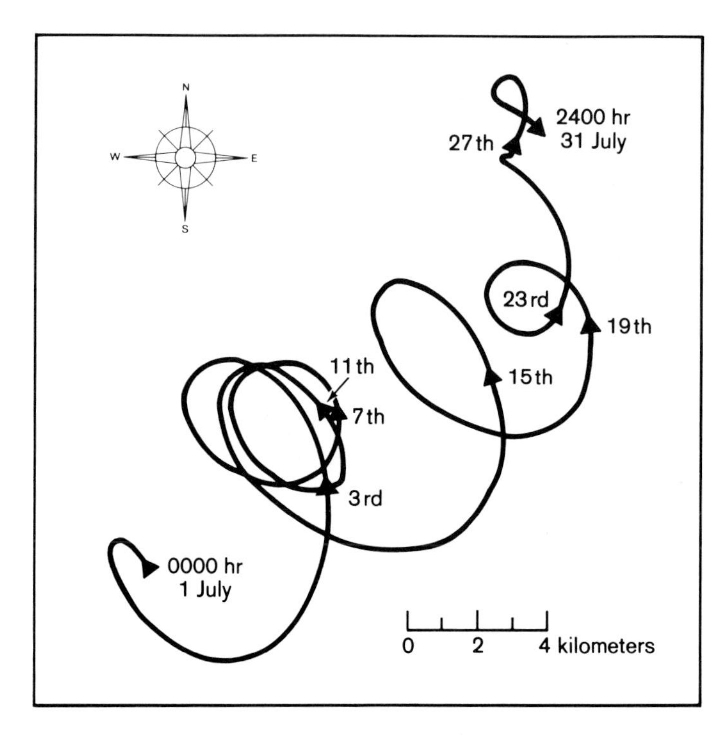

Fig. 5.16. Progressive vector diagram of low-pass filtered currents observed near the center of the southern basin of Lake Michigan. From Saylor *et al.* (1980).

At stations nearshore the same oscillation was present, but the north-south (longshore) current was 180° out of phase, and the current vector rotated anticyclonically. Saylor *et al.* (1980) point out that the character of this rotational mode is very much like Lamb's model for a circular paraboloid (see Section 4.5.3 above), although the observed period is 90 hr in place of the theoretical 123 hr. A circular cone basin has about the same vorticity-wave frequency as observed, although it is no closer to the topography of the actual basin than a paraboloid. The vorticity waves were observed to be excited by north-south wind stress impulses, approximately according to the scenario envisaged in the topographic gyre-vorticity wave models discussed in Chapter 4.

5.5. CLIMATOLOGY OF COASTAL CURRENTS

A general characteristic of various inertial model responses to forcing may be said to be a clear difference between some 'nearshore' region and the rest of a shallow sea. Wind driven transient currents were generally found to be 'trapped' within some distance l of the coast due to the interplay of various physical factors. Repeated application of impulsive forcing by wind, as it occurs at mid-latitudes, is therefore certain to give rise to a coastal current climatology which differs sharply from mid-basin current climatology. This distinction has in fact been clearly and conclusively made as the end result of various studies of current climatology in the Great Lakes.

The first clear distinction between current regimes observed near and far from shore has apparently been drawn by Verber (1966). In the course of a large scale experiment conducted by the Federal Water Quality Administration in Lake Michigan, a number of current meters were deployed, covering the whole lake in a more or less even grid pattern. Some of the meters were near shore in shallow water. In classifying the types of current regimes observed, Verber notes that 'straightline flow' is always found near shore, while far from shore the currents generally oscillate in all directions. Figure 5.17 taken from

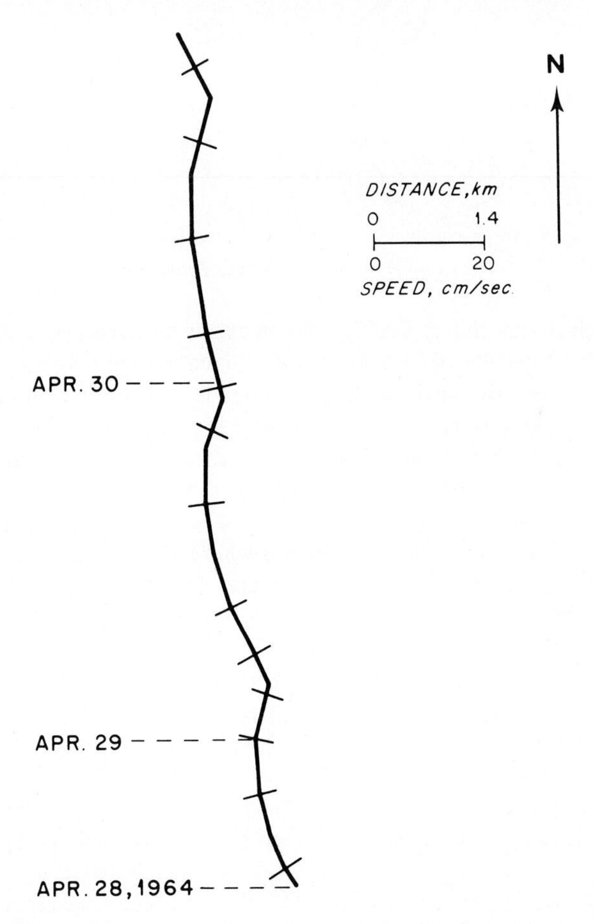

Fig. 5.17. Progressive vector diagram of currents observed in Lake Michigan at a nearshore mooring. From Verber (1966).

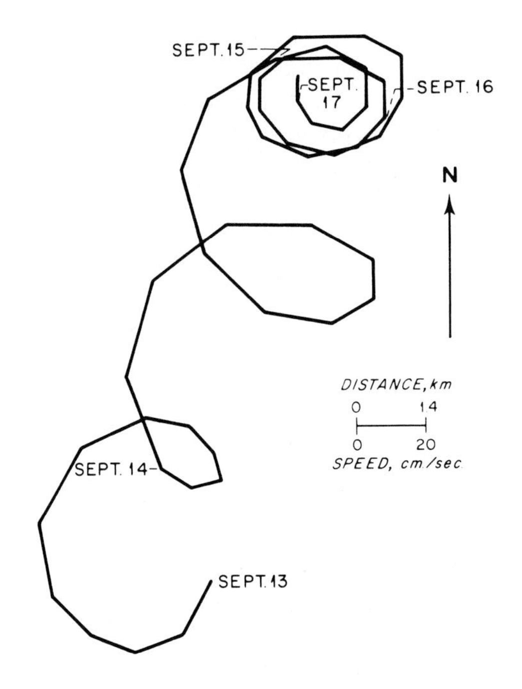

Fig. 5.18. As previous figure, but at a mid-lake station.

Verber's paper illustrates the typical current behavior by means of a progressive vector diagram near shore: typical offshore currents are shown in Figure 5.18.

The same data were later analyzed in greater detail by Birchfield and Davidson (1967) and Malone (1968). Spectra at stations far from shore were dominated by a large peak at frequencies slightly above inertial, while close to shore most of the energy was in low-frequency motions. According to Malone (1968), there is a "large current component nearly parallel to the coast" at nearshore meters. The percentage of energy in the near-inertial peak is three times less near shore than offshore.

All this early evidence on the distinct character of the coastal zone became consolidated and placed into perspective by work during and in preparation for IFYGL. Moored current meter observations (Weiler, 1968) have shown that the kinetic energy of surface currents peaks some 8 km from shore. Blanton (1974) has corroborated this finding and showed that the percentage of energy in long period motions falls dramatically between 8 and 10 km from shore (Figure 5.19). Blanton (1975) has also demonstrated that the onshore-offshore component of the surface currents nearshore is much smaller than the longshore current, typically by a factor of 5. Sato and Mortimer (1975) note that in Lake Michigan, as in Lake Ontario, the coastal zone is the "repository of most of the lake's kinetic energy".

The statistical evidence of fixed point (Eulerian) current measurements is also supported by studies of the movement of Lagrangian tracers, notably flourescent dye. Wind-driven longshore currents have been shown to carry tracers alongshore for considerable

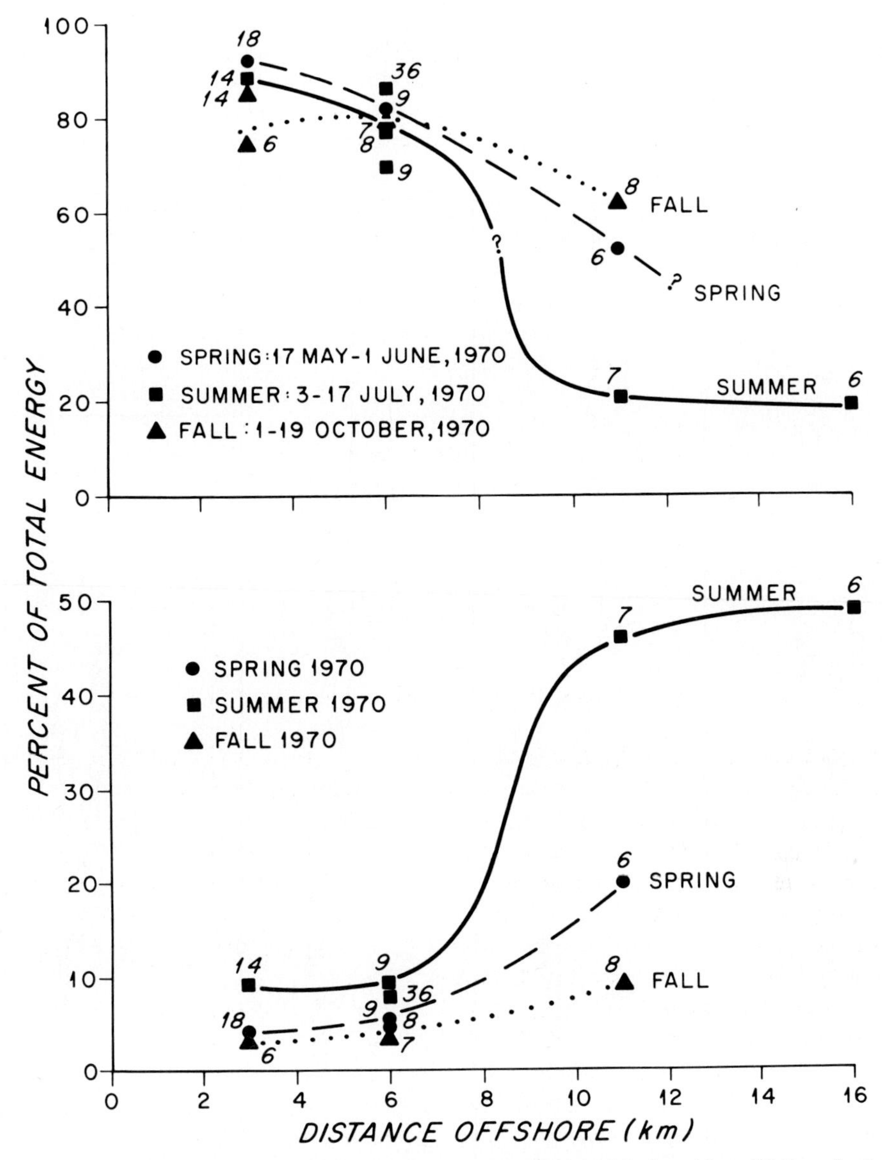

Fig. 5.19. Proportion of energy in long period currents (top) and in inertial oscillations (bottom) versus distance from shore. Lake Ontario observations from Blanton (1974).

distances in the Great Lakes, if only they are released nearshore (Csanady, 1970, 1974c). Release at an offshore location results in much more erratic movement (Murthy, 1970). Figure 5.20 here, redrawn from data of Pritchard and Carpenter (1965) shows long dye plumes generated by a continuous source placed in the coastal current along the south shore of Lake Ontario.

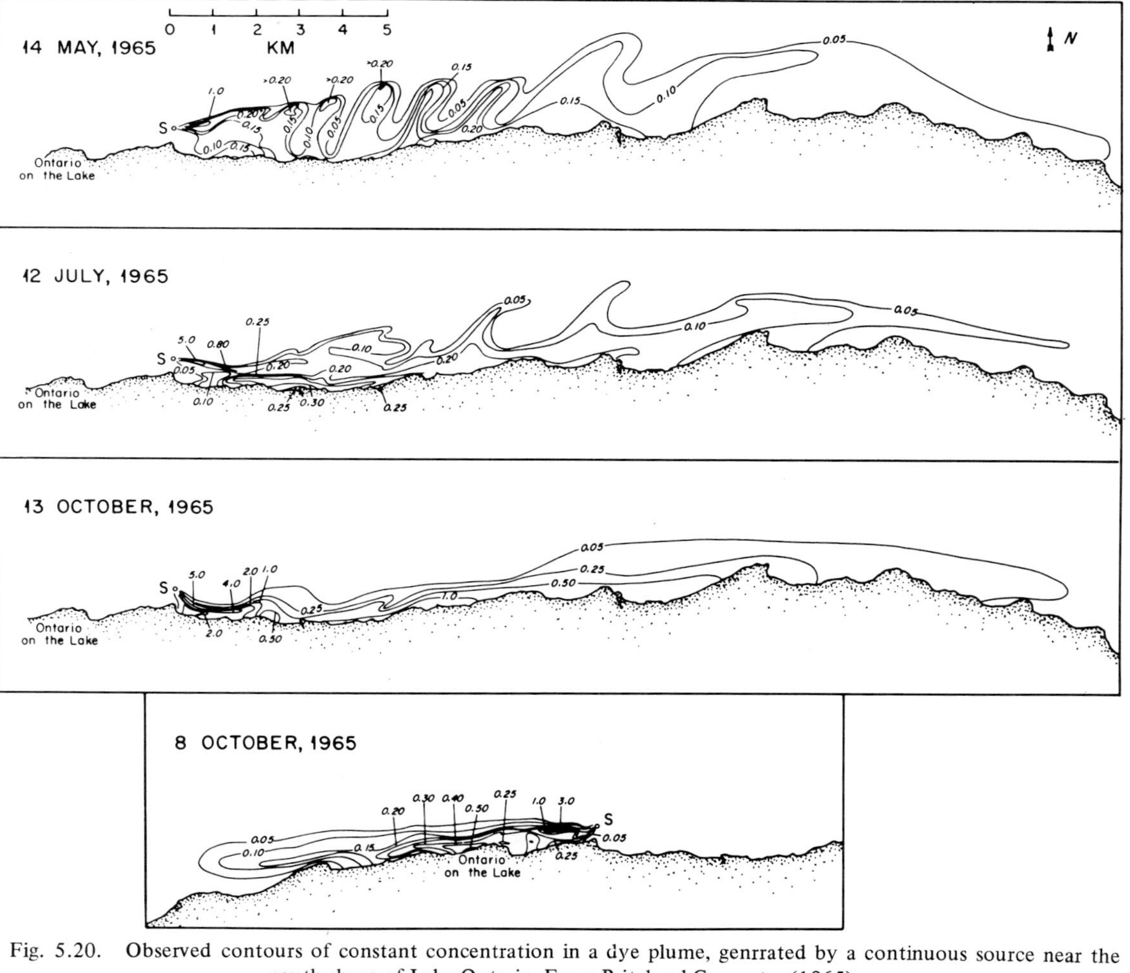

Fig. 5.20. Observed contours of constant concentration in a dye plume, genrrated by a continuous source near the south shore of Lake Ontario. From Pritchard-Carpenter (1965).

Similar Lagrangian tracer studies also show the more or less complete disappearance of a dye-plume on the reversal of the coastal current due to an opposing wind impulse (Csanady, 1974c). The adjustment drift is thus seen to perform the very important practical task of renewing the coastal water mass. For a strong enough wind impulse, the renewal is more or less complete.

Simple as the inertial models of flow in the coastal ocean are, they clearly account for an impressive array of observed facts. Most importantly, perhaps, they allow one to conceptualize in simple ways a dynamical behavior which is quite complex in the aggregate.

CHAPTER 6

Flow Controlled by Bottom Friction

6.0. INTRODUCTION

The foregoing four chapters have all been concerned with the physics of developing or transient coastal currents. In most of the discussion so far, an important part of the problem was the acceleration of the fluid, accompanying the development of the flow, geostrophic adjustment, or wave propagation. As could be seen from the discussion of the observational evidence, transient models account for many important features of circulation in various shallow seas. There are, however, other important phenomena involved in circulation in which the fluid accelerations do not play a significant role. Such steady flow phenomena arise when external forcing comes to be balanced by dissipative processes, primarily bottom friction.

The potential importance of bottom friction was already touched upon on several occasions before. In Section 5.1, for example, a simple model of a developing longshore current was compared with observation. At the shore, the frictionless model predicted substantial longshore transport V, which is clearly unrealistic in very shallow water, because it could only develop if the velocity became very high. As the longshore velocity increases, so does bottom stress, until the latter balances the applied wind stress and any longshore pressure gradient. The longshore current, and with it the transport, is thus limited in intensity by bottom friction. Important questions are, how large the limiting velocity and transport become, how long it takes for the flow to adjust to frictional equilibrium, and how the frictional adjustment time varies with depth.

The above example merely suggests that the results of frictionless theory should be amended in a fairly obvious way to take into account bottom friction. There have been earlier indications of more subtle frictional effects. In Section 2.2.1, for example, a wind setup calculation was carried out on the hypothesis that the cumulative effects of bottom friction remain negligible in the vorticity balance for the time period of interest. Without this hypothesis a different steady flow pattern could have been calculated. In general, when the curl of bottom stress is a significant term in the balance of the vorticity tendencies, one should not be surprised to find a flow pattern completely different from what applies in an analogous frictionless case.

In the present chapter both the simpler and the more complex aspects of frictionally controlled flow are explored. In order to minimize complexity, a homogeneous fluid will be again postulated. The earlier results suggest that such an approach may apply also to a fluid with the usual small density variations found in shallow seas, except in regions of limited extent where the constant density surfaces develop significant inclinations. In the next chapter some problems of steady flow in a nonhomogeneous fluid are explored, while the final chapter of this monograph is devoted to a survey of experimental evidence on steady flow, relating to the conceptual models developed in this chapter and the next.

6.1. FRICTIONAL ADJUSTMENT

A convenient starting point is a reconsideration of the problem discussed in Section 4.4.2, sudden longshore wind applied over a coastal strip of variable depth, but now with bottom friction taken into account. The transport equations are

$$\frac{\partial U}{\partial t} - fV = -gH\frac{\partial \zeta}{\partial x} + F_x - B_x,$$

$$\frac{\partial V}{\partial t} + fU = -gH\frac{\partial \zeta}{\partial y} + F_y - B_y, \tag{6.1}$$

$$\frac{\partial U}{\partial x} + \frac{\partial V}{\partial y} = -\frac{\partial \zeta}{\partial t}.$$

The coast will be supposed very long, coincident with the y-axis, the domain of interest the $x \geqslant 0$ half-plane. The depth distribution will be taken to be a function of x only, $H(x)$, so that there is no reason why gradients along y should exist. The discussion will at first be confined to a nearshore region where the coastal constraint holds to a satisfactory approximation:

$$U = 0, \qquad (x \text{ suitably small}). \tag{6.2}$$

Subject to later discussion of the local problem the longshore component of the bottom stress will be parameterized by means of a drag law involving the depth-average velocity:

$$B_y = c_{da}\left(\frac{V}{H}\right)^2, \tag{6.3}$$

where c_{da} is a drag coefficient referred to this velocity. Intuitively, one expects the simplifications (6.2) and (6.3) to be reasonable in a well stirred water column close to shore and to give a first order answer to questions raised above regarding frictional adjustment to equilibrium.

Writing $F_y = u_*^2$ for $t > 0$, supposing that the motion starts from rest, and with the simplifications (6.2) and (6.3) introduced, the second of Equations (6.1) is easily solved by elementary methods, with the result:

$$V = \frac{u_* H}{\sqrt{c_{da}}}\left(\frac{1 - \exp(-2u_* t\sqrt{c_{da}}/H)}{1 + \exp(-2u_* t\sqrt{c_{da}}/H)}\right). \tag{6.4}$$

The x-dependence in this result is contained entirely in the variable depth, $H(x)$. For a given depth, at suitably short times, Equation (6.4) reduces to the inertial response result:

$$V = u_*^2 t, \qquad (t << t_f), \tag{6.4a}$$

where

$$t_f = \frac{H}{2u_* c_{da}^{1/2}} \tag{6.4b}$$

is a 'frictional adjustment' time scale. At times long compared to t_f, on the other hand, one finds

$$V = u_* H c_{da}^{-1/2}, \qquad (t >> t_f), \tag{6.4c}$$

i.e., a constant depth-average velocity of $u_* c_{da}^{-1/2}$, which is just large enough to evoke a bottom stress exactly equal to the applied wind stress.

For a 'typical' value of the wind stress of 0.1 Pa, $u_* = 0.01$ m s^{-1}, and supposing $c_{da} = 2 \times 10^{-3}$, one finds t_f a little over 30 hr in 100 m. The frictional adjustment time varies directly with depth, and inversely with the square root of the wind stress and of the bottom drag coefficient. Under a hurricane, for example, both u_* and c_d are high and frictional adjustment time is short.

Equation (6.4b) may also be written down in terms of the surface Ekman depth $D = 0.1\, u_* f^{-1}$:

$$t_f = f^{-1} \frac{H}{10\sqrt{c_d}\, D} \tag{6.4d}$$

which is typically $2(H/D)f^{-1}$. Thus for H/D large, ft_f is also large so that frictional equilibrium flow is confined to depths of the order of D, given that typical storm durations are a few times f^{-1}.

These order of magnitude estimates may be compared with the observations shown in Figure 5.2, which was quoted above in support of the frictionless inertial response theory. The agreement with that theory is in fact confined to water deeper than about 30 m. The vanishing transport at the shore and its slow increase with distance shows a behavior consistent with Equation (6.4), the asymptotic solution $V \sim H$ being a more or less realistic description of the observations in water up to about 30 m in depth. This is exactly what one would expect from Equation (6.4d) for a storm of about 16 hr duration, which preceeded the observations shown in Figure 5.2.

The simplicity of the result (6.4) hides a rather complex depth-time dependence. Very close to thore the asymptotic longshore velocity $c_{da}^{-1/2} u_*$ given by (6.4c) is reached in a vanishingly short time. In greater depth, the longshore velocity increases more slowly, so that a velocity gradient $\partial v/\partial x$ develops between the faster moving water over the shallow depth and the slower water offshore. Maximum vorticity occurs over the depth $H_m = c_{da}^{1/2} u_* t$, i.e., along an isobath moving constantly offshore. This isobath may be considered to be a boundary between a coastal frictional flow regime and a deepwater inertial one. An intense storm lasting for about a day causes this boundary to move to a depth of the order of 100 m.

6.2. INTERIOR VELOCITIES

The simplicity of the result in Equation (6.4) rests on the parameterization scheme of (6.3) which now needs to be justified. In the absence of a longshore pressure gradient the local problem is subject to the equation

$$\frac{\partial u}{\partial t} - fv = -g\frac{\partial \zeta}{\partial x} + \frac{\partial}{\partial z}\left(K\frac{\partial u}{\partial z}\right),$$
$$\frac{\partial v}{\partial t} + fu = \frac{\partial}{\partial z}\left(K\frac{\partial v}{\partial z}\right). \tag{6.5}$$

As in Section 2.7, the velocities are resolved into pressure field induced (subscript 1) and frictional (subscript 2) components. The non-oscillatory parts of these are taken to be subject to, on dropping $\partial u/\partial t$ to filter out inertial oscillations:

$$-fv_1 = -g\frac{\partial \zeta}{\partial x},$$
$$\frac{\partial v_1}{\partial t} + fu_1 = 0. \tag{6.6}$$

The frictional component is

$$-fv_2 = \frac{\partial}{\partial z}\left(K\frac{\partial u_2}{\partial z}\right),$$
$$fu_2 = \frac{\partial}{\partial z}\left(K\frac{\partial v_2}{\partial z}\right). \tag{6.7}$$

Top and bottom boundary conditions on the frictional velocity are

$$K\frac{\partial u_2}{\partial z} = 0, \qquad K\frac{\partial v_2}{\partial z} = u_*^2, \qquad (z = 0)$$
$$K\frac{\partial u_2}{\partial z} = B_x, \qquad K\frac{\partial v_2}{\partial z} = B_y, \qquad (z = -H). \tag{6.8}$$

The bottom stresses are also given by a drag law (Equation (1.26)):

$$B_x = c_d(u_1 + u_2)q, \qquad B_y = c_d(v_1 + v_2)q, \qquad (z = -H), \tag{6.9}$$

where $q = [(u_1+u_2)^2 + (v_1+v_2)^2]^{1/2}$ is the *total* velocity magnitude. With these relationships substituted into (6.8) the boundary conditions allow the determination of the four integration constants arising in the solution of (6.7). These are expressed in terms of u_1 and v_1, or if one takes note of Equation (6.6), $\partial\zeta/\partial x$ and $\partial v_1/\partial t$. The solution of the global problem, i.e., of the depth integrated equations, then renders the interior velocities determine. These equations give for v_1 and $\partial v_1/\partial t$:

$$\frac{\partial v_1}{\partial t} = -fu_1 = \frac{1}{H}(u_*^2 - B_y),$$
$$v_1 = \frac{g}{f}\frac{\partial \zeta}{\partial x} = \frac{V}{H} - \frac{B_x}{fH}. \tag{6.10}$$

Consistency of Equations (6.10) requires a vanishingly slow rate of change of B_x. This is because the frictional velocities were effectively postulated to develop on a slow time scale, when time-dependent terms were not included in (6.7). As will be shown below, the cross-isobath component of the bottom stress plays a subordinate dynamical role in any case, and its time variation is safely neglected.

The general problem of solving the above set of equations leads to complex calculations. To gain some insight, limiting cases will first be considered.

6.2.1. The Shallow Water Limit

As earlier examples have shown, the solutions of Equations (6.7) are scaled by the Ekman depth $D = (2Kf^{-1})^{1/2}$. Where the total depth H is small compared to the Ekman depth D, the shear stress changes only by an order H/D quantity from the top of a water column to its bottom. This implies at once that, to zeroth order in H/D:

$$B_y \cong u_*^2, \qquad B_x \cong 0, \qquad (H \ll D). \tag{6.11}$$

The relationships (6.10) now also show $u_1 \cong 0$, $v_1 = V/H$. This is clearly the case of well-stirred steady flow, with no significant rotational effects, the type of flow that can be produced in a laboratory flume by applying shear stress along the axis of the flume of the upper surface. The velocity distribution observed in such a case is illustrated in Figure 6.1. The bottom stress equals the imposed wind stress, the stress being in fact independent

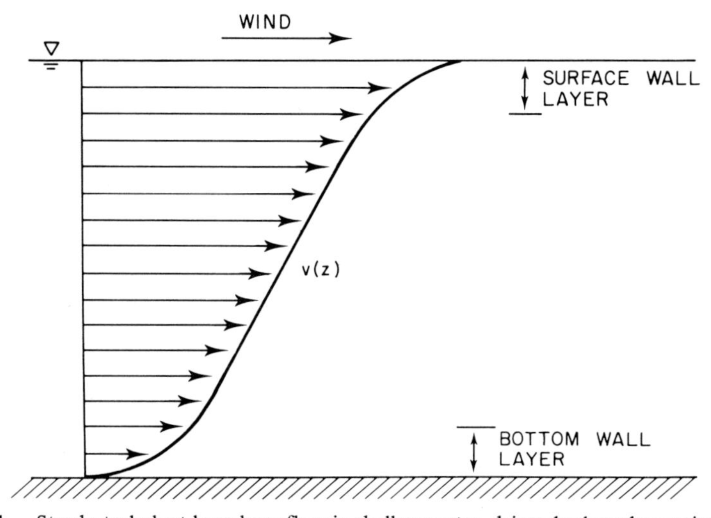

Fig. 6.1. Steady turbulent longshore flow in shallow water, driven by longshore wind stress.

of depth. Two thin, highly sheared layers form at the free surface and at the bottom respectively. These are the 'inner' or 'wall' layers within which the length scale of eddies varies rapidly, in direct proportion to distance from the free surface and bottom respectively: they have generally been ignored in this monograph. The region in between the wall layers is occupied by an 'outer' layer, within which the eddy length scale is approximately constant, and proportional to the depth H, the eddy velocity scale to the friction velocity u_*. The eddy viscosity is then approximately constant over the outer layer, characterized by an eddy Reynolds number Re (Equation (1.28)) of about 16. Correspondingly, the shear in the outer layer is

$$\frac{dv_2}{dz} = 16\,\frac{u_*}{H}. \tag{6.12}$$

Thus if the outer layer occupies 80% of the depth H, the velocity difference from just below the free surface wall layer to just above the bottom wall layer is about $13u_*$. The average velocity V/H over the section coincides approximately with the velocity at mid-depth, since most of the volume is transported within the outer layer. Extrapolating the constant velocity gradient to the bottom one finds

$$v_b = \frac{V}{H} - 8u_*. \tag{6.13}$$

Applying the bottom drag law (6.9) to this extrapolated velocity one arrives at

$$\frac{V}{H} - 8u_* = u_*\, c_d^{-1/2}, \tag{6.14}$$

where c_d is the drag coefficient referred to the extrapolated or 'slip' velocity. Equation (6.14) may also be expressed as:

$$B_y = u_*^2 = \frac{(V/H)^2}{\left(8 + c_d^{-1/2}\right)^2} \tag{6.14a}$$

which is equivalent to (6.3), justifying the analysis of the previous section for the shallow water limit. The drag coefficient referred to the depth-average velocity is seen to be:

$$c_{da} = \left(8 + c_d^{-1/2}\right)^{-2}. \tag{6.15}$$

As a typical example, if c_d is 2×10^{-3}, $c_{da} = 1.1 \times 10^{-3}$. The drag coefficient referred to the depth-average velocity is somewhat less than that referred to the slip velocity, but the relationship in this case of the well stirred water column is a straightforward one.

6.2.2. Deep Water Limit

When the water depth H is large compared to the Ekman depth D, it is convenient to divide the water column into top and bottom Ekman layers (each with their own wall layers) and a frictionless interior. The velocity distribution in the bottom Ekman layer is illustrated in Figure 6.2. Across the bottom wall layer the velocity increases rapidly in

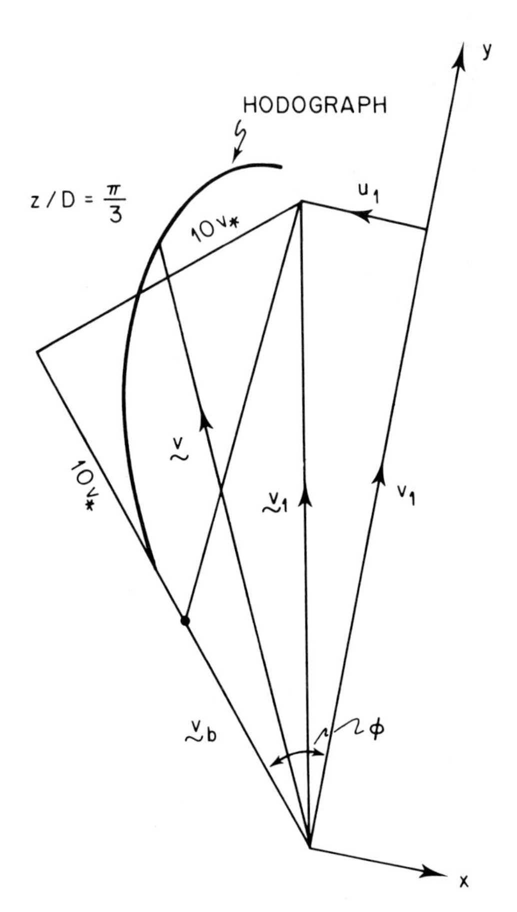

Fig. 6.2. Bottom boundary layer of developing longshore flow, driven by wind stress pointing to positive y. The geostrophic velocity vector $\mathbf{v}_1$ above the Ekman layer has components u_1, v_1, the cross-shore component u_1 being adjustment drift. The velocity vector $\mathbf{v}_b$ is found just above the bottom wall layer and is directed along the line of action of bottom stress. The velocity changes from $\mathbf{v}_b$ to $\mathbf{v}_1$ in the outer Ekman layer, characterized by the usual spiral (see hodograph and the sample velocity vector at $z/D = \pi/3$).

magnitude from zero, but remains directed along the bottom stress, the latter pointing at some angle ϕ to the left of the wind stress (supposed applied along positive y). Within the outer Ekman layer the velocity increases further both in the direction of the bottom stress and perpendicular to it. By a calculation similar to that in Section 1.11.3 the increase may be shown to be v_*/fD in both these directions, where v_* is bottom friction velocity:

$$v_*^2 = B_x^2 + B_y^2 = c_d(u_b^2 + v_b^2), \tag{6.16}$$

where $\mathbf{v}_b(u_b, v_b)$ is the extrapolated or slip velocity at the bottom.

Given the empirical value $D = 0.1\, v_* f^{-1}$ for bottom Ekman depth one also has $v_*/fD = 10$ for the increments of both velocity components across the outer bottom Ekman layer, see Figure 6.2. The magnitude of the extrapolated bottom velocity is $|v_b| = c_d^{-1/2} v_*$, in virtue of the definition of the drag coefficient c_d.

The velocities above the bottom Ekman layer are the pressure field-induced contributions, u_1, v_1. The geometry of the vectors shown in Figure 6.2 allows the pressure field-induced velocity components to be expressed in terms of v_* and ϕ:

$$\begin{aligned} u_1 &= -10 v_* \sin\phi + 10 v_* \cos\phi - c_d^{-1/2} v_* \sin\phi, \\ v_1 &= 10 v_* \sin\phi + 10 v_* \cos\phi + c_d^{-1/2} v_* \cos\phi, \end{aligned} \tag{6.17}$$

where the empirical value $v_*/fD = 10$ has been used. These are implicit equations from which the bottom stress v_*^2 and its orientation ϕ may be determined for given u_1, v_1. For example, when u_1 vanishes (i.e., when steady state flow has been reached, see the second of Equation (6.6)), one finds

$$\begin{aligned} \tan\phi &= (1 + 0.1 c_d^{-1/2})^{-1}, \\ v_* &= (c_d^{-1/2} + 10 \sin\phi + 10 \cos\phi)^{-1}\, v_1. \end{aligned} \tag{6.18}$$

A 'geostrophic' drag coefficient c_g is sometimes introduced in turbulent boundary layer theory, defined for the present case by:

$$v_*^2 = c_g\, v_1^2. \tag{6.19}$$

The value of this coefficient follows from the second of (6.18). The relationship of the the bottom stress to the depth-average velocity may be determined from (6.10), where now $B_x = v_*^2 \sin\phi$. The term involving bottom stress in Equation (6.10) is, however, at most a few percent of fv_1/g, so that to a good approximation:

$$v_1 \cong \frac{V}{H}. \tag{6.20}$$

Physically, the principal cross-shore force balance (second of Equation (6.10)) is between the Coriolis force of longshore flow and the cross-shore pressure gradient, the cross-shore bottom stress component playing a subordinate role. With this approximation, the longshore bottom stress may be written:

$$B_y = c_g \cos\phi \left(\frac{V}{H}\right)^2$$

which is again of the form of (6.3) and contains the expression for c_{da}. To the approximation of Equation (6.20), the parameterization scheme of Equation (6.3) has thus been justified for the deep water limit.

Given a 'typical' bottom drag coefficient $c_d = 2 \times 10^{-3}$, referred to the extrapolated bottom velocity, one finds from the above results:

$$\begin{aligned} \phi &= -17.172^\circ, \\ c_g &= 0.8226 \times 10^{-3}, \\ c_{da} &= 0.7859 \times 10^{-3}, \\ B_x &= -0.2429 \times 10^{-3}\, v_1^2. \end{aligned}$$

The cross-shore force per unit mass arising from bottom stress is also 0.2429×10^{-3} (v_1/fH) times $\partial\zeta/\partial x$, or $2^{1/2}$ percent of the cross-shore pressure gradient force for H = 100 m, v_1 = 1 m s^{-1}, $f = 10^{-4}$ s^{-1}. The appropriate value of c_{da} in deep water differs from the shallow water value (Equation (6.15)), although both are of the same order as c_d.

6.2.3. Intermediate Depths

Similar calculations for intermediate depths are more involved, but not in principle different. The steady flow case with equilibrium between forcing and bottom stress and subject to the coastal constraint (6.2), has been discussed in the literature many times, but usually on the basis of an arbitrary constant eddy viscosity, and mostly with the unrealistic bottom boundary condition of zero velocity. With a realistic parameterization of turbulent flow and a quadratic bottom friction law the velocity distribution in wind-driven flow with zero longshore pressure gradient may be determined in principle as follows.

As was pointed out in Section 1.7, interior stresses in a homogeneous fluid are realistically described by a gradient momentum transport relationship and an eddy viscosity K scaled by the total depth H in shallow water, or by the Ekman depth $D = 0.1u_*/f$ in deep water, as well as by the friction velocity u_*. In the steady flow case $(\partial v_1/\partial t = 0)$ with zero longshore pressure gradient top and bottom friction velocities are (nearly) the same and the variation of K with depth between the limiting case of shallow and deep water has the general form:

$$K = \text{func}\,(u_*, H, f) \tag{6.22}$$

which is also, in terms of nondimensional variables

$$\frac{K}{u_*H} = \text{func}\left(\frac{fH}{u_*}\right). \tag{6.22a}$$

Limiting values of this functional relationship have been specified before as

$$\begin{aligned} \frac{K}{u_*H} &= \frac{1}{16}, && \left(\frac{fH}{u_*} \to 0\right) \\ \frac{K}{u_*H} &= \frac{1}{200}\frac{u_*}{fH}, && \left(\frac{fH}{u_*} \to \infty\right). \end{aligned} \tag{6.22b}$$

Suitable interpolation formulae could clearly be designed, but a simple step is to take the first of these relationships to apply for $fH/u_* < 0.08$ (where the two yield equal K/u_*H) the second at higher fH/u_*. It is then easy to see from the structure of the equations (6.6) to (6.10) that the solution for v_1/u_* will be of the form

$$\frac{v_1}{u_*} = \text{func}\left(\frac{fH}{u_*}, c_d\right). \tag{6.23}$$

As pointed out before, v_1 equals V/H to a good approximation, so that (6.23) is effectively again a drag law of the form of Equation (6.3). However, the reciprocal drag coefficient appearing on the right of (6.23) is not a constant, but varies with nondimensional depth, and with the drag coefficient referred to the bottom velocity (which is a function of bottom roughness and may also vary in the cross-shore direction).

The solution (6.4a) remains valid if c_{da} is a function of nondimensional depth, or varies in an arbitrary manner with x, the only change compared to a c_{da} = constant case being a more complex dependence of V on cross-shore distance. The approach of Section 6.1 is thus adequately justified: a detailed discussion of bottom drag in a developing flow would merely belabor the point. Note, however, that only the specific case of forcing by longshore wind stress has been explicitly considered here: cross-shore wind and longshore pressure gradient complicate the problem further.

Beyond justifying the parameterization scheme of Equation (6.7), the considerations of this section also demonstrate some general properties of cross-shore flow in a coastal region. Interior nondimensional velocities u/u_*, v/v_* were seen to be a function of nondimensional depth z/D (or of fz/u_*, which is equivalent), as well as of fH/u_* and c_d. The cross-shore velocity was seen to vanish in very shallow water and to be significant in deep water only in top and bottom Ekman layers, where this velocity gives rise to equal and opposite Ekman transport. At intermediate depths were the transition between the above two extremes must take place, the algebra is a little more involved, but the same steady-flow calculations as indicated above may be carried out without difficulty, using (6.22b) for eddy viscosities and stress boundary conditions at top and bottom.

Consider specifically a coastal region with an arbitrary depth distribution $H(x)$ in function of offshore distance x. At each x, the cross-shore velocity distribution may be calculated by the above recipe for the local depth. From the calculated distribution of cross-shore velocities $u(x, z)$ it is then possible to infer a streamline pattern $\psi(x, z)$ in the cross-shore plane, such that

$$u = -\frac{\partial \psi}{\partial z}, \qquad w = \frac{\partial \psi}{\partial z} \tag{6.23a}$$

taking the surface and bottom as the $\psi = 0$ streamline. This calculation also determines the small vertical velocities which are otherwise neglected in the quasi-horizontal flow approximation. Two examples of such a pattern calculated for an east coast example are shown in Figure 6.3 for a 5 m s^{-1} and 25 m s^{-1} longshore wind respectively ($u_* = 0.775$ and 3.873 cm s^{-1}, from Csanady, 1975).

The region where the streamlines close in these illustrations may be regarded as a frictional coastal (horizontal) boundary layer. The width of this can vary considerably according to the magnitude of the surface stress exerted: in the strong wind example of Figure 6.3b the frictional coastal boundary layer occupies the entire continental shelf.

These illustrations answer a question raised in earlier chapters, what happens to cross-shore flow in the Ekman layer as a coast is approached. In a constant depth (laboratory) container a complex boundary layer structure occurs adjacent to the vertical side wall (Greenspan, 1968). In a coastal zone of realistic topography the slow variation of depth dominates and gives rise to a basically much simpler structure: vertical velocities remain small enough everywhere for the above approach to be adequate.

The pattern shown in Figure 6.3 was calculated from linearized equations of motion.

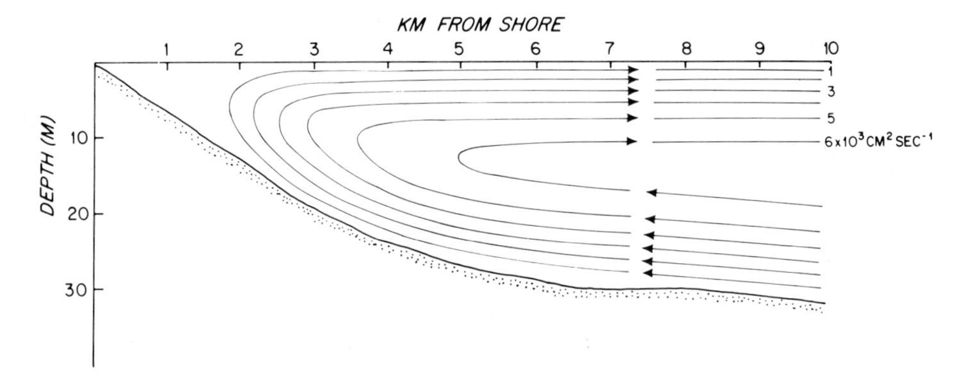

Fig. 6.3a. Circulation (transport streamlines) in a cross-shore transect accompanying steady longshore wind of 5 cm s^{-1} with vanishing longshore pressure gradient. Topography is that off south coast of Long Island. From Csanady, 1975.

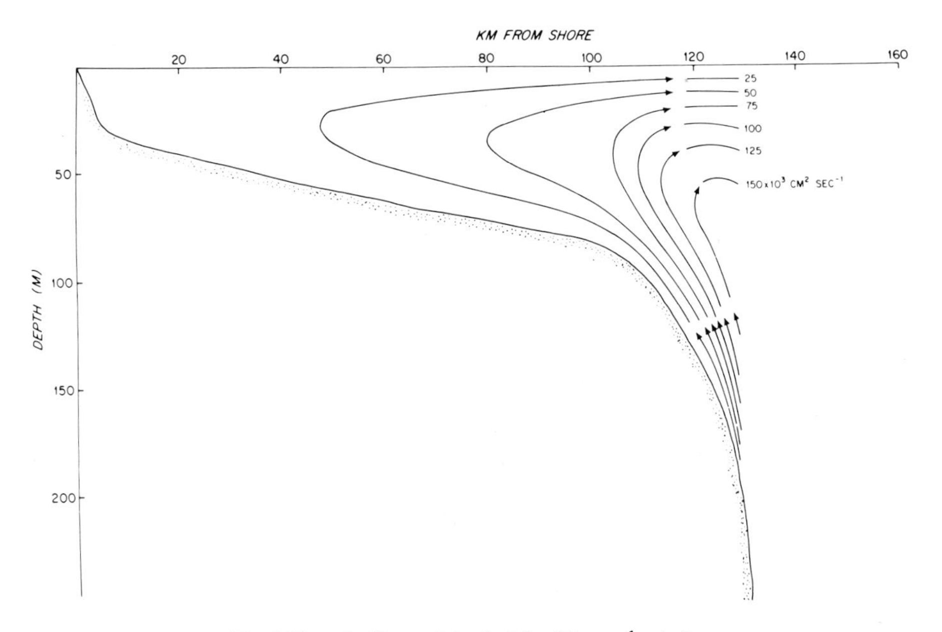

Fig. 6.3b. As Figure 6.3a, but for 25 m s^{-1} wind.

In a more accurate approach it is necessary to take into account cross-shore momentum flux by mean flow and by turbulence. The neglect of the mean flow momentum flux divergence, in particular, is not very accurate within the frictional coastal boundary layer, as some tentative estimates show (Csanady, 1975). However, the general character of the flow field is not altered by the likely momentum advection effects. Empirical knowledge of the frictional coastal boundary layer is inadequate at present to pursue this problem further.

6.3. SIMPLIFIED PARAMETERIZATION OF INTERIOR AND BOTTOM STRESS

Applied to the circulation problem the discussion of frictional effects in the previous sections has a serious weakness. Given a quadratic bottom friction law, and an interior momentum exchange coefficient which increases with the friction velocity, oscillatory components of motion interact with circulation and cannot be simply filtered out. Consider first the interior stresses and write for the total velocity $u + u'$, eddy viscosity $K + K'$ etc., where the unprimed part is the circulation component (i.e., a suitably filtered value of the variable concerned, such as a moving average) and the primed part is the oscillation. The typical interior stress component, averaged over, say a full tidal or inertial period, is

$$\frac{\tau_x}{\rho} = K\frac{\partial u}{\partial z} + \left\langle K'\frac{\partial u'}{\partial z}\right\rangle, \tag{6.24}$$

where the circumflex brackets indicate an average value. According to previous discussion, the average value of the momentum exchange coefficient K is determined by the average magnitude of the boundary stress, wind stress or bottom stress. When the typical oscillatory velocity u' is large compared to the circulation velocity u, the mean exchange coefficient is determined mainly by u' and is more or less independent of u. Interior friction is then linear in the mean (circulation component) velocity. This is a welcome simplification of the problem, as long as the second term on the right of Equation (6.24) is negligible. That second term is an exchange coefficient-velocity gradient correlation. This term indicates the possibility that a significant portion of the interior stress is unrelated to the mean velocity gradient. In an example discussed in detail elsewhere (Csanady, 1976a) some 20% of the mean sea-surface stress was estimated to be carried by such a fluctuating product correlation.

Similar considerations apply to the bottom stress. With fluctuating velocities considered, Equation (6.9) gives for the typical mean bottom stress component:

$$B_x = c_d\, uq + c_d\, \langle u'q'\rangle, \qquad (z = -H). \tag{6.25}$$

The correlation $\langle u'q'\rangle$ may be (but is not necessarily) small compared to the mean product uq. However, when it is negligible and when u is typically small compared to u' and v', q is essentially independent of u and linear bottom friction law emerges:

$$B_x = ru, \qquad B_y = rv, \qquad (z = -H), \tag{6.26}$$

where $r = c_d\, q$ is a resistance coefficient of the dimension of velocity. Such a linearized drag law appears to give good results in describing the mean circulation component in the Mid-Atlantic Bight, as well as frictional equilibrium storm currents in the same location, or hurricane driven currents in the Gulf of Mexico (Scott and Csanady, 1976; Bennett and Magnell, 1979; Forristal *et al.*, 1977). The appropriate value of r in all the quoted cases was empirically found to be of the order of 10^{-3} m s^{-1}. Such a value of r may be interpreted physically as an average bottom velocity magnitude of order 0.3 m s^{-1}, times a drag coefficient c_d close to 3×10^{-3}.

It should be clear that there are considerable inherent inaccuracies in parameterizing interior and bottom stresses as they affect the circulation component of the flow, when strong oscillatory components are known to be present and are yet ignored. Consequently, the complex calculations discussed in the previous sections relating to relatively minor details of interior velocity in turbulent flow for shallow and deep water become a rather meaningless exercise. The main point those calculations have demonstrated was that for an important special case of forcing a bottom friction law can be cast in terms of the depth-average velocity or the bottom geostrophic velocity, just as well as in terms of the velocity above the bottom wall layer. In a well-stirred water column, i.e., where the total depth is of the order of the Ekman depth and less, a corresponding linear parameterization scheme is for the longshore bottom stress component:

$$B_y = r_* \frac{V}{H}. \tag{6.27}$$

If no oscillatory flow components are present, the resistance coefficient may be interpreted as $r_* = c_{da}(V/H)$ with reference to Equation (6.3).

In deeper water a more realistic parameterization scheme is to set bottom stress proportional to near bottom geostrophic velocity:

$$B_y = r \frac{g}{f} \frac{\partial \zeta}{\partial x} = r v_1. \tag{6.28}$$

With reference to Equation (6.21), this resistance coefficient may be interpreted for the circulation component of the flow as $r = (c_g \cos \phi) q$, with q an average near-bottom fluctuating geostrophic velocity magnitude, c_g a geostrophic drag coefficient and ϕ the angle the bottom stress vector includes with the y-axis.

The cross-shore component of the bottom stress has not been written down here, because previous calculations have shown that it contributes negligibly to the cross-shore momentum balance in a coastal region with relatively large longshore transport V. It is well within the accuracy of any bottom stress parameterization scheme for the circulation component of the flow field to neglect B_x in such circumstances. The parameterization, $B_x = 0$, and B_y as given by (6.27) or (6.28) leads to a considerably simplified theoretical approach, while representing bottom friction affecting the circulation component of the flow as accurately as possible without explicit consideration of transient events. This approach will be used in the developments below. Its inherent shortcomings should not be forgotten, however.

6.4. STEADY CIRCULATION NEAR A STRAIGHT COAST

Previous sections should have placed the problem of a 'steady' i.e., frictionally controlled circulation component in perspective and established a suitably simple theoretical framework for its further discussion. The remainder of this chapter will be devoted to a discussion of simple models of steady circulation in a homogeneous coastal ocean, beginning with a long, straight coast model. The transport equations for steady flow are, with the simplification of neglecting the cross-shore component of the bottom stress:

$$- fV = - gH \frac{\partial \zeta}{\partial x} + F_x,$$

$$fU = - gH \frac{\partial \zeta}{\partial y} + F_y - B_y, \tag{6.29}$$

$$\frac{\partial U}{\partial x} + \frac{\partial V}{\partial y} = 0,$$

where B_y is parameterized according to either Equation (6.27) or (6.28). Isobaths are supposed straight and parallel to a long straight coast coincident with the y-axis, $H = H(x)$.

By taking curl on the first two of Equations (6.29) a depth-integrated vorticity tendency balance is obtained:

$$g \frac{dH}{dx} \frac{\partial \zeta}{\partial y} = W - \frac{\partial B_y}{\partial x}, \tag{6.30}$$

where $W = (\partial F_y / \partial x) - (\partial F_x / \partial y)$ is the wind-stress curl. The operation resulting in Equation (6.30) may also be regarded as expressing the non-divergent nature of transport, substituting for U and V from the first two equations (6.29) into the third. Correspondingly, the terms of Equation (6.30), after division by the Coriolis parameter f, may also be viewed from left to right, as divergence of geostrophic transport, divergence of surface Ekman transport, and divergence of bottom Ekman transport, the three of them having to balance.

As in deep ocean dynamics, the vorticity tendency balance will be found particularly illuminating for gaining insight into the physics of steady currents in the coastal ocean. It is therefore important to reexamine how far the neglect of the cross-shore component of the bottom stress affects this balance. The contribution of B_x to Equation (6.30), if it had been included, would have been $\partial B_x / \partial y$. This will be taken to be of order B_x / L_y, where L_y is the scale of longshore variations of the flow field. Where longshore variations will be considered below, they will be of a relatively large scale, such as are associated with major changes in the orientation of continental coasts, or with wind-stress variations on the scale of weather systems. By contrast, observations show that cross-shore variations of the flow take place on a much shorter offshore scale $L_x \ll L_y$. Therefore $\partial B_x / \partial y$ is small compared to $\partial B_y / \partial x$ = order B_y / L_x on two counts, (1) because B_y is generally substantially larger than B_x, see previous remarks, and (2) because of the discrepancy of scales $L_y \gg L_x$, by one or two orders of magnitude.

6.4.1. Parallel Transport Model

Consider first flow subject to the coastal constraint, Equation (6.2), so the transport is along the isobaths, parallel to a long, straight coast. Such a simple distribution of transport may be expected to arise if forcing by the wind stress is uniform in space, F_x = constant, F_y = constant. For steady flow, the coastal constraint and the continuity equation (third of (6.29)) imply $V = V(x)$, longshore transport a function of distance from shore only, so that V remains constant along isobaths between upstream and downstream transects. This idealization can clearly be valid for certain portions of a shallow sea only and then to a certain degree of approximation, so that its limitations will have to be explored later.

The aysmptotic distribution of transport produced by uniform longshore wind, discussed in Section 6.1 is of this kind, but applies to the case of vanishing longshore pressure gradient. More generally, a cross-shore wind and a longshore sea level gradient $\partial\zeta/\partial y$ may also be present. However, the first of (6.29) shows that with $V = V(x), F_x =$ constant,

$$\frac{\partial^2 \zeta}{\partial x \partial y} = 0 \tag{6.31}$$

so that the longshore pressure gradient is constant with distance from shore. The magnitude of this gradient is arbitrary and in a sense parameterizes the interaction of a limited coastal region, to which the parallel transport model is thought to apply, with the flow outside (this point of view is discussed further in Csanady, 1976a).

The depth-integrated force balance (second of Equation (6.24)) between a longshore pressure gradient constant with distance from shore, wind stress and bottom stress, for a variable depth water column is similar to what was discussed in Chapter 4 for transient flow over variable depth, with 'setup' opposing the wind stress. In the general case, however, the pressure gradient need not oppose the local wind. Where the pressure gradient supports the wind in generating longshore flow in a given direction, bottom stress must be strong enough to balance both. If the two are in opposition, the bottom stress must make up the difference. In shallow water, the integrated pressure gradient force $gH(\partial\zeta/\partial y)$ is small and bottom stress must always oppose wind stress. Where the depth is large enough for $gH(\partial\zeta/\partial y)$ to overwhelm the wind stress, bottom stress opposes the pressure gradient force. Thus, if wind stress and longshore pressure gradient are in opposition, bottom stress vanishes along a critical isobath (where $gH(\partial\zeta/\partial y) = F_y$) and changes sign on crossing this isobath. This is analogous to the change of acceleration in the transient, developing current along the same isobath.

The cross-shore force balance (first of (6.29)) shows that the pressure gradient force must balance both the cross-shore wind and the Coriolis force of longshore flow. In the vorticity balance (Equation (6.30)) the wind stress curl vanishes, so that the vortex stretching term balances bottom stress curl. In a shallow, well-stirred water column the stress gradient and the velocity may be supposed constant in the vertical so that $v = V/H$, $u = 0$, all fluid particles move along depth contours. The appearance of a 'vortex-stretching' term in the vorticity balance under such circumstances is puzzling. However, dividing Equations (6.20) by the depth H and *then* taking curl, one also finds

$$\frac{\partial}{\partial x}\left(\frac{F_y - B_y}{H}\right) - \frac{\partial}{\partial y}\left(\frac{F_x}{H}\right) = -\frac{fU}{H^2}\frac{dH}{dx} \tag{6.32}$$

or that the curl of the net external force acting in the fluid interior vanishes with the cross-isobath transport. The fluid particles in a well stirred column then move along isobaths. Equation (6.32) is simply another version of the vorticity tendency balance, equivalent to (6.30) but more directly relevant to a well-stirred water column.

In deeper water, it is instructive to state the depth-integrated longshore momentum balance in terms of Ekman transports, simply dividing the second of Equations (6.29) by the Coriolis parameter:

$$\frac{F_y}{f} - \frac{B_y}{f} = \frac{gH}{f}\frac{\partial \zeta}{\partial y}\,. \tag{6.33}$$

The left-hand side is Ekman transport in the surface layer plus Ekman transport in the bottom layer (due note being taken of signs), while the right-hand side is geostrophic cross-shore transport associated with the longshore pressure gradient. In the interior of the water column frictional influences are negligible, and only the geostrophic cross-shore flow is in evidence.

In the parallel transport model, where $\partial \zeta / \partial y$ is constant (Equation (6.31)), the geostrophic cross-shore transport is proportional to the local depth H. This has important consequences for the *longshore* flow driven by such a pressure gradient. In the simple case when the longshore wind stress vanishes, $F_y = 0$, the longshore bottom stress B_y must then also be proportional to depth to balance the longshore pressure gradient. This can happen only if the longshore velocity of the pressure-gradient driven flow increases with depth H, in direct proportion to H if a linear drag law such as Equation (6.28) applies. The divergence of the geostrophic cross-shore transport is in this case balanced by a convergence of the bottom Ekman transport. Further consideration of Equation (6.33) shows that with F_y nonzero and in opposition to the pressure gradient, bottom layer Ekman transport changes sign at some critical isobath, where the wind stress is balanced by the pressure gradient. This isobath thus becomes a line of divergence or convergence for near-bottom flow.

Returning to the vorticity tendency equation (6.30), this was seen to express the balance between vortex stretching associated with cross-isobath flow and bottom stress curl, both as affecting the whole (deep) water column. Focussing on the frictionless interior above the bottom Ekman layer, vortex squashing would arise from the kinematic effect of the geostrophic velocity running into reducing depth, which would induce upward motion at the bottom. However, vortex stretching through bottom Ekman layer 'suction' exactly cancels this effect and the frictionless interior is unaffected, because the net vertical velocity above the Ekman layer vanishes. Note that it is not necessary to postulate that Ekman layers be limited to some small fraction of the water column: the balance of vorticity tendencies or of suitably defined horizontal transports (Equation (6.33)) remains the same for whatever ratio of Ekman depth to total depth. All that changes between shallow and deep water is the detailed method of satisfying the various integral balances in the fluid interior. Figure 6.4 illustrates the properties of the parallel transport model ($U = 0$ everywhere) of flow in a coastal region of variable depth.

6.4.2. Coastal Boundary Layer Model of Shelf Circulation

The parallel transport model yielded some illuminating insights into the role of the bottom stress, but of course the model is overidealized, because the coastal constraint is not valid everywhere, and the longshore pressure gradient is no 'deus ex machina'. In a more

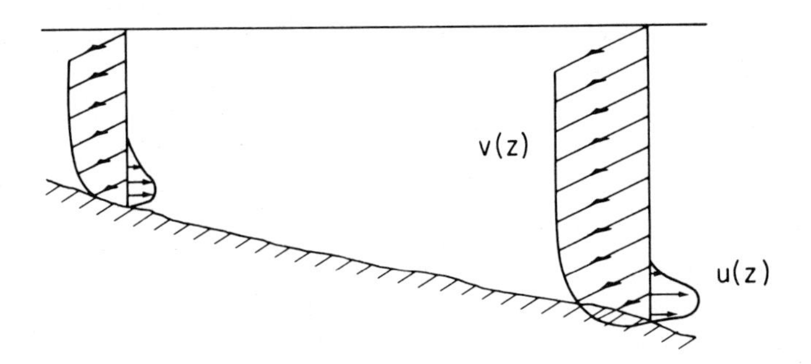

Fig. 6.4. Parallel transport model: structure of pressure gradient driven flow over variable depth continental shelf. Same gradient acting in deeper water adds up to greater total force, which must be balanced by larger bottom stress, implying more intense longshore flow.

general approach the distribution of transports and sea level has to be calculated from Equations (6.29) in order to discover by what physical mechanism a steady longshore pressure gradient is generated along an open coastline and under what circumstances (if any) it is reasonable to rely on the results of the parallel transport model.

On substituting the parameterization scheme of Equation (6.28) into the vorticity tendency equation (6.30) a simple equation for sea level distribution emerges:

$$\frac{\partial^2 \zeta}{\partial x^2} + \frac{f}{r}\frac{\mathrm{d}H}{\mathrm{d}x}\frac{\partial \zeta}{\partial y} = \frac{f}{rg} W. \tag{6.34}$$

The coastal constraint, $U = 0$, is now applied as a boundary condition at $x = 0$. If, as is realistic, also the depth H is taken to vanish at the coast, the second of Equations (6.29) show that

$$F_y = B_y, \quad (x = 0) \tag{6.35}$$

or that the longshore component of the wind stress is balanced by bottom stress, the longshore pressure gradient force vanishing with depth. Given the parameterization scheme of Equation (6.28), the last result is equivalent to

$$\frac{\partial \zeta}{\partial x} = \frac{fF_y}{rg}, \quad (x = 0). \tag{6.35a}$$

This is clearly a convenient form to apply in conjunction with Equation (6.34).

Equation (6.34) is a parabolic equation for the sea level ζ analogous to the one dimensional heat conduction equation, with thermometric conductivity $\kappa = rf^{-1}\,(\mathrm{d}H/\mathrm{d}x)^{-1}$ and negative y playing the role of time (for positive f). In the coordinate system used here (the domain of interest being $x \geqslant 0$), negative y is the direction of propagation of Kelvin or topographic waves, called before the 'cyclonic' direction. In the absence of any forcing, an arbitrary $\zeta(x)$ distribution at a chosen transect ($y = 0$, say) spreads out along x, in the time-like negative y direction, much as a hot spot spreads out in a conducting rod. The heat conduction analogy at once makes apparent what conditions are necessary to find a solution: boundary conditions at $x = 0$ and at some 'outer' boundary $x = l$ of the coastal region of interest, as well as 'initial' conditions at $y = 0$ or other chosen transect.

In view of the linearity of Equation (6.34) it is convenient to discuss separately component solutions due to different kinds of forcing. One may, for example, specify an 'inert' ocean, $\zeta = 0$ at $x = l$ (or even at $x \to \infty$), zero inflow from a 'backward' ($y > 0$) portion of the coast, $\zeta = 0$ at $y = 0$, and calculate the response of a coastal ocean model to some specified local forcing alone. One may then consider an ocean-driven case, and a pattern caused by some specified inflow. All such component solutions are in principle easily superimposed.

Suppose that a solution for (6.34) has been found in which the two key homogeneous terms on the left-hand side are of the same order of magnitude. Let such solutions be characterized by an offshore length-scale L_x and a longshore length-scale L_y. The terms on the left are of the same order if

$$L_x^2 = \kappa L_y, \tag{6.36}$$

where $\kappa = r/(f\ \mathrm{d}H/\mathrm{d}x)$ is the equivalent thermometric conductivity, as before. Typical values of κ are of the order of 3 km supposing r = order 10^{-1} cm s^{-1}, $f = 10^{-4}$ s^{-1}, $\mathrm{d}H/\mathrm{d}x = 3 \times 10^{-3}$. The longshore scale L_y is imposed by variations of forcing or of geography and has already been supposed to be fairly large, say 300 km or more. It follows then that L_x is of order 30 km, i.e., small compared to L_y. This same assumption was made on an essentially empirical basis earlier, in simplifying the equations of motion as they apply to steady coastal currents. The result means that, for typical quantitative parameters characteristic of continental shelves and other coastal seas, Equation (6.34) yields solutions of the same general character as suggested by observation.

In virtue of these scale relationships, Equation (6.34) may be referred to as the "coastal boundary layer model of shelf circulation". The boundary layer character of the equation may be formally established by expressing all variables in a nondimensional form. The externally impressed length scale is that of the forcing, L_y. Elevations ζ are proportional to u_*^2/gs, where s is a typical value of $\mathrm{d}H/\mathrm{d}x$, as one finds by equating the vortex stretching term to the forcing term. Introducing the nondimensional variables:

$$x^* = \frac{x}{L_y}, \qquad y^* = \frac{y}{L_y}$$

$$\zeta^* = \frac{gs\zeta}{u_*^2},$$

$$W^* = \frac{WL_y}{u_*^2},$$

$$\frac{\mathrm{d}H^*}{\mathrm{d}x^*} = \frac{1}{s}\frac{\mathrm{d}H}{\mathrm{d}x},$$

one arrives at the following nondimensional form of Equation (6.34), after dropping the stars again:

$$\frac{r}{sfL_y}\frac{\partial^2\zeta}{\partial x^2}+\frac{dH}{dx}\frac{\partial\zeta}{\partial y}=W. \tag{6.34a}$$

The highest order term in this equation is multiplied by the nondimensional number r/sfL_y. For typical values of r, f, and s this number is small compared to unity even for $L_y = 100$ km, and smaller yet for forcing on the scale of extratropical storms. Hence the solutions are expected to have a boundary layer character, with the first term in the equation (the bottom stress curl term) being significant only in a coastal boundary layer of scale width $L_x = (r/sfL_y)^{1/2} L_y = (rL_y/fs)^{1/2}$.

6.4.3. Coastally Trapped Flow Fields

Boundary-layer type solutions of Equation (6.34), characterized by a small scale ratio L_x/L_y, may also be described as coastally 'trapped', as some transient features of the flow were described earlier, for example topographic waves. Far from the coast such solutions vanish:

$$\zeta = 0, \qquad (x\to\infty). \tag{6.37}$$

This is effectively a boundary condition at the outer limit of the coastal sea of interest and represents physically the case of an 'inert' deep ocean, i.e., no dynamical interaction between shallow and deep portions of the ocean.

For continental shelves, a realistic idealization is to postulate very great depths beyond the shelf-edge, i.e.,

$$H\to\infty, \qquad (x\to\infty). \tag{6.38}$$

Some important general properties of coastally trapped solutions of (6.34) follow from Equations (6.37) and (6.38) (Csanady, 1978a). The transport equations (Equation (6.29)) imply:

$$U=\frac{F_y}{f}, \qquad V=-\frac{F_x}{f}, \qquad (x\to\infty) \tag{6.39}$$

which is the familiar Ekman transport.

An integration of the continuity equation leads to

$$\frac{d}{dy}\int_0^\infty V\,dx=-\frac{F_y}{f} \tag{6.40}$$

independently of the depth distribution. Thus the total longshore transport in a y = constant cross-section is independent of bottom topography. A second integration with respect to y shows moreover:

$$\int_0^\infty V\,dx=\int_y^\infty \frac{F_y}{f}\,dy' \tag{6.40a}$$

or that the integrated longshore transport is the sum of all 'backward' ($y' > y$) Ekman transport to or from the coast. Furthermore, the boundary condition at the shore (6.35), effectively determines the longshore velocity at small x. The cross-shore transport at large x is Ekman transport, given by (6.38). These constraints determine the general character of the flow field to the point where one can almost sketch in transport streamlines.

The flow field may conveniently be described in terms of a transport stream function ψ defined by

$$U = \frac{\partial \psi}{\partial y}, \qquad V = -\frac{\partial \psi}{\partial x}. \tag{6.41}$$

With the elevation field known from a solution of (6.34), $\psi(x, y)$ may be easily calculated from the first two of (6.29). The constraints on the flow just discussed imply that, while the streamlines may be distorted in various ways over complex bottom topography, the total range of ψ-variation is fixed by the wind-stress field. Also, the asymptotic spacing of streamlines both close to the shore and at large distances from it remains the same for a given wind-stress field. Even a highly idealized coastal ocean model should therefore give a realistic description of trapped flow and pressure fields.

6.5. INCLINED PLANE BEACH MODEL

A convenient simple coastal ocean model is the inclined plane beach, with the depth distribution

$$H = sx, \qquad (s = \text{constant}). \tag{6.42}$$

For this simple case the equivalent thermometric conductivity $\kappa = r/fs$ is constant and standard solutions of Equation (6.34) become available to represent trapped flow fields.

6.5.1. Flow and Pressure Field of a Coastal Mound

The simplest case to consider is that of vanishing wind stress everywhere, all flow being due to the 'initial' condition imposed at the section $y = 0$. Let the level distribution in this section be

$$\zeta = \zeta_0(x), \qquad (y = 0). \tag{6.43}$$

Physically this may be viewed as a coastal 'mound' of water maintained near $y = 0$ by some unspecified effect, e.g., by shoreward Ekman transport over the backward sector $y > 0$, or by freshening of coastal waters. With the aid of the first of (6.29) and $F_x = 0$, Equation (6.43) may also be interpreted as a given distribution of inflowing longshore transport across the initial section.

Substituting for V from the first of (6.29) one arrives at, after one integration:

$$\int_0^\infty \zeta \, dx = \int_0^\infty \zeta_0 \, dx = \text{constant} = M \tag{6.24}$$

showing conservation of total excess volume in the coastal mound. Interpreted in terms of the heat conduction analogy, this is conservation of total heat content. The result is consistent with the boundary condition at $x = 0$ (Equation (6.35a)) for $F_y = 0$, which translates into a zero heat flux (insulating) boundary. The full solution is simply transcribed from the equivalent heat conduction problem (Carslaw and Jaeger, 1959):

$$\zeta = \frac{1}{2(-\pi\kappa y)^{1/2}} \int_0^\infty \zeta_0(x') \left[\exp \frac{(x-x')^2}{4\kappa y} + \exp \frac{(x+x')^2}{4\kappa y} \right] dx'. \tag{6.45}$$

If significant ζ_0 variations are confined to a strip $x \leqslant l$, then at distances $(-y)$ large compared to l^2/κ Equation (6.45) approaches the simple source solution which is sufficiently instructive to be considered in detail here. The pressure and transport fields of this solution are given by:

$$\begin{aligned} \zeta &= \sqrt{\frac{2}{\pi}} \frac{M}{L_x} \exp\left(-\frac{x^2}{2L_x^2}\right), \\ U &= \frac{1}{\sqrt{2\pi}} \frac{gsM}{fy} \left(\frac{x}{L_x}\right)^3 \exp\left(-\frac{x^2}{2L_x^2}\right), \\ V &= -\sqrt{\frac{2}{\pi}} \frac{gsM}{fL_x} \left(\frac{x}{L_x}\right)^2 \exp\left(-\frac{x^2}{2L_x^2}\right), \end{aligned} \tag{6.46}$$

where

$$L_x = \left(-\frac{2ry}{fs}\right)^{1/2}$$

is the cross-shore length scale ('trapping width') of the field. The solution is illustrated in Figures 6.5 and 6.6.

The contours of constant surface elevation are typically long and cigar-shaped (if the same scale is used along the x and y axes) resembling effluent plume concentration isopleths. The cross-shore pressure gradient vanishes at the coast but it has the relatively high peak value at $x/L_x = 1$ of:

$$\frac{\partial \zeta}{\partial x} = -\sqrt{\frac{2}{\pi}} \frac{M}{L_x^2}, \quad (x = L_x). \tag{6.47}$$

This gradient is proportional to the longshore velocity (V/H). The longshore pressure gradient is

$$\frac{\partial \zeta}{\partial y} = -\frac{M}{\sqrt{2\pi}\, y L_x} \left(1 - \frac{x^2}{L_x^2}\right) \exp\left(-\frac{x^2}{2L_x^2}\right). \tag{6.48}$$

This is seen to be positive (for $y < 0$) at the coast, i.e., to drive the flow toward negative y. At $x = L_x$, where the longshore velocity peaks, the pressure gradient $\partial\zeta/\partial y$ changes

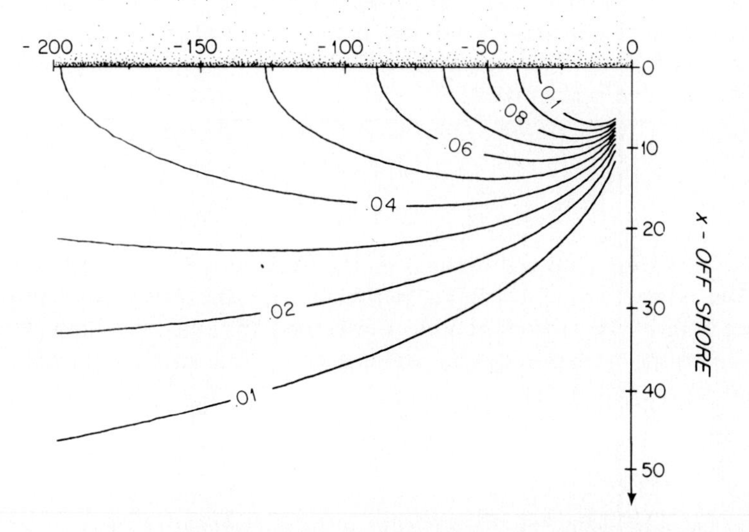

Fig. 6.5. Pressure field of a coastal 'mound', in nondimensional units $\kappa\zeta/M$, with $\kappa = r/fs$, which is also the unit of distance along x and y. Note that the x coordinate axis is stretched by a factor of 2.5. From Csanady (1981b).

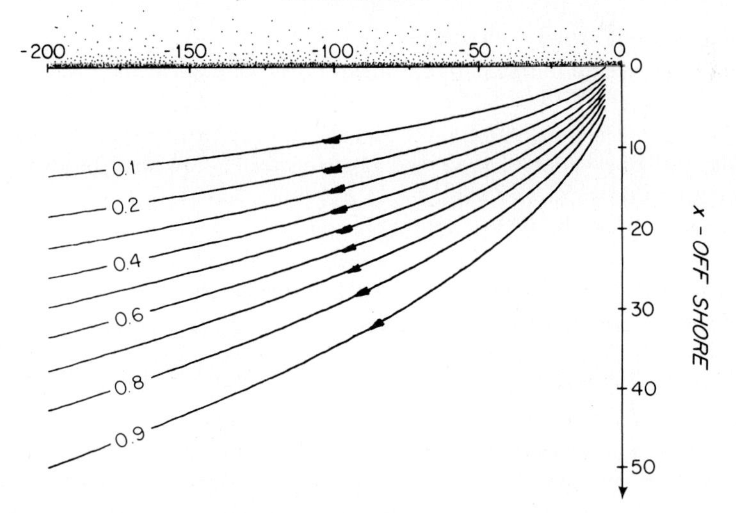

Fig. 6.6. Transport streamlines ($f\psi/gsM$ = constant lines) of a coastal mound, distance scales as in previous figure. From Csanady (1981b).

sign and begins to oppose the flow. On analyzing the longshore momentum balance (second of (6.29)) on finds that at small x/L_x longshore pressure gradient and bottom friction balance to order $(x/L_x)^2$. However, at $x = L_x$ the Coriolis force of cross-shore transport balances the bottom stress and becomes, at still greater distances from the shore, the largest term in the longshore momentum balance, opposing both longshore pressure gradient and bottom stress.

The transport stream function (Equation (6.41)) may be calculated to be

$$\psi = \frac{gsM}{f}\left[\operatorname{erf}\left(\frac{x}{\sqrt{2}\,L_x}\right) - \sqrt{\frac{2}{\pi}}\,\frac{x}{L_x}\,\exp\left(-\frac{x^2}{2L_x^{\,2}}\right)\right]. \tag{6.49}$$

Because the stream function depends on the single variable x/L_x, its distribution (in the far field, where (6.46) are valid) is 'self-similar', i.e., the same at any transect, except for a stretching of the cross-shore scale. Correspondingly, the streamlines ψ = constant connect points of constant x/L_x, and are parabolae of the form x = constant $y^{1/2}$. One also observes from (6.49) that

$$\frac{U}{V} = \frac{1}{2}\frac{x}{y}$$

which gives an easy rule for the construction of the transport streamlines, U/V being their tangent inclination. The total longshore transport is

$$Q = \int_0^\infty V\,\mathrm{d}x = -\frac{gsM}{f} \tag{6.50}$$

which is independent of the y-coordinate.

The transport pattern in this simple model is thus characterized by constant total longshore flow. However, the width of the flow field grows as L_x, i.e., as $y^{1/2}$. The longshore transport V has a maximum at $x = \sqrt{2}\,L_x$, the cross-shore transport U at $x = \sqrt{3}\,L_x$, both being negligible close to the coast. The longshore flow may be thought of as a relatively narrow stream centered at $x = \sqrt{2}\,L_x$.

As was pointed out above, the cross-shore transport plays a dominant role in the longshore momentum balance beyond $x = L_x$. Supposing that the water at these distances is much deeper than Ekman depth, the part of the cross-shore transport balancing bottom stress is carried in the bottom Ekman layer. It is interesting to reflect, however, that the broadening of the flow and pressure field in the negative y-direction necessarily implies an adverse pressure gradient on the outer edges of the field, and an associated geostrophic cross-shore flow. The partition of cross-shore transport between bottom Ekman layer and interior geostrophic flow changes with distance from shore: at the inner edge of the main stream cross-shore transport is all within the bottom Ekman layer, on the other fringes it is all geostrophic flow. The relative simplicity of the transport streamline pattern thus hides some intricate details.

The above results remain valid for negative M, i.e., for a sea level depression at $y = 0$

instead of a mound. The signs of ζ, U, and V are then all reversed. Longshore transport, in particular, is in this case toward positive y, so that at $y = 0$ a quantity of water M is flowing *out* of the domain of interest, $y < 0$. To represent this situation, the arrows in Figure 6.6 on the streamlines should be reversed. The physical interpretation is that the fluid removed at $y = 0$ is gathered from a coastal boundary layer of increasing width (increasing toward negative y, as before). At the outer fringes of the field there is now shoreward flow, first mainly in the frictionless interior, then closer to shore mainly in the bottom Ekman layer.

The narrowness of the flow and pressure fields arises from the typical parameters characterizing continental shelves. On a broad, flat shelf at mid-latitudes the key parameters are of order: $r = 0.05$ cm s^{-1}, $f = 10^{-4}$ s^{-1}, $s = 3 \times 10^{-3}$. These give $L_x = 18$ km at $-y = 100$ km, $L_x = 60$ km at $-y = 1000$ km, i.e., very slow growth in the trapping width of the flow field with increasing longshore distance, in the 'time-like' or negative y direction. A typical 'mound' might contain enough water to raise sea level by 10 cm over a coastal strip 100 km wide, i.e., $M = 10^4$ m^2. Then with the above numbers and at $-y =$ 500 km, $\partial\zeta/\partial y$ at the coast is 2×10^{-7}, or quite significant, even at such a fairly large distance from the mound. The maximum cross-shore elevation gradient here is about -5×10^{-6}, corresponding to a longshore geostrophic velocity of 0.5 m s^{-1} toward negative y, and it occurs at a distance from the coast of $L_x \approx 40$ km, in water 120 m deep. The longshore transport peaks even further from shore, at $x = \sqrt{2}\, L_x$ or close to 60 km.

This example shows that the maintenance of a nearshore mound at some location has some surprisingly far-reaching effects in the forward direction (negative y), affecting at such large distances primarily the outer portion of the shelf, however.

6.5.2. Wind Stress Along Portion of Coast

The discussion of the previous section leaves open the question, by what physical agency a 'mound' of water can be maintained in steady state at some initial section $y = 0$. One plausible possibility is longshore wind blowing toward negative y over a finite 'backward' sector of the coast $0 \leqslant y \leqslant Y$. From the general result (6.40), such wind is known to have a tendency to generate raised coastal sea levels.

Consider therefore again an inclined plane beach and a steady longshore wind, distributed in space as follows:

$$\left.\begin{array}{ll} F_y = -u_*^2, & (0 \leqslant y \leqslant Y) \\ F_y = 0, & y \leqslant 0 \end{array}\right\} \quad \text{(all } x\text{)}. \tag{6.51}$$

For the portion $y \leqslant 0$ the results of the previous section will apply: here the main interest is to see what distribution of sea level arises at $y = 0$ and how this builds up over the backward portion of the coast where the wind acts.

The boundary condition at $x = 0$ is for the sector of interest, from (6.35a):

$$\frac{\partial\zeta}{\partial x} = -\frac{fu_*^2}{rg}. \tag{6.52}$$

Interpreted in terms of the heat conduction analogy this is constant heat flux *into* the domain of interest, which tends to raise the temperature, i.e., sea level. The solution of the relevant one-dimensional heat conduction problem is known and is illustrated in Figures 6.7 and 6.8. The equations describing the fields are, for $0 \leqslant y \leqslant Y$:

$$\zeta = \frac{f u_*^2}{rg} L_x \operatorname{ierfc}\left(\frac{x}{L_x}\right),$$

$$\psi = \frac{u_*^2}{f} (Y-y)\left[1 - \operatorname{erfc}\left(\frac{x}{L_x}\right) - 2\frac{x}{L_x} \operatorname{ierfc}\left(\frac{x}{L_x}\right)\right], \tag{6.53}$$

$$\frac{V}{H} = -\frac{u_*^2}{r} \operatorname{erfc}\left(\frac{x}{L_x}\right),$$

where now

$$L_x = 2\left[\frac{r}{fs}(Y-y)\right]^{1/2}$$

is the cross-shore length scale of the flow. At the coast, the longshore velocity is u_*^2/r, to negative y, in accordance with the boundary condition (6.35a). Far from the coast transport is shoreward, with the transport streamlines evenly spaced, representing surface layer Ekman transport in the quantity u_*^2/f. In between, the average longshore velocity (V/H) distribution is self-similar, monotonically reducing from the maximum value at the coast as its range grows with the scale L_x. The total longshore transport correspondingly grows as it absorbs the arriving Ekman transport:

$$Q = \int_0^\infty V \, dx = -\psi(\infty) = \frac{u_*^2}{f}(Y-y). \tag{6.54}$$

The pressure field associated with this flow pattern is characterized by a relatively large cross-shore gradient, in geostrophic balance with the longshore flow. The longshore gradient is smaller but its distribution is only a little more intricate:

$$\frac{\partial \zeta}{\partial y} = -\frac{2u_*^2}{gsL_x}\left[\operatorname{ierfc}\left(\frac{x}{L_x}\right) + \frac{x}{L_x}\operatorname{erfc}\left(\frac{x}{L_x}\right)\right]. \tag{6.55}$$

Close to the coast the first term dominates, the gradient right at the coast being

$$\left.\frac{\partial \zeta}{\partial y}\right|_0 = -1.1284 \frac{u_*^2}{gsL_x}.$$

For typical values of the parameters, $r = 0.05$ cm s^{-1}, $f = 10^{-4}$ s^{-1}, $s = 3 \times 10^{-3}$, at a distance $Y - y = 100$ km from the beginning of the wind-stress square wave the distance scale L_x calculates to 26 km, the longshore gradient $\partial\zeta/\partial y|_0$ with $u_*^2 = 1$ cm^2 s^{-2} to about 1.4×10^{-7} or 1.4 cm in 100 km. The longshore gradient retains the same sign, opposing the applied wind stress everywhere, and remains of the same order of

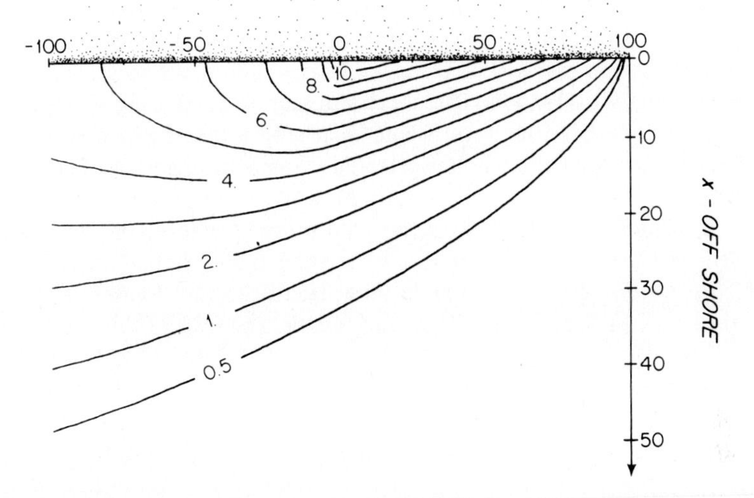

Fig. 6.7. Pressure field of a square wave of wind-stress, u_*^2 = constant for $0 \leqslant y/\kappa \leqslant 100$. Contours shown are lines of constant $\zeta gs/u_*^2$, units of distance are κ. From Csanady (1981b).

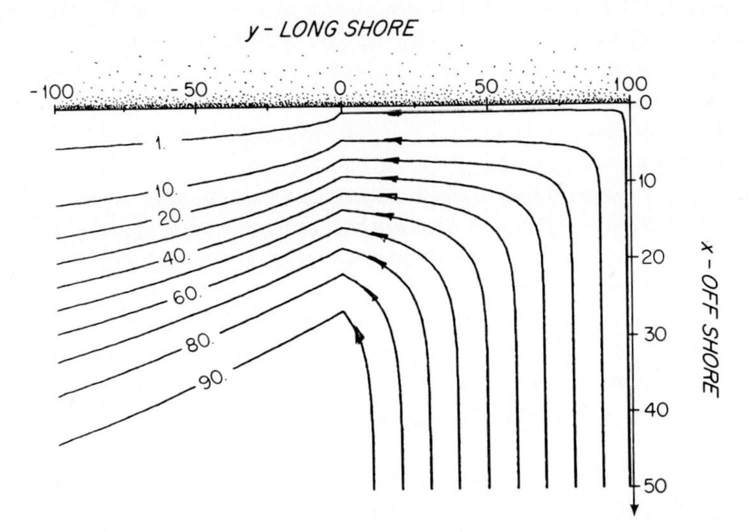

Fig. 6.8. Transport streamlines corresponding to previous figure, in units of $u_*^2\kappa/f$. From Csanady (1981b).

magnitude at the coast to a little beyond $x = L_X$. Physically, the existence of this gradient is a consequence of the buildup of coastal sea level, necessary to accomodate the arriving Ekman transport in a continually broadening longshore current. Note that the 'buildup' occurs not in time, but in the time-like negative y-direction of the coast.

For wind pointing into the opposite, positive y direction the signs change in all three of Equations (6.53), so that a coastal sea level depression develops gradually, *still in the negative y or time-like direction*. The physical reason is exactly the same as before, the coastal sea level drop reflecting geostrophic balance of a longshore current of increasing width, directed now to positive y, however. The arrows on the streamlines in Figure 6.8 are reversed.

At $y = 0$, where the wind stops acting, the pressure gradient at the coast abruptly reverses sign, and for $y < 0$ the conditions discussed in the previous section prevail. Near the coast the pressure gradient drives the flow against bottom friction, but the velocity drops to zero in a nearshore band of increasing width. The scale of this is

$$L_X{}^* = \left(-\frac{2ry}{fs}\right)^{1/2}$$

or the same as the scale of the field of a mound, defined in (6.46). Over this forward sector $L_X{}^* < L_X$, so that a second or 'internal' boundary layer may be said to form alongshore, growing in the negative y direction. As discussed in the previous section, this internal boundary layer growth forces the longshore current further and further offshore. The present example shows a fuller circulation pattern, specifying where those transport streamlines come from (Ekman transport between $y = 0$ and $y = Y$) that cross the $y = 0$ transect and later, over the forward sector, become slowly forced offshore by bottom friction. Or, reversing the wind stress, the pattern shows where all the Ekman transport originates that leaves the coast in the segment $0 \leqslant y \leqslant Y$: the supply is in that case from the forward section, $y < 0$, to the backward one.

A generalization of this model allows one to answer immediately the question, what happens along a coastline subject to a piecewise constant longshore wind stress distribution. The contribution of each piece of coastline may be calculated as above. At any given location, the flow is the sum of contributions from all 'backward' sections. Recalling that at $y < 0$ in Figures 6.7 and 6.8 an internal boundary layer appears, growing toward negative y as $(-2ry/fs)^{1/2}$, one infers that with many active backward sections there are many internal boundary layers. The effect of distant sections (those lying toward large positive y) is, however, increasingly less significant. These backward sections also affect increasingly the outer (high x) portions of the shelf. In each internal boundary layer the flow is forward or backward according to whether the section where it originated was subject to inward or outward Ekman transport.

6.5.3. Spacewise Periodic Longshore Wind Stress

To expand on the effects of wind stress variations along the coast, a fairly complex coastline along which the wind blows both in the forward and the backward direction may be modelled by periodic forcing:

$$F_y = u_*^2 \cos ky. \tag{6.56}$$

Periodic solutions of Equation (6.34) are readily obtained by the method of the separation of variables, as for topographic waves. Writing

$$\zeta = Z_1(x)\,Z_2(y) \tag{6.57}$$

one finds on substitution into Equation (6.34) that separation is possible if

$$-\frac{Z_2'}{Z_2} = \frac{r}{fs}\frac{Z_1''}{Z_1} = \text{constant}, \tag{6.58}$$

where primes denote differentiation. Sinusoidal longshore variation, consistent with the character of the forcing, Equation (6.56), requires an imaginary separation constant, ik, writing e^{iky} for $Z_2(y)$.

Therefore the distribution Z_1 is subject to the equation

$$\frac{d^2 Z_1}{dx^2} + \frac{ikfs}{r} Z_1 = 0 \tag{6.59}$$

and the boundary conditions

$$\left.\begin{aligned} \frac{dZ_1}{dx} &= \frac{fu_*^2}{rg}, \qquad (x = 0)\\ Z_1 &= 0, \qquad (x \to \infty). \end{aligned}\right\} \tag{6.60}$$

After some routine calculations one finds

$$\left.\begin{aligned}
\frac{gs}{u_*^2}\zeta &= \frac{2}{kL} e^{-x/L} \sin\left(ky + \frac{x}{L} - \frac{\pi}{4}\right),\\
\frac{fU}{u_*^2} &= \cos ky - e^{-x/L}\cos\left(ky + \frac{x}{L}\right) -\\
&\quad - \sqrt{2}\,\frac{x}{L}\, e^{-x/L}\cos\left(ky + \frac{x}{L} - \frac{\pi}{4}\right),\\
\frac{rV}{u_*^2 H} &= e^{-x/L}\cos\left(ky + \frac{x}{L}\right),\\
\frac{fk}{u_*^2}\psi &= \sin ky - e^{-x/L}\sin\left(ky + \frac{x}{L}\right) -\\
&\quad - \sqrt{2}\,\frac{x}{L}\, e^{-x/L}\sin\left(ky + \frac{x}{L} - \frac{\pi}{4}\right),
\end{aligned}\right\} \tag{6.61}$$

where

$$L = \left(\frac{2r}{fks}\right)^{1/2} \tag{6.62}$$

this being a cross-shore length scale of the distribution. The transport streamfunction ψ has been defined in Equation (6.41). From the solution it may be seen that, at the shore ($x = 0$), the longshore velocity distribution is in phase with the wind stress. The physical reason is that, by the second of Equations (6.29), in very shallow water the wind stress must be balanced by the bottom stress, both longshore pressure gradient and Coriolis force of offshore flow being negligible in comparison. The bottom stress is, of course, in phase with the longshore velocity. Figures 6.9 and 6.10 illustrate the solution.

The longshore pressure gradient leads the longshore velocity by $3\pi/4$, which means that at the shore the pressure gradient force opposes the wind stress over 75% of the shoreline (Figure 6.9). Constant phase lines of surface elevation and longshore velocity have an inclination of $\tan^{-1} kL$ against the x axis. Both ζ and $V = V/H$ decay to negligible values within a distance of order L from the coast.

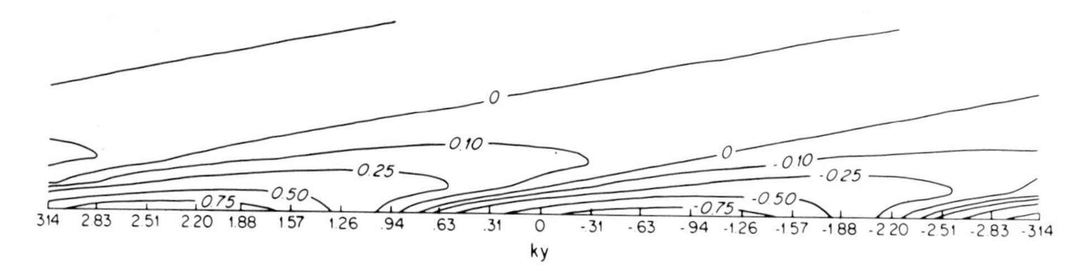

Fig. 6.9. Pressure field due to sinusoidally varying longshore wind over inclined plane beach. Contours shown are lines of constant $\zeta gs/u_*^2$, the distance unit alongshore is k^{-1}, cross-shore. From Csanady (1978a).

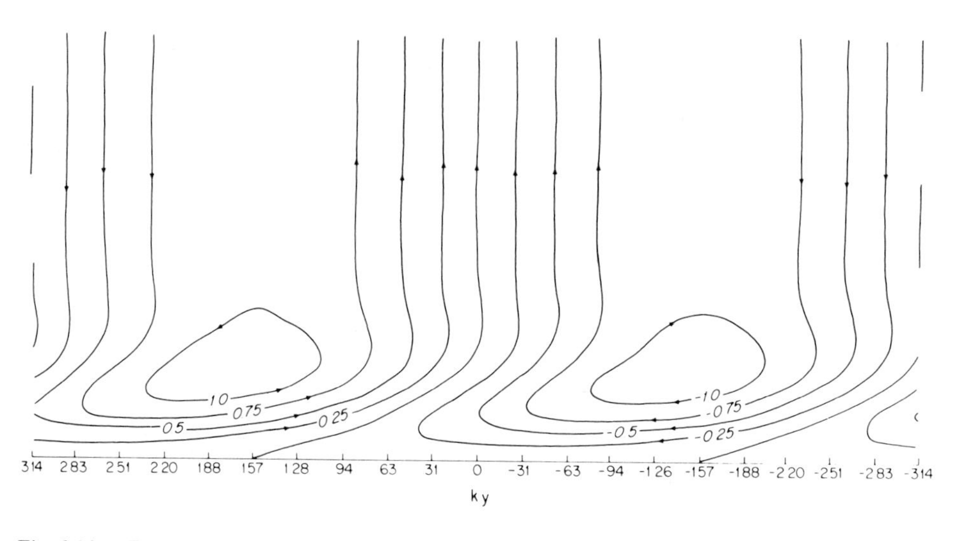

Fig. 6.10. Transport streamlines corresponding to previous figure, in units of u_*^2/fk. From Csanady (1978a).

The variation of offshore velocity and transport streamfunction is more complex. Far from the coast the transport becomes the offshore Ekman transport corresponding to the local longshore wind stress.

The x and y length scales in Equations (6.61) are different, and their ratio kL determines to what extent the streamlines are squeezed together near the coast. Once kL is fixed, however, Equations (6.61) may be illustrated by nondimensional contours, valid for any specific case characterized by the same kL. In Figures 6.9 and 6.10 calculated results are given for $kL = 0.183$, which is fairly typical, with $L = 10$ km and $2\pi/k = 350$ km or so, the latter being the longshore wavelength.

The figures reveal how the flow satisfies the various integral constraints and boundary conditions discussed earlier. At the shore the flow is downwind, longshore velocity and wind stress being in phase. Over much of the coastline, a 'return' flow region appears further offshore, where the flow is against the wind. This is driven by the longshore pressure gradient, which opposes the wind over much of the coastline. Far from shore the transport distribution becomes the Ekman drift, *cross-shore* velocity being in phase with wind stress.

The figures also illustrate how further and further backward sections influence progressively outer domains of the shelf, although at rapidly reducing intensity, see earlier remarks.

6.6. DEEP OCEAN INFLUENCE

The solutions of Equation (6.29) discussed in the previous few sections were all coastally trapped, satisfying the condition $\zeta = 0$ at large distances from the shore. Physically, these solutions represent effects of forcing acting near shore, which generate characteristic flow patterns in a coastal band, without interaction with the flow in the deeper ocean.

However, elevation gradients of the same order of magnitude as observed or inferred over continental shelves also exist in the deep ocean. They are caused by forces of much greater total magnitude owing to the great depth of the ocean and are a consequence of larger scale phenomena, e.g., of wind stress acting over an entire oceanic basin. They may legitimately be regarded as extraneous factors from the point of view of a limited coastal zone model. To discover their effects on the coastal zone appropriate boundary conditions at the outer edge of the continental shelf are imposed in place of the simple hypothesis of constant or vanishing elevation. Such boundary conditions may be regarded as modeling the influence of an 'active' deep ocean on the coastal zone.

Because the idealizations of the simple coastal zone model discussed here become unrealistic at depths much in excess of 100 m, it is convenient to limit the model to some nearshore strip $0 < x < l$. An arbitrary distribution of surface elevation $\zeta(l, y)$ along the $x = l$ line may then be used to represent the influence of the open ocean on the shelf. This constitutes a simple replacement for the previous boundary condition at infinity, and is clearly only one of several possible choices in this regard.

For a coastal zone of constant bottom slope the heat conduction analogy may again be invoked. The zone $x = 0$ to l corresponds to a slab of constant conductivity. A specific $\zeta(l, y)$ distribution at the outer edge corresponds to a prescribed variation of temperature at one surface of the slab. At the other surface, $x = 0$, one may prescribe finite or zero heat flux, according to whether or not a longshore wind is present.

As an example of oceanic influence, consider the case of a constant longshore pressure gradient at the shelf edge, acting at $y < 0$:

$$\left.\begin{aligned} \frac{\partial \zeta}{\partial y} &= \gamma = \text{constant}, \qquad && (y \leqslant 0) \\ \zeta &= 0, && (y > 0) \end{aligned}\right\} \quad (x = l). \tag{6.63}$$

Wind is supposed absent so that the boundary condition at the shore is $\partial\zeta/\partial x = 0$. In the heat conduction analogy the corresponding problem is a slab of zero initial temperature, no heat flow across $x = 0$, the temperature at $x = l$ changing at a constant rate in time after $t = 0$. The appropriate solution of the heat conduction equation is (Carslaw and Jaeger, p. 104).

$$\zeta = \gamma y - \frac{\gamma(x^2 - l^2)}{2} - \frac{16\gamma l^2}{\kappa\pi^3} \sum_{n=0}^{\infty} \frac{(-1)^n}{(2n+1)^3} \exp\left[(2n+1)^2 \pi^2 \kappa y/4l^2\right] \cos\left[(2n+1)\pi x/2l.\right. \tag{6.64}$$

The largest of the exponential factors multiplying the cosine series corresponds to $n = 0$, and becomes negligible for

$$-y \gg 4l^2/\pi^2\kappa.$$

All others vanish at shorter $-y$. At such large distances the ζ distribution with cross-shore distance does not change in shape, the level rising or dropping uniformly at all x in the y direction. The corresponding longshore velocity distribution is

$$\begin{aligned} v &= \frac{g}{f}\frac{\partial \zeta}{\partial x} = -\frac{g\gamma x}{f\kappa} = -g\frac{\gamma s x}{r} \\ &= -\frac{gH}{r}\frac{\partial \zeta}{\partial y}, \qquad (-y \gg 4l/\pi^2\kappa) \end{aligned} \tag{6.65}$$

having noted that

$$\frac{\partial \zeta}{\partial y} \cong \left.\frac{\partial \zeta}{\partial y}\right|_l = \gamma$$

at all x at these distances $-y$. Equation (6.65) expresses balance between bottom stress rv and longshore pressure gradient force $-gH(\partial\zeta/\partial y)$. The offshore transport U is zero everywhere. The simple physical situation is then that the 'active' ocean impresses its longshore gradient on the entire coastal zone, the water flowing alongshore, down the gradient. The longshore velocity increases with depth, because the total pressure-gradient force does, and this has to be balanced by bottom stress. In other words, the flow field is exactly as discussed under the parallel transport model (Section 6.4.1.).

One should note, however, that this asymptotic state is only reached after sufficient water has moved onto the shelf to build up (or deplete) coastal sea levels so as to allow the geostrophic balance demanded by (6.65). The deep ocean can only drive shelf circulation against bottom friction if it 'pumps' water onto the shelf. To show this formally, multiply the first and second of Equations (6.29) by U and V respectively and add the two to arrive at:

$$g\left[\frac{\partial}{\partial x}(U\zeta) + \frac{\partial}{\partial y}(V\zeta)\right] = F_x\frac{U}{H} + (F_y - B_y)\frac{V}{H}. \tag{6.66}$$

An area integral over a finite coastal domain A with boundary B now yields

$$-g\oint_B V_n\,\zeta\,\mathrm{d}s = \iint_A \left[F_x\frac{U}{H} + (F_y - B_y)\frac{V}{H}\right]\mathrm{d}x\,\mathrm{d}y. \tag{6.67}$$

The line integral on the left is the 'flow work' term, or local pressure times inward transport. The right-hand side represents energy input by wind and energy dissipation by bottom stress. In the problem considered in the present section only the energy dissipation term is present. This must be balanced by flow work, or the 'pumping' of fluid onto the shelf in the region of high pressure.

6.7. PERIODIC CROSS-SHORE WIND

Another instance of a pressure field not trapped at the coast is encountered when the forcing term on the right of (6.29), the wind stress curl, is non-zero at $x \to \infty$. The problem arises in considering what the effects of cross-shore winds arc, when these vary in the longshore direction. A periodic distribution of cross-shore windstress is a sufficiently general example. It is described by:

$$F_x = u_*^2 \cos ky. \tag{6.68}$$

As discussed earlier (Section 4.2), cross-shore winds over realistic beach topography give rise to a logarithmic singularity at $x = 0$ in the solutions of the linearized equations of motion. In order to calculate realistic longshore gradients, the coastal zone model is conveniently cut off at some depth H_0, at $x = x_0$, representing physically the outer edge of the wave-dominated littoral zone.

An inclined plane beach is considered again, of bottom slope s, $H = sx$, with $H_0 = sx_0$, so that offshore distance is measured from a virtual origin, $x = x_0$ being the coast. With cross-shore wind acting over shallow water, the parameterization scheme of (6.27) is more realistic than (6.28) and will be used. With the aid of the second of (6.29), with (6.27) substituted, one finds that the boundary condition at the coast, $U = 0$, is equivalent to:

$$\frac{\partial\zeta}{\partial x} + \frac{fH_0}{r}\frac{\partial\zeta}{\partial y} = \frac{F_x}{gH_0}, \quad (x = x_0). \tag{6.69}$$

With (6.27) used to represent bottom stress the forcing term on the right is missing:

in water a few meters deep this would be unrealistic. Equation (6.30), with (6.27) substituted for bottom stress results in

$$\frac{\partial^2 \zeta}{\partial x^2} + \frac{f}{r}\frac{dH}{dx}\frac{\partial \zeta}{\partial y} = \frac{f}{rg} W + \frac{1}{gH}\left(\frac{\partial F_x}{\partial x} - \frac{dH}{dx}\frac{F_x}{H}\right), \tag{6.70}$$

where now $W = -\,\partial F_x/\partial y$. A comparison with (6.24) reveals an additional forcing term on the right, related to the cross-shore wind stress. The new forcing term is present only in shallow enough water for the depth-average velocity to determine bottom stress. This is the case where the depth is of the order of the Ekman depth or less.

As in dealing with coastally trapped fields earlier, it is convenient to nondimensionalize cross-shore distances with the length scale

$$L = (2r/fks)^{1/2}. \tag{6.71}$$

Elevations are proportional to the forced amplitude

$$\zeta_a = u_*^2/gs. \tag{6.72}$$

Longshore distances are best nondimensionalized by the reciprocal wave number k^{-1}. The following nondimensional variables thus emerge:

$$\zeta^* = \frac{\zeta}{\zeta_a}, \qquad x^* = \frac{x}{L}, \qquad y^* = ky. \tag{6.73}$$

Dropping the stars on these variables now, (6.70) takes on the nondimensional form,

$$\frac{\partial^2 \zeta}{\partial x^2} + 2\frac{\partial \zeta}{\partial y} = -\frac{\cos y}{x^2} + 2 \sin y. \tag{6.74}$$

The boundary condition (6.69) similarly transforms into

$$\frac{\partial \zeta}{\partial x} + 2x_0 \frac{\partial \zeta}{\partial y} = \frac{\cos y}{x_0}, \qquad x = x_0. \tag{6.75}$$

Solutions are now readily calculated (for details see Csanady, 1980). Figures 6.11 and 6.12 illustrate these, obtained for the following typical parameters:

$x_0 = 10$ km, $\quad L = 40$ km

$H_0 = 10$ m, $\quad s = 10^{-3}$, $\quad f = 10^{-4}\ \mathrm{s}^{-1}$,

$r = 0.1\ \mathrm{cm\ s}^{-1}$, $\quad \dfrac{2\pi}{k} = 500$ km,

$k = 1.25 \times 10^{-7}\ \mathrm{cm}^{-1}$.

The stream function has been introduced as defined in (6.41), or in nondimensional form

$$\psi^* = \frac{f\psi}{u_*^2 L} \tag{6.76}$$

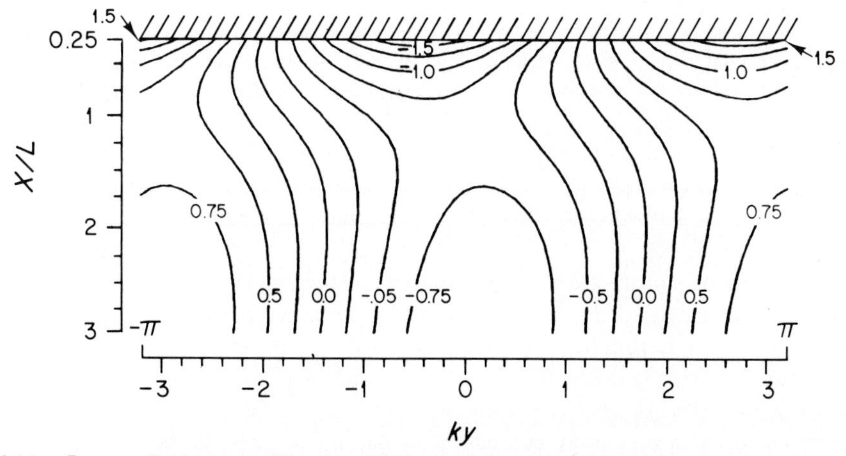

Fig. 6.11. Pressure field generated by sinusoidally varying cross-shore wind over inclined plane beach, contours of $\zeta gs/u_*^2$. From Csanady (1980).

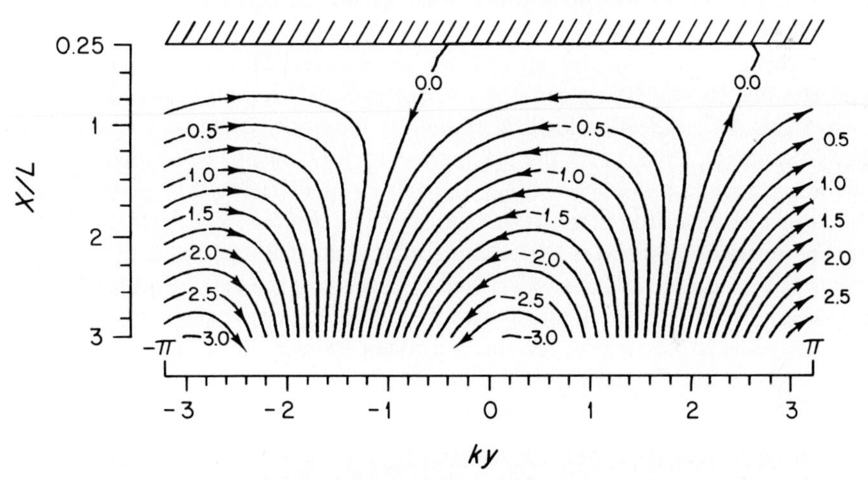

Fig. 6.12. Transport streamlines correspond to previous figure, contours of $f\psi/u_*^2 L$, $L = (2\kappa k^{-1})^{1/2}$. From Csanady (1980).

which is subject to the equation, after dropping the stars on the nondimensional variables:

$$\frac{\partial \psi}{\partial x} = x \frac{\partial \zeta}{\partial x} - \cos y. \tag{6.77}$$

This is readily integrated to yield

$$\psi = x\zeta(x) - x_0\zeta(x_0) - \int_{x_0}^{x} \zeta(x)\, \mathrm{d}x - (x-x_0)\cos y, \tag{6.78}$$

where $\psi = 0$ is the coast, $x = x_0$.

The surface elevation map, Figure 6.11, shows how the larger hills and valleys of the 'trapped' part of the pressure field at $x - x_0 < L$ merge smoothly into the somewhat smaller undulations far offshore. At the coast there is a slight phase shift between the maximum offshore wind (acting at $ky = 0$) and the minimum elevation, which occurs at $ky = -0.4$. Nevertheless, the effect on coastal elevations is more or less as one intuitively expects, setup where the wind blows onshore, setdown where it is offshore. At larger distances the hills and valleys shift to $\pm \pi/2$, i.e., they occur *between* the cells of onshore and offshore wind.

The pattern of surface elevation contours is understood better when the streamline picture (Figure 6.12) is also considered. Far from shore the Ekman transport to the right of the wind is supplied by an inflow/outflow pattern which is strongest where the cross-shore wind vanishes. The function of the long cross-shore hills and valleys far offshore in the $\zeta(x, y)$ map is to provide geostrophic balance for this inflow-outflow pattern. Close to shore a more complex flow pattern prevails, as the flow has to adjust to the coastal constraint, and this requires the somewhat larger, skewed hills and valleys. Right at the coast the longshore momentum balance is between pressure gradient and bottom stress.

The calculated flow pattern far offshore is characterized by the balance of vorticity tendencies due to wind stress curl and to cross isobath flow, in an essentially frictionless manner. This is the direct analog of the Sverdrup interior of larger scale ocean circulation models (Stommel, 1965), with the planetary vorticity tendency being replaced by vortex stretching through cross-isobath flow. The theory predicts this flow pattern for depths of the order of 50 m and more. A certain degree of skepticism is in order in connection with this prediction: the vortex stretching effect can only operate if the flow actually extends to the bottom, i.e., in a well-mixed case. Much of the year, stratification insulates layers below about 30 m from direct wind stress effects. Frictionally controlled, stratified flow over realistic topography remains a difficult problem, however, the understanding of which will require further theoretical progress.

6.8. CIRCULATION IN A CIRCULAR BASIN WITH A PARABOLIC DEPTH PROFILE

As in earlier chapters, it remains to relate infinite coast results to a finite basin model. A remarkable study of frictionally controlled circulation in a closed basin has been published by Birchfield (1967). A discussion of his results will serve the present purpose very well. The basin is taken to have circular outlines and a parabolic depth distribution:

$$H = H_0 \left(1 - \frac{r^2}{r_0^2}\right). \tag{6.79}$$

Birchfield has succeeded in finding explicit analytical solutions for both the pressure distribution and the transport streamlines.

Polar coordinates (r, ϕ) are used in this section, as in Sections 2.8.1, 3.9, and 4.5.3. To avoid confusion with the bottom friction coefficient, the latter will in this section be labelled r_f. The transport equations have been written down in Section 6.4, Equations (6.29). On introducing the parameterization scheme of Equation (6.28) for the *tangential* component of the bottom stress, neglecting the radial component, and taking curl on the steady-flow transport equations, one finds the circular-basin equivalent of Equation (6.34).

$$\frac{\partial^2 \zeta}{\partial r^2} + \frac{1}{r}\frac{\partial \zeta}{\partial r} + \frac{f}{r_f}\frac{\mathrm{d}H}{\mathrm{d}r}\frac{\partial \zeta}{r\partial \phi} = \frac{f}{r_f g}W. \tag{6.80}$$

The boundary condition at the coasts is $V_r = 0$, which translates to, since $H = 0$:

$$F_\phi = B_\phi = r_f \frac{g}{f}\frac{\partial \zeta}{\partial r}, \quad (r = r_0). \tag{6.81}$$

The case to be considered in detail is uniform wind along the x-azis, hence

$$F_\phi = -u_*^2 \sin \phi. \tag{6.82}$$

The solution can be expressed in terms of Kelvin functions, for details see Birchfield (1967). The resulting surface elevation and stream function maps are shown in Figure 6.13. The surface elevations are proportional to

$$\zeta_a = \frac{u_*^2 r_0}{g}\left(\frac{f}{r_f H_0}\right)^{1/2}. \tag{6.83}$$

A typical value of this is about 2 cm, using $u_*^2 = 1$ cm s^{-1}, $r_0 = 50$ km, $f = 10^{-4}$ s^{-1}, $r_f = 0.05$ cm s^{-1}, and $H_0 = 100$ m. The stream function amplitude is proportional to

$$\psi_a = \frac{u_*^2 r_0}{f} \tag{6.84}$$

a typical value of which is 5×10^{10} cm^3 s^{-1}.

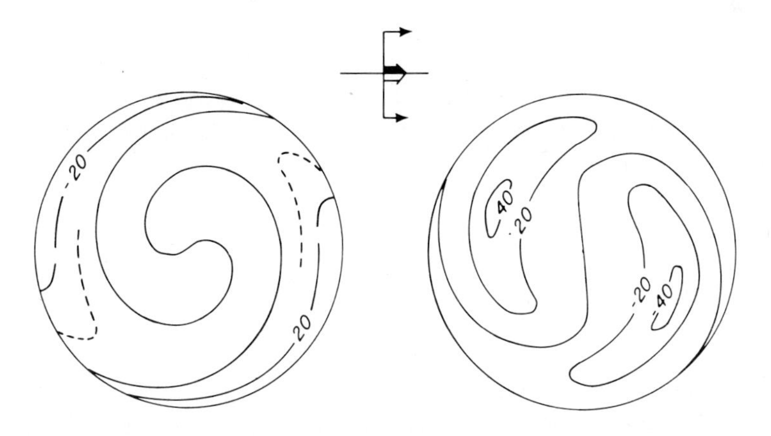

Fig. 6.13. Elevation contours (left) and transport streamlines in circular paraboloid under uniform wind stress. One unit of the elevation is typically 0.15 cm, of the stream function 1.5×10^3 m^3 s^{-1}.. From Birchfield (1967). Copyright American Geophysical Union.

The boundary layer length scale in this problem is $(2r_f \, r_0/fs)$ or typically 22 km, which is not very small compared to the radius of 50 km. Nevertheless, the pattern of elevation contours and streamlines is pretty much as one would infer from wrapping around the solutions due to sinusoidally varying wind stress. At the center of the basin, transport is to the right of the wind and represents Ekman transport. Coastal boundary layers (relatively broad ones, in this case) transport the Ekman transport arriving at the coast around the perimeter, their width-scale increasing in a cyclonic direction around the basin. The surface elevation field is flat at basin center, with levels rising or falling toward the coasts, in such a way as to provide the cross-shore pressure gradients necessary for the geostrophic balance of the coastal boundary layer transports.

The reader will find it rewarding to compare this pattern with the frictionless solution for a flat bottomed basin, Figure 2.3. There is a basic similarity in the sense that both patterns show a boundary layer structure. The details are different, because the boundary layers are due to different physical causes, stratification in one case, variable depth and bottom friction in the other.

CHAPTER 7

Thermohaline Circulation

7.0. INTRODUCTION

In coastal waters several thermodynamic processes combine to generate horizontal gradients of density, due to differences in temperature or salinity, or both. During early spring, solar radiation heats up nearshore shallow waters rapidly and generates a warm wedge of water at the shores separated from the deeper waters by an inclined density interface (Figure 7.1) known as the 'spring thermocline'.* As was discussed in Section 1.3, horizontal density gradients imply the presence of horizontal pressure gradients, which may then drive fluid moths. When the density gradients cause the motion in this manner (and are not themselves a consequence of the internal adjustment of fluid masses to geostrophic equilibrium in e.g., a wind-driven current) the resulting circulation is described as 'thermohaline'.

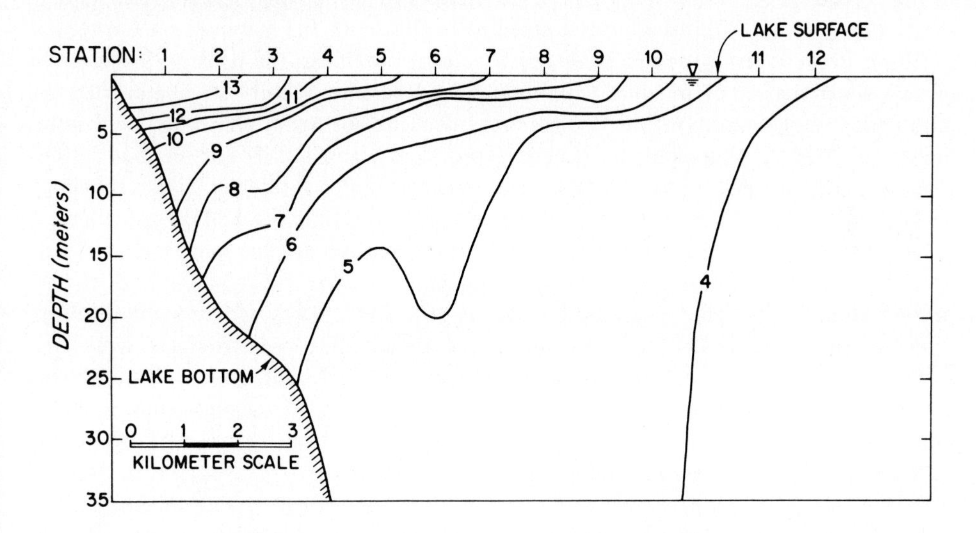

Fig. 7.1. Typical spring temperature distribution nearshore, observed in Lake Ontario. Contours of constant temperature, °C. From Csanady (1974b).

---

* There is an unfortunate tendency to confuse the spring thermocline in the Great Lakes with the 'thermal bar' phenomenon described by Tikhomirov (1964) and Rodgers (1965), see further remarks below on cabbeling.

The effects of freshwater runoff from land are broadly similar to those of nearshore heating, but they often involve larger coastal water masses and are thus more important in generating thermohaline circulation. This is the case over continental shelves with relatively large river water-input, for example over the Alaskan continental shelf or off the North American east coast. On these shelves, in winter, turbulence due to wind-driven and tidal motions effectively mixes the water column from shore to some limiting depth of 80–100 m. At this depth an *S*-shaped density interface often forms, similar to that due to nearshore heating, compare Figures 7.1 and 7.2. The horizontal density gradient between the coast and the density interface near the 100 m isobath does not vanish, the isopycnals merely becoming nearly vertical (Figure 7.3). Horizontal salinity gradients are a necessary consequence of the considerable freshwater flux which must cross the shelf and eventually find its way to a freshwater 'sink' or oceanic desert, where evaporation exceeds precipitation.

Over a coastal-ocean desert, such as the Arabian Sea, the converse occurs and salinity is increased by excess evaporation, again generating horizontal density gradients between such a coastal sea and the deep waters offshore, in the opposite sense to freshwater run-off. The converse of spring heating, autumn cooling, also causes appropriate density difference and may generate a noticeable thermohaline circulation component. Further such possibilities are connected with the freezing and melting of sea ice.

To complete this catalog of thermodynamic processes of importance in shallow seas, the phenomenon of 'cabbeling' should be mentioned (from the German 'Kabbelung', see Horne *et al.* (1978), usually pronouned in English with the emphasis on the second syllable). Because the equation of state of water is nonlinear, the density of a mixture of two water masses of unequal temperature (and possibly salinity) is greater than the arithmetic average density of the two components. Thus surface mixing of adjacent water masses of differing characteristics generally produces water locally denser than its surroundings, which tends to sink. Sinking motion so generated is referred to as 'cabbeling'.

Tikhomirov (1964) has described the cabbeling process in Lake Ladoga and thermed it the 'thermal bar', a rather unfortunate choice apparently implying some sort of separation between the water masses involved in cabbeling. Rodgers (1965, and a series of later papers) studied the same process in detail in the Great Lakes and brought the term 'thermal bar' into wide currency. However, as is already clear from Rodgers' work, and as later work during the International Field Year on Lake Ontario has demonstrated beyond doubt (Csanady, 1974b) the cabbeling circulation associated with the 4 °C isotherm is more of a mixer than a bar to mixing. The barrier between nearshore and offshore waters is the spring thermocline forming after the 4 °C isotherm has moved further offshore. It is undesirable to retain the term thermal bar in place of the standard one now in wide use in oceanography for the same thermodynamic process, which is cabbeling.

Given the variety of thermodynamic processes generating density differences in a fluid, dynamical models of thermohaline circulation, verified by observation, are at present few in number and generally overidealized. Whatever the underlying thermodynamic process, the magnitude of the horizontal density gradients produced depends on the intensity of vertical and horizontal mixing. Thus in the above illustrations regions of small and large horizontal density gradients occur, as already pointed out. Two opposite idealizations

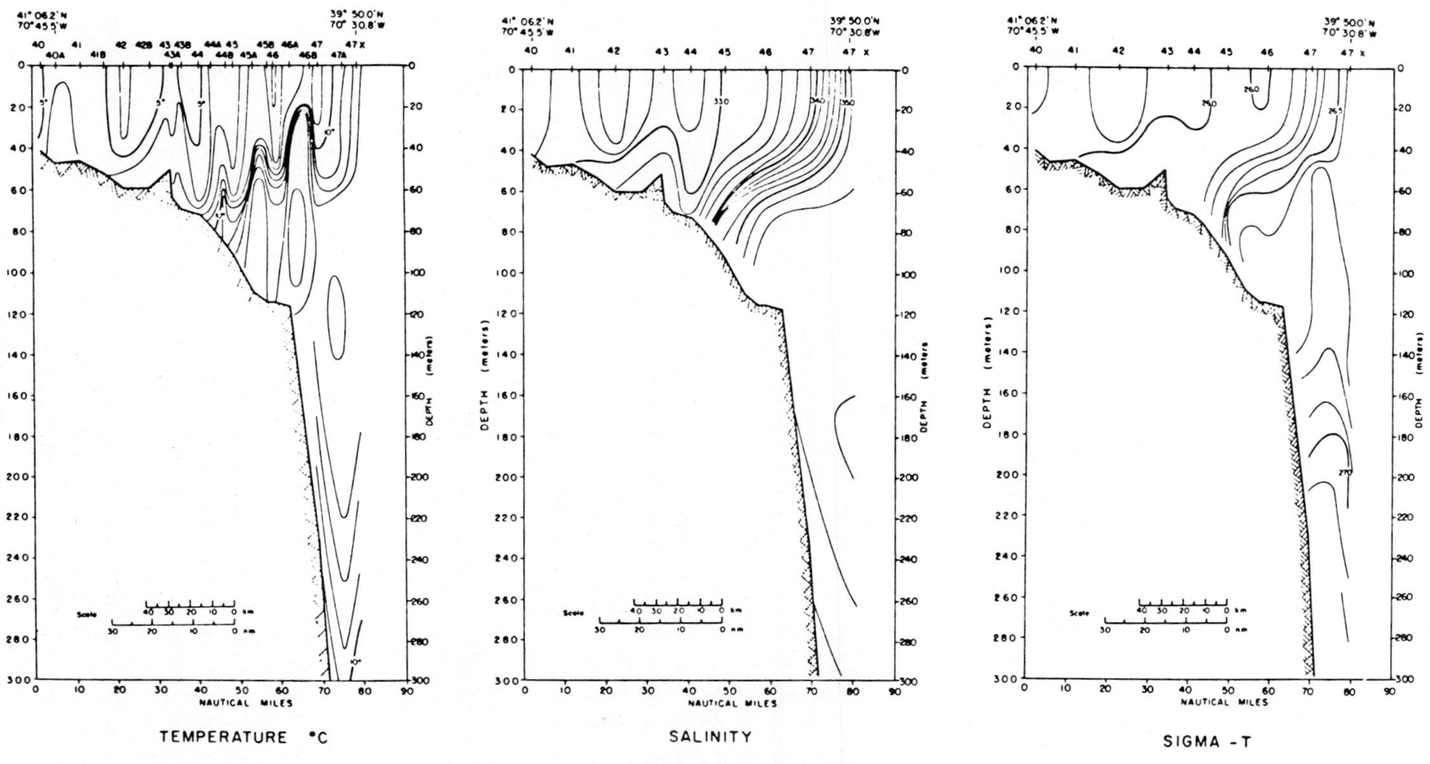

Fig. 7.2. Distribution of temperature (°C), salinity (‰) and density (σ_t ‰) at the edge of the continental shelf in winter. From Beardsley and Flagg (1976).

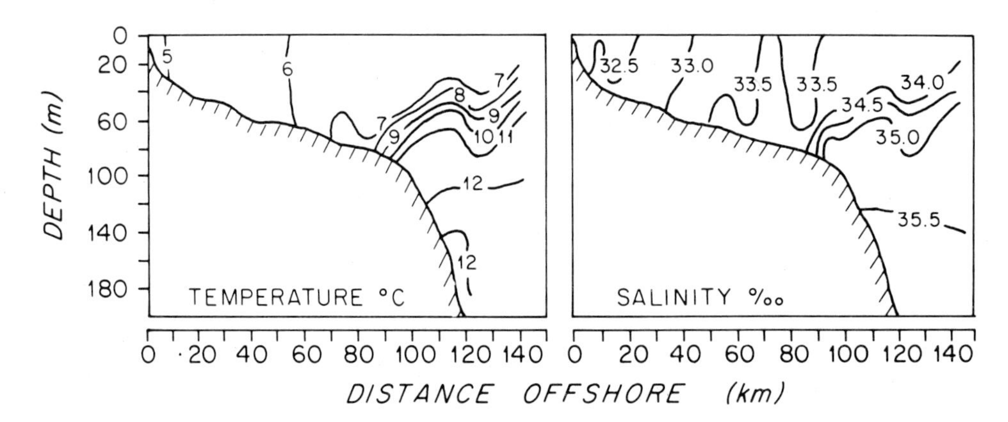

Fig. 7.3. Distribution of temperature (°C) and salinity (‰) in a transect of the continental shelf south of Long Island in early April 1975, showing well mixed condition shoreward of the shelf-edge front. From Walsh *et al.* (1978).

correspondingly suggest themselves: one, of regarding the density interface as a single surface or 'front' similar to an upwelled pycnocline, subject to negligible mixing only, or two, of supposing the horizontal and vertical density gradients weak and to vary suitably slowly in the horizontal direction.

The thermodynamic processes producing horizontal density gradients operate on a long, essentially seasonal time scale. Steady flow models are therefore most appropriate in representing thermohaline circulation. However, a time-dependent, inertial adjustment model provides useful physical insight into the generation of thermohaline circulation, and even models some observed phenomena, which on account of hydrodynamic instability never reach a true steady state. The next section will briefly discuss this model, which is in effect the thermohaline counterpart of inertially controlled, wind induced frontal adjustment discussed earlier in Chapter 3. Later, steady models of thermohaline circulation will be analyzed in greater detail.

7.1. FRONTAL ADJUSTMENT

The preferential heating of nearshore waters is depth dependent and is appropriately modelled as a two-dimensional process, independent of the longshore coordinate y. The freshening of coastal waters is mostly due to rivers, i.e., many point sources, but in a first approximation may also be viewed as if the freshwater input were uniformly distributed along the coastline. In the present section the question will be discussed, what flow pattern such a two-dimensional density distribution, suddenly imposed, will generate from rest.

Consider the spring thermocline shown in Figure 7.1 or the analogous thermohaline front in Figure 7.2. Idealized as an S-shaped density interface, how could such a frontal structure arise and what is the accompanying flow field? A relevant simple thought experiment is to imagine a membrane inserted into the fluid at $x = 0$ (Figure 7.4), and the region on the shore-side, $-l \leqslant x \leqslant 0$, heated or freshened to reduce its density by the small amount $\epsilon\rho$. The depth of the fluid in this region then increases from H to $(1+\epsilon)H$. Although the higher rate of heating of coastal waters is originally due to their shallow depth, depth variations are of secondary importance in the dynamics of frontal adjustment so that H = const. will be supposed here. Also, the process is taken to be two-dimensional, independent of the y-coordinate. The imaginary membrane at $x = 0$ is withdrawn at $t = 0$, and the water is allowed to move freely afterwards, but without mixing or friction between 'heavy' and 'light' columns.

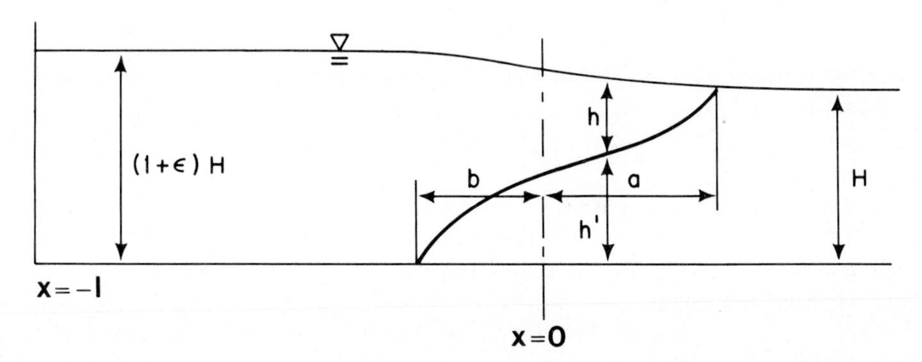

Fig. 7.4. Simple theoretical model of nearshore warming or freshening, with a frontal region near $x = 0$. separating slightly lighter nearshore water (density defect $\epsilon\rho$) from slightly heavier offshore water.

The heated water stands higher on one side to the membrane and has excess potential energy per unit mass, of amount $\frac{1}{2}\epsilon gH$. When the membrane is withdrawn, offshore motion sets in at the surface, while return flow occurs at depth. A characteristic velocity c for these 'intrusive' motions may be calculated by assuming conversion of potential into kinetic energy:

$$\frac{c^2}{2} = \frac{1}{2}\,\epsilon gH, \tag{7.1}$$

i.e., $c = (\epsilon gH)^{1/2}$, called the 'densimetric velocity' by Keulegan (1966) in a discussion of similar stratified flow problems.

The offshore moving surface waters and the shoreward moving deeper waters are deflected in opposite directions by the Coriolis force within a period of order f^{-1}. One may expect the fluid then to adjust to geostrophic equilibrium, in which state the horizontal pressure gradients associated with the density field come to be balanced by the Coriolis force of longshore flow. Any oscillatory motions which accompany the adjustment process are of no interest in the circulation problem and will not be discussed.

In the absence of interface and bottom friction potential vorticity is conserved. After the supposed adjustment the cross-shore (i.e., cross-front) momentum balance is geostrophic. Longshore acceleration is generated by cross-shore adjustment drift. These principles are the same that underlie the finite amplitude pycnocline displacement theory discussed in Section 3.10, except for the absence of an external wind-impulse. The equations of motion and potential vorticity are therefore as written down there, in this case applying to the region where two layers of fluid are present after adjustment:

$$\frac{f}{(1+\epsilon)H} = \frac{f + \mathrm{d}v/\mathrm{d}x}{h}, \qquad \frac{f}{H} = \frac{f + \mathrm{d}v'/\mathrm{d}x}{h'}, \qquad (-b \leqslant x \leqslant a)$$

$$v = \frac{g}{f}\frac{\mathrm{d}}{\mathrm{d}x}(h+h'), \qquad v' = \frac{g}{f}\frac{\mathrm{d}}{\mathrm{d}x}(h+h'-\epsilon h), \tag{7.2}$$

$$\frac{\mathrm{d}v}{\mathrm{d}t} = -fu, \qquad \frac{\mathrm{d}v'}{\mathrm{d}t} = -fu',$$

where the symbols are as defined in Figure 7.4. Integrating the third set of two equations (7.2) one also finds

$$v = -\xi f, \qquad v' = -f\xi',$$

where ξ and ξ' are cross-frontal displacements of light and heavy columns respectively in the course of the adjustment process. The columns initially on either side of the imaginary membrane at $x = 0$ move to $\xi = a$ and $-b$ respectively (see Figure 7.4) and become compressed to negligible height:

$$\left.\begin{aligned} h' &= 0, \\ v' &= fb, \end{aligned}\right\} \quad (x = -b)$$
$$\left.\begin{aligned} h &= 0, \\ v &= -fa, \end{aligned}\right\} \quad (x = a). \tag{7.4}$$

Outside the two-layer region, at $x \geqslant a$ and $x \leqslant -b$ the right and left hand columns of Equations (7.2) apply, with h and h' respectively set equal to zero. Boundary conditions at locations distant from the frontal zone are

$$\begin{aligned} \xi &= 0, \qquad (x = -l) \\ h' &\to H, \qquad (x \to \infty). \end{aligned} \tag{7.5}$$

At $x = -b$ and $x = a$ the single layer solutions have to be matched to the two-layer ones, i.e.,

$$\left.\begin{aligned} h_- &= h_+ \\ \xi_- &= \xi_+ \end{aligned}\right\} \quad (x = -b)$$
$$\left.\begin{aligned} h'_- &= h'_+ \\ \xi'_- &= \xi'_+ \end{aligned}\right\} \quad (x = a). \tag{7.6}$$

The differential equations governing surface and interface elevations can be obtained from the first two pairs of Equation (7.2) as follows:

$$\left[D^2 - (1+\epsilon)^{-1}\right]\frac{h}{H} = -1, \qquad (x \leqslant -b)$$

$$\left.\begin{aligned}&\left[\epsilon D^2 - \frac{2+\epsilon}{1+\epsilon}D^2 + \frac{1}{1+\epsilon}\right]\frac{h}{H} = 1,\\ &\frac{h'}{H} = \left[\frac{1}{1+\epsilon} - \epsilon D^2\right]\frac{h}{H},\end{aligned}\right\} \qquad (-b \leqslant x \leqslant a) \qquad (7.7)$$

$$(D^2 - 1)\frac{h}{H} = -1, \qquad (x \geqslant a)$$

where $D = (gH)^{1/2} f^{-1}$ (d/dx) is a non-dimensional differential operator. The solutions of these equations contain a total of eight integration constants. Two further unknown constants are the positions a and b where the front comes to intersect the free surface and the bottom respectively. All of the constants may be determined from the ten boundary conditions written down in Equations (7.4) to (7.6).

Particular solutions of Equations (7.7) are simply the initial depths, $h = (1+\epsilon)H$ and $h' = H$, while the homogeneous equations have solutions of the form $\exp(mx/R)$, with $R = (gH)^{1/2} f^{-1}$, the familiar radius of deformation in the barotropic mode. On substituting the exponential solution into the homogeneous equations one finds the following algebraic equations for m:

$$m^2 - (1+\epsilon)^{-1} = 0, \qquad (x \leqslant -b)$$

$$\epsilon m^4 - \frac{2+\epsilon}{1+\epsilon}m^2 + (1+\epsilon)^{-1} = 0, \qquad (-b \leqslant x \leqslant a) \qquad (7.8)$$

$$m^2 - 1 = 0, \qquad (a \leqslant x).$$

Let the roots of the first of these equations be designated m_1 and m_2, of the second m_3 to m_6, of the third, m_7 and m_8. To the lowest order in the small quantity ϵ these roots are:

$$\begin{aligned}m_{1,2} &= 1,\\ m_{3,4} &= (2)^{-1/2},\\ m_{5,6} &= (2/\epsilon)^{1/2},\\ m_{7,8} &= 1.\end{aligned} \qquad (7.9)$$

Most of the roots are of order one, except m_5 and m_6 which are much larger. The corresponding exponential solutions are scaled by $m^{-1}R$, so that those solutions containing m_5 and m_6 have a short spatial scale and represent a response akin to the internal mode response in linear theory. The other solutions are scaled by the barotropic radius of deformation.

The determination of the ten constants from the ten boundary conditions involves some tedious algebra which tends to obscure the important physical features of the solution. It proves more practical to focus first on those aspects of the solution which vary rapidly across the frontal zone, i.e., the interface shape and the velocity differences between light and heavy layers.

Surface elevation changes remain of order ϵH after the withdrawal of the membrane. Hence the matching conditions (7.6) may be replaced approximately by:

$$\begin{aligned} h' &= H + 0(\epsilon H), \qquad (x = a) \\ h &= H + 0(\epsilon H), \qquad (x = -b). \end{aligned} \tag{7.10}$$

These conditions, together with (7.4), supply six equations involving only the solution within the band $-b \leqslant x \leqslant a$. Consequently, the four integration constants appearing in the solution of the second of (7.7) *and* the location constants a, b may be determined. The results are

$$\begin{aligned} \frac{h}{H} &= \frac{1}{2}\left[1 - \frac{\sinh x/R_i}{\sinh a/R_i}\right] + 0(\epsilon), \\ \frac{h'}{H} &= \frac{1}{2}\left[1 + \frac{\sinh x/R_i}{\sinh a/R_i}\right] + 0(\epsilon), \end{aligned} \tag{7.11}$$

where R_i is the internal mode radius of deformation,

$$R_i = f^{-1}(\epsilon g H/2)^{1/2}. \tag{7.12}$$

The equations imply that the interface is symmetrical about the origin, $a = b$. The velocity distributions to the lowest order in ϵ are

$$\begin{aligned} \frac{v}{fR_i} &= -\frac{1}{2}\left[\frac{x}{R_i} + \frac{\cosh (x/R_i)}{\sinh (a/R_i)}\right] + 0(\epsilon^{1/2}), \\ \frac{v'}{fR_i} &= -\frac{1}{2}\left[\frac{x}{R_i} - \frac{\cosh (x/R_i)}{\sinh (a/R_i)}\right] + 0(\epsilon^{1/2}), \end{aligned} \tag{7.13}$$

as is easily verified from (7.11) and the *first* of (7.2) by integration.

The frontal zone width may now be obtained by substituting (7.13) into (7.4):

$$a/R_i = \coth(a/R_i) \tag{7.14}$$

the root of which is $a/R_i = 1.20$, so that $a = b = 1.2R_i$. Figure 7.5 illustrates the frontal shape and the velocity distribution, neglecting $0(\epsilon^{1/2})$ terms and smaller.

As in similar adjustment problems, the width of the frontal zone is scaled by R_i, the along-front velocities by fR_i, which is $(2)^{-1/2}$ times the densimetric velocity. The simple physical principles which underlie this result may be summarized as follows: if the cross-frontal displacement of the fluid column furthest from the initial position is a, its along-front velocity is fa, having been generated in the course of geostrophic adjustment by Coriolis force. Geostrophic balance also gives $(\epsilon g/f)\partial h/\partial x$ for this velocity, the bottom layer being here at rest. The geometry of the front after adjustment is such that the slope is of order $H/2a$, i.e.,

$$fa = \text{order}\ \frac{\epsilon g}{f}\,\frac{H}{2a}$$

or a = order R_i, as is of course also obtained by more accurate calculations.

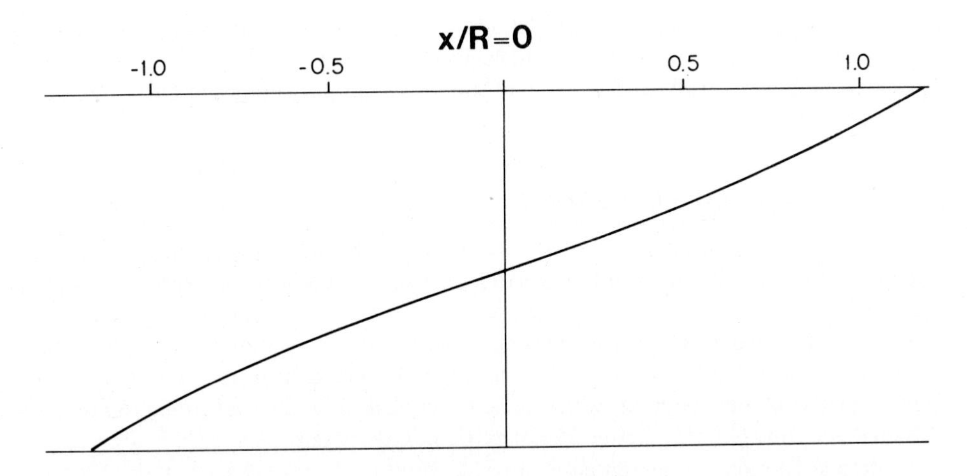

Fig. 7.5a. Calculated shape of density interface after adjustment to geostrophic equilibrium. From Csanady (1978b).

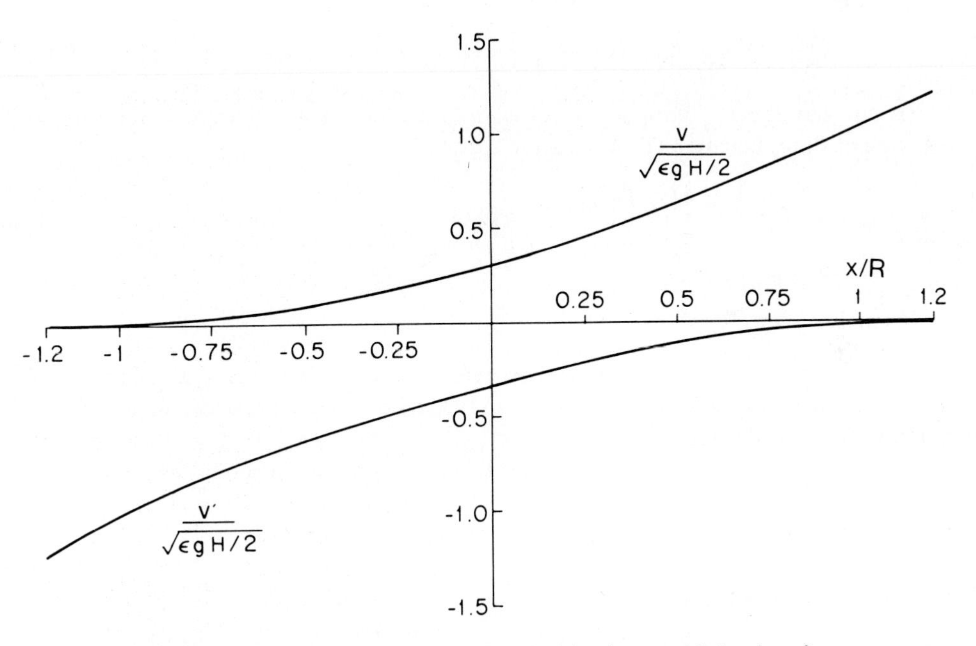

Fig. 7.5b. Calculated after adjustment velocities above and below interface.

Actually, the bottom layer is not quite at rest at $x = a$, nor is the top layer at $x = -b$. Some of the fluid originally between $x = -l$ and $x = -b$ escapes in the adjustment process, so that ξ at $x = -b$ is positive, and v by Equation (7.3) negative. The magnitude of this 'barotropic' velocity may be estimated from a mass balance argument. If $l >> b$, the maximum amount of fluid that can escape is of order ϵHl, giving rise to ξ of order ϵl, and by Equation (7.3), to v of order $f\epsilon l$. Over a typical continental shelf, for example, $f = 10^{-4}\ \text{s}^{-1}$, $\epsilon = 10^{-3}$, $l = 100$ km, which gives a barotropic velocity of order 1 cm s^{-1}. By contrast fR_i is typically 0.5 m s^{-1} and more, so that neglecting the barotropic velocity in comparison with fR_i is justified.

7.2. GRAVITATIONAL CONVECTION

While the inertial adjustment problem treated in the previous section yields a certain amount of physical insight into the dynamical effects of density variations, it is clearly over-idealized. Density changes observed in shallow seas are continuous not stepwise, and internal friction and mixing have important effects on the long time scales over which any thermohaline circulation component develops. A more general treatment of stratified fluid dynamics is thus required, which is best approached by first returning to Equation (1.6) and adopting (for this section alone) suffix notation again.

If internal friction is represented by an eddy viscosity K, and by a generalized form of the gradient-transport relationship written down in Equation (1.27), on substituting into Equation (1.6) one obtains

$$\frac{\partial(\rho u_i)}{\partial t} + \frac{\partial}{\partial x_j}(\rho u_i u_j) = -\frac{\partial p}{\partial x_i} - g\rho\delta_{ij} + 2\rho\epsilon_{ijk}\, u_j\, \Omega_k + \rho K \nabla^2 u_i. \tag{7.15}$$

Because the density differences involved are generally small, a linearized form of the equation of state, Equation (1.14), is often adequate:

$$\rho = \rho_0(1 - \alpha\theta + \beta\sigma), \tag{7.16}$$

where

$$\theta = T - T_0,$$

$$\sigma = S - S_0,$$

and (ρ_0, T_0, S_0) is some suitably chosen reference state. The excess temperature and salinity, θ and σ, give rise to a deficiency and excess of density respectively, or using the notation of Equation (1.11):

$$\epsilon = \beta\sigma - \alpha\theta,$$

where the thermodynamic definitions of the coefficients are:

$$\alpha = -\frac{1}{\rho}\frac{\mathrm{d}\rho}{\mathrm{d}T}\bigg|_{T_0,\, S_0}, \qquad \beta = \frac{1}{\rho}\frac{\mathrm{d}\rho}{\mathrm{d}S}\bigg|_{T_0,\, S_0}. \tag{7.17}$$

Both of these coefficients have been defined so as to be positive under most circumstances (except that α is negative for freshwater between 0 °C and 4 °C).

When one wishes to model cabbeling, it is also necessary to retain at least the quadratic terms in the equation of state, (7.16). These are of consequence especially for freshwater near the temperature of maximum density, 4 °C. Models of cabbeling in freshwater lakes have been discussed by Elliott (1971), Bennett (1971, and Huang (1971), for example.

The distribution of temperature and salinity is subject to the conservation laws written down in Equations (1.15) and (1.17). If eddy diffusion of heat and salt is also expressible in terms of eddy coefficients K_T and K_S, substitution into the conservation laws yields:

$$\frac{\partial \theta}{\partial t} + \frac{\partial}{\partial x_i}(u_i\theta) = K_T \nabla^2 \theta,$$
$$\frac{\partial \sigma}{\partial t} + \frac{\partial}{\partial x_i}(u_i\sigma) = K_S \nabla^2 \sigma. \tag{7.18}$$

Fluxes are usually imposed through the boundary conditions: heat flux at the surface (Equation (1.15a)), freshwater flux at the coast, see the discussion below, Section 7.3. In using eddy coefficients the same caveats should be observed that were mentioned in connection with eddy viscosity in Chapter 1.

In substituting a linearized equation of state into the equation of motion (7.15) it is sufficiently accurate to take $\rho \cong \rho_0$ in all terms, except where the excess density is multiplied by the gravitational acceleration (Boussinesq approximation). The resulting equation is

$$\frac{\partial u_i}{\partial t} + \frac{\partial}{\partial x_j}(u_i u_j) = -\frac{\partial}{\partial x_i}\left(\frac{p}{\rho_0} + g x_3\right) + (\alpha\theta - \beta\sigma) g \delta_{i3} + $$
$$+ 2\epsilon_{ijk}\, u_j\, \Omega_k + K \nabla^2 u_i. \tag{7.19}$$

The term containing $\epsilon = (\alpha\theta - \beta\sigma)$ (ϵ being the proportionate density defect) is the acceleration due to the buoyancy of a fluid element warmer and/or fresher than its environment. Motions forced by this term are described as gravitational convection. Such motions are often due to heating of a fluid from below or cooling from above. Buoyant plumes and thermals are particularly well known manifestations of gravitational convection.

The derivation of Equation (7.19) in this section differs from the approach of Chapter 1 in that the approximation of quasi-horizontal motion has not been made here. In buoyant or heavy plumes, for example, vertical and horizontal motions are of about the same intensity. Density driven motions of this kind have been extensively treated in the literature, see e.g., Chandrasekhar (1961), Scorer (1978), or with an orientation to large-scale meteorology, the anthology edited by Saltzman (1962).

Density driven motions involving relatively large vertical velocities are also important in oceanography, and specifically in the circulation of the coastal ocean. This is true of convective overturn caused by surface cooling, or of the cabbeling process already referred to. Another example is the field of motion within a frontal zone intersecting the free surface, such as shown in Figures 7.1 and 7.2 above. The idealization of the previous sections of regarding a pycnocline as a single inclined density interface is not useful where the isopycnals actually intersect the free surface. Given negligible surface heat and salt flux, the isopycnals must be vertical here, instead of being inclined at a small angle against the horizontal. The horizontal spacing of the isopycnals correspondingly becomes relatively wide. In such a region, wind action may force denser fluid over lighter fluid and initiate gravitational convection. Even in the absence of wind, surface frontal zones are known to be regions of intense convergence, where vertical velocities are considerable. Cabbeling may be of importance in such a region. The precise dynamical and mixing processes in frontal zones are not well understood at present. Some illuminating numerical calculations of thermohaline motions in surface fronts have, however, recently been published by Kao *et al.* (1978).

An important general result concerning such motions may be stated in terms of vorticity, the three components of which are

$$\omega_i = \epsilon_{ijk} \frac{\partial u_k}{\partial x_j} \,. \tag{7.20}$$

The vorticity tendency equation is obtained from Equation (7.19) on taking curl. For the present purposes it is sufficient to write down the linearized result, ignoring vortex tilting terms arising from the curl of the momentum advection term in (7.19):

$$\frac{\partial \omega_i}{\partial t} = \epsilon_{ijk}\, g\delta_{k3} \left(\alpha \frac{\partial \theta}{\partial x_j} - \beta \frac{\partial \sigma}{\partial x_j} \right) + 2\Omega_j \frac{\partial u_i}{\partial x_j} + K\nabla^2 \omega_i. \tag{7.21}$$

The bracketed term is simply $\partial\epsilon/\partial x_j$ (with ϵ the proportionate density defect), and it is seen to be part of a vorticity source term associated with horizontal variations of density (note that $\epsilon_{ijk}\delta_{k3}$ is nonzero only for $k = 3, j = 1$ or 2). That horizontal density variations generate vorticity is an important general result in the theory of nonhomogeneous fluids (Yih, 1965). Without rotation, $\underset{\sim}{\Omega} = 0$, if horizontal density variations are present, either $\partial\omega_i/\partial t$ is non-zero and motion is being generated, or $K\nabla^2\omega_i$ is non-zero, in which case a circulation is already in existence and vorticity diffusion balances vorticity generation. Rotation introduces the additional possibility that the second term on the right of (7.21) balances the vorticity tendency associated with nonuniform density. This second term represents the interaction of planetary vorticity with the velocity field. Part of a fluid element's (absolute) vorticity is in this case $2\underset{\sim}{\Omega}$. As this vector is tilted by a sheared velocity field, the vorticity components change at a rate given by the second term in (7.21). From a postulated balance between the first and second terms on the right of (7.21) the shear $\partial u_i/\partial x_j$ may be calculated, yielding what is known in meteorology as the 'thermal wind equation'.

This general result may be applied to a surface frontal zone such as illustrated in Girues 7.1 and 7.2 as follows. Let the front coincide with the x_2 axis, so that v_1 and v_3 are horizontal and vertical velocities in the plane of the illustrated cross-frontal sections. The along-front component of the vorticity is

$$\omega_2 = \frac{\partial u_1}{\partial x_3} - \frac{\partial u_3}{\partial x_1}$$

and Equation (7.21) reduces to, with horizontal density variations being present only along the x_1 (not the x_2) axis:

$$\frac{\partial \omega_2}{\partial t} = g \frac{\partial \epsilon}{\partial x_1} + f \frac{\partial u_2}{\partial x_3} + K\nabla^2 \omega_2. \tag{7.21a}$$

Given the density distributions shown in Figures 7.1 and 7.2 the tendency in the first instance is to generate cross-frontal circulation, with the light fluid intruding over the heavier layers, much as the inertial adjustment model has already demonstrated. When the density term and the Coriolis force term on the right balance, geostrophic equilibrium prevails. Given the presence of some friction, this is never quite the case, a slow cross-frontal circulation remaining to reestablish continually near-geostrophic balance. Kao *et al.* (1978) discuss this in greater detail in connection with their numerical model of a surface front, illustrated here in Figure 7.6.

The main source of complexity in density-driven motions is the coupling of the velocity and density fields. The nonuniform density distribution, which induces the circulation, itself depends on the velocity field, being shaped by advection as well as by diffusion, according to Equations (7.18). In the example illustrated in Figure 7.6, the sharpness of the front is largely a consequence of density advection by the cross-frontal circulation. Indeed advection generally dwarfs diffusion: the relative importance of these two phenomena is measured by the eddy Péclet and Lewis numbers

$$\frac{uL}{K_T} \quad \text{and} \quad \frac{uL}{K_S},$$

where u and L are velocity and length scales of the flow. In turbulent layers these are generally of the same order of magnitude as eddy Reynolds numbers, i.e., are large compared to unity rather than small. In the interior of a stratified fluid eddy coefficients are small and eddy Péclet and Lewis numbers may be very large.

Theoretical models fully coming to grips with the 'self-advection' of density-driven flow have yet to be developed. In some instances it is possible to skirt this problem: this is what the models to be discussed in the remainder of this chapter effectively do. One approach is to represent the aggregate effect of density advection as horizontal diffusion, a realistic step under some special circumstances. Another point of view is to ignore the advection-diffusion equations (7.18), and solve the equations of motion and continuity with a density field supposed known from observation. This is known as a 'diagnostic' calculation, in which the fluid is supposed to have solved its own equations, but the velocity field is inferred from the density field instead of being directly observed, observations of velocity being much more difficult. While one gains some physical insight in this manner, the theory of thermohaline circulation is clearly in need of further development.

7.3. THE 'NEARLY HOMOGENEOUS' FLUID IDEALIZATION

Over mid-latitude continental shelves in winter turbulent mixing is particularly vigorous on account not only of tidal friction, but also of gravitational convection due to surface cooling, and of frequent, intense storms. As a result, the water column becomes nearly homogeneous to depths of the order of 100 m, except for weak horizontal density

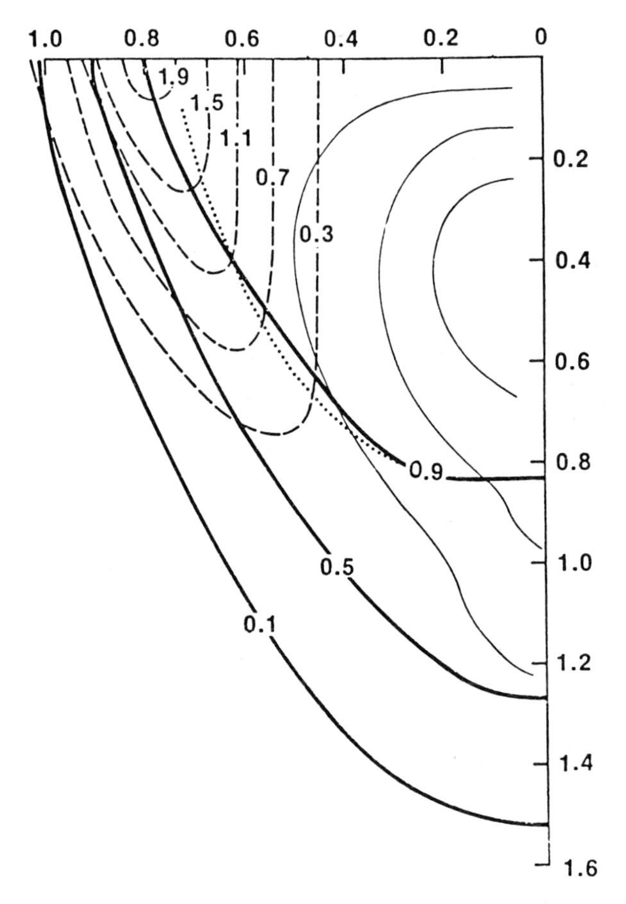

Fig. 7.6. Structure of flow in frontal zone, according to numerical model of Kao *et al.* (1978). Heavy full lines are contours of constant density, given in fraction of the total density defect, the contours 0.1 and 0.9 effectively enclosing the frontal zone. Broken lines are constant alongfront velocity contours, expressed as multiples of the densimetric velocity. The alongfront velocity is seen to peak at the surface, on the inner edge of the front. Light full lines are streamlines of acrossfront circulation, with flow directed toward the front at the surface. This circulation serves to maintain the front against friction and mixing. Vertical mixing is unrealistically intense in this model, as shown by the wide spacing of the isopycnals where these turn horizontal.

gradients caused by freshwater inflow at the coast. Differential advection of top versus bottom layers then also causes weak density gradients in the vertical. Figure 7.3 above has illustrated an observed instance of such a winter density field. The complex interplay between the thermohaline circulation and the density field simplifies somewhat in the 'nearly homogeneous' case, The idealization of weak density gradients and consequent slow thermohaline circulation is almost the logical opposite to the frontal adjustment model discussed earlier in this chapter.

It is now convenient to return to xyz notation, with the x-axis laid cross-shore, the

z-axis vertical, as before. For the present, the winter shelf density distribution will be supposed two-dimensional, i.e., the alongshore (y) variations to be negligible. Such a distribution is taken to be produced by uniform inflow of freshwater at the coast, at the constant rate of Q m^3 s^{-1} per m of coastline.

Typical winter shelf density gradients are: one part in one thousand in a horizontal distance of 100 km, or $(\partial/\partial x)(\beta\sigma)$ = order 10^{-8} m^{-1}, and one part in ten thousand in a vertical water column of 100 m depth, or $(\partial/\partial z)(\beta\sigma)$ = order 10^{-6} m^{-1}. If the horizontal density gradient term in the vorticity tendency balance is matched by the thermal wind, the vertical gradient of longshore velocity is of order 10^{-3} s^{-1}, corresponding to a surface-bottom longshore velocity difference of 0.1 m s^{-1} in water 100 m deep. Based on a horizontal length scale of L = 100 km, the Rossby number v/fL is then of order 10^{-2}, so that one may expect linearized equations of motion to be sufficiently accurate.

In shallow seas large-scale thermohaline motions are quasi-horizontal, no less than wind-driven ones, and the idealizations discussed in Chapter 1 apply. The shallow water equations for steady flow, with dynamic height terms added to allow for the effects of variable density, thus govern the dynamics of these motions. These equations are, from (1.12) and (1.32):

$$\begin{aligned} -fv &= -g\frac{\partial\zeta}{\partial x} + g\frac{\partial\zeta_d}{\partial x} + K\frac{\partial^2 u}{\partial z^2}, \\ fu &= -g\frac{\partial\zeta}{\partial y} + g\frac{\partial\zeta_d}{\partial y} + K\frac{\partial^2 v}{\partial z^2}, \end{aligned} \tag{7.23}$$

where

$$\zeta_d(z) = -\int_z^0 \epsilon\,dz = -\beta\int_z^0 \sigma\,dz \tag{7.24}$$

is the dynamic height, in the present problem due to freshening of nearshore waters. Given the supposed absence of longshore salinity variations, the gradient $\partial\zeta_d/\partial y$ vanishes. The distribution of excess salinity is governed by the salt diffusion equation, which for the two-dimensional field to be considered is:

$$\frac{\partial\sigma}{\partial t} + \frac{\partial}{\partial x}(u\sigma) = K_S\left(\frac{\partial^2\sigma}{\partial x^2} + \frac{\partial^2\sigma}{\partial z^2}\right). \tag{7.25}$$

The ratio of the horizontal advection and diffusion terms in this equation is the Lewis number vL/K_S, which for typical winter shelf conditions (K_S = 0.01 m^2 s^{-1}) is 10^6, with the above used values of v and L, so that the diffusion term $K_S\,\partial^2\sigma/\partial x^2$ is safely neglected. The vertical diffusion term is, on the other hand, proportional to L^2/H^2 times the horizontal diffusion term (H = depth), so that its ratio to the advection term is of order one, with H of the order of 100 m. The importance of this is that the effects of density advection come to be limited by vertical mixing, a phenomenon that may be examined independently of the precise velocity distribution, $u(x, z)$.

7.3.1. Advection and Diffusion of Freshwater

Consider therefore Equation (7.25) in steady state, with the terms $\partial\sigma/\partial t$ and $\partial^2\sigma/\partial x^2$ dropped, and subject to the condition of a uniform inflow rate Q of freshwater per unit length of coastline. Such a model is somewhat artificial: in the strict two-dimensional case the salinity σ would keep decreasing at any given x, and the fresh water would ultimately displace saline water altogether over the shelf, or in any finite nearshore band. Physically, the freshwater must eventually be transported to a sink, or oceanic 'desert', where evaporation exceeds precipitation, and allowed to re-enter the atmospheric leg of the hydrologic cycle.

Although a steady, two-dimensional solution of (7.25) is thus not strictly possible, the necessary calculation of salinity gradients can be carried out with adequate accuracy in a coordinate system moving slowly offshore with a parcel of freshwater, at a cross-shore velocity Q/H. An alternative physical interpretation of the same calculation is to consider the cross-shore transport U strictly zero (instead of $U = Q$) and to suppose a certain amount of salt to be removed at the coast, equal to the total salt deficiency of the entering freshwater. Defining therefore:

$$u' = u - \frac{Q}{H} \tag{7.26}$$

one has

$$\int_{-H}^{0} u' \, dz = 0 \tag{7.27}$$

and the problem is to solve

$$u' \frac{\partial\sigma}{\partial x} = K_S \frac{\partial^2\sigma}{\partial z^2} \tag{7.28}$$

subject to the boundary conditions

$$\frac{\partial\sigma}{\partial z} = 0, \qquad (z = 0, -H)$$
$$\int_{-H}^{0} u'\sigma \, dz = QS_e, \tag{7.29}$$

where S_e is the (total) salinity at the outer edge of the shelf, and QS_e is the salt transport needed to make up the salt deficiency of the freshwater influx Q.

It is convenient to define the depth averaged salinity anomaly σ_a:

$$\sigma_a = \frac{1}{H} \int_{-H}^{0} \sigma \, dz \tag{7.30}$$

and write $\sigma' = \sigma - \sigma_a$. A solution for (7.28) may now be found of the following form:

$$\sigma = \sigma_a(x) + \sigma'(z) \tag{7.31}$$

the second contribution to which is

$$\sigma'(z) = \frac{1}{K_S} \frac{\mathrm{d}\sigma_a}{\mathrm{d}x} \int_{-H}^{z} U(z)\,\mathrm{d}z, \tag{7.32}$$

where

$$U(z) = \int_{-H}^{z} u\,\mathrm{d}z$$

is cross-shore transport over a *variable* portion of the water column. It is now also easy to show that the total horizontal salt transport is

$$QS_e = -\frac{1}{K_S} \frac{\mathrm{d}\sigma_a}{\mathrm{d}x} \int_{-H}^{0} U^2(z)\,\mathrm{d}z. \tag{7.33}$$

This transport is proportional to the horizontal salinity gradient and is directed down that gradient. An effective horizontal diffusivity K_H may consequently be defined by

$$K_H = \frac{1}{K_S H} \int_{-H}^{0} U^2(z)\,\mathrm{d}z \tag{7.34}$$

in terms of which the horizontal salinity gradient may be expressed as

$$\frac{\mathrm{d}\sigma_a}{\mathrm{d}x} = -H \frac{QS_e}{K_H}. \tag{7.35}$$

This result shows that under certain conditions advection by cross-shore flow results in a salinity distribution very similar to what vigorous horizontal mixing would produce. This kind of advection-diffusion process ('shear diffusion') has first been identified by G. I. Taylor (1954) in a pioneering study. The essence of the process is a form of equilibrium between the tendency of differential advection to distort the salinity distribution in the vertical, and the equalizing effect of vigorous vertical mixing. Formally the key criterion for the validity of the above development is that $\partial\sigma'/\partial x$ should be negligible compared to $\mathrm{d}\sigma_a/\mathrm{d}x$. This is a reasonable hypothesis if K_S is suitably large, and the shelf depth and the cross-shore velocities do not vary too rapidly with offshore distance, see Equation (7.32). Taylor's (1954) method was first applied to shelf circulation by Stommel and Leetmaa (1972) and was further discussed recently by Fischer (1980).

7.3.2. Parallel Flow Model of Thermohaline Circulation

The net result of the preceding analysis is that, in the nearly homogeneous fluid idealization, a horizontal density gradient appears, independent of the vertical coordinate z. For a given value of this gradient, and supposing that surface elevation gradients are known, it is then possible to solve the local problem (Equations (7.23)), determine the velocity distribution, and in the end return to Equation (7.35) to close the problem, by calculating K_H and the actual value of the horizontal density gradient.

Instead of specifying both components of the horizontal surface level gradient as externally imposed parameters, the problem may be made determinate also by imposing the coastal constraint and an arbitrary longshore surface gradient. This leads to what was described in Chapter 6 as the parallel transport model. The extension of the same model to the nearly homogeneous fluid case is subject to the conditions:

$$\begin{aligned}
&\frac{\partial \zeta}{\partial y} = \gamma = \text{constant}, \qquad U = 0 \\
&\beta \frac{\partial \sigma}{\partial x} = \beta \frac{\sigma_0}{L} = \text{constant}, \\
&K \frac{\partial u}{\partial z} = F_x, \qquad K \frac{\partial v}{\partial z} = F_y, \qquad (z = 0) \\
&K \frac{\partial u}{\partial z} = ru, \qquad K \frac{\partial v}{\partial z} = rv, \qquad (z = -H).
\end{aligned} \tag{7.36}$$

The constants are understood to be constant with respect to the vertical coordinate, and σ_0 is the salinity excess over a shelf of width L. An arbitrary wind stress is applied at the surface. The dynamic height (Equation (7.4)) is linear with depth in a constant density gradient, so that the equations of motion reduce to:

$$\begin{aligned}
-fv &= -g \frac{\partial \zeta}{\partial x} + g\beta\sigma_0 \frac{z}{L} + K \frac{\partial^2 u}{\partial z^2}, \\
fu &= -g\gamma + K \frac{\partial^2 v}{\partial z^2}.
\end{aligned} \tag{7.37}$$

Because Equations (7.37), as well as the various constraints and conditions imposed, (7.36), are all linear in the velocities u and v, the resulting solution may be regarded as a linear combination of components due separately to F_x, F_y, γ, and $\beta\sigma_0/L$. Each component may be calculated by setting the other three forcing factors zero, and satisfying separately the constraints and conditions imposed. Specifically, the circulation component forced by the cross-shore density ($\beta\sigma_0/L$ nonzero) is calculated with zero stress boundary conditions, zero longshore gradient and zero cross-shore transport, bottom stress as per (7.36), where bottom velocities are taken to be those due to the cross-shore density gradient alone. Setting $\beta\sigma_0/L$ equal to zero in calculating the other circulation components one encounters the same problem that was discussed in detail in Section 6.4.1.

The present model differs only in that another component of shelf circulation has been added, generated by the nonuniform density field. Here it is only necessary to calculate this additional component.

The velocities may be split into a geostrophic component and frictional components u_f, v_f, where

$$v_g = \frac{g}{f}\left(\frac{\partial \zeta}{\partial x} - \beta\sigma_0 \frac{z}{L}\right), \qquad u_f = u, \qquad v_f = v - v_g. \tag{7.38}$$

The frictional velocities are subject to the Ekman layer equations, Equations (1.48), and in deep water may be expected to be significant only within surface and bottom Ekman layers of scale depths $D = (2K/f)^{1/2}$. The zero-stress surface boundary condition yields

$$\frac{\partial u_f}{\partial z} = 0, \qquad \frac{\partial v_f}{\partial z} = -\frac{\partial v_g}{\partial z} = \frac{g\beta\sigma_0}{fL}, \qquad (z = 0) \tag{7.39}$$

so that the surface Ekman layer is the same as that due to a wind stress of magnitude $Kg\beta\sigma_0/fL$. Below the surface Ekman layer the gradient of the geostrophic velocity is $(-g\beta\sigma_0/fL)$, with which there is associated an interior shear stress of magnitude $(-Kg\beta\sigma_0/fL)$. This is brought to zero within a surface Ekman layer, the structure of which is the same as of a wind-driven Ekman layer, across which the stress changes by the same amount.

A depth integration of the second of Equations (7.37) results in, given the coastal constraint (7.36):

$$\frac{\partial v}{\partial z} = 0, \qquad (z = -H) \tag{7.40}$$

which also implies $v = 0$ at the bottom, given (7.36). Frictional velocities within the bottom Ekman layer may be written as

$$v_f = \exp\left(-\frac{z+H}{D}\right)\left[A \cos\left(\frac{z+H}{D}\right) + B \sin\left(\frac{z+H}{D}\right)\right],$$
$$u_f = \exp\left(-\frac{z+H}{D}\right)\left[A \sin\left(\frac{z+H}{D}\right) - B \cos\left(\frac{z+H}{D}\right)\right], \tag{7.41}$$

where A, B are integration constants. The conditions (7.36) allow the determination of the three constants, A, B and the cross-shore elevation gradient $\partial\zeta/\partial x$. The calculations are straightforward and yield

$$A = -\frac{g\beta\sigma_0 D}{fL}\,\frac{1+rD/K}{2+rD/K},$$
$$B = -\frac{g\beta\sigma_0 D}{fL}\,\frac{1}{2+rD/K}, \tag{7.42}$$
$$\frac{\partial \zeta}{\partial x} = -\frac{\beta\sigma_0 H}{L}\left[1 - \frac{D(1+rD/K)}{H(1+rD/K)}\right].$$

Thus the geostrophic velocity distribution is

$$v_g = -\frac{g\beta\sigma_0(z+H)}{fL} + \frac{g\beta\sigma_0 D}{fL}\frac{1+rD/K}{2+rD/K}. \tag{7.43}$$

With $D/H \ll 1$, as supposed in these calculations, the second term is a small correction. Over most of the water column the geostrophic velocity is as calculated from the dynamic height relative to the bottom.

Fig. 7.7 illustrates the distribution of the thermohaline circulation velocity for $D/H = 0.2$. The cross-shore velocity is to positive x or offshore within the surface Ekman layer, shoreward near the bottom. The transport in each is $Kg\beta\sigma_0/f^2L$, or the same as produced by wind stress of a magnitude equal to the interior stress associated with the shear or the geostrophic velocity.

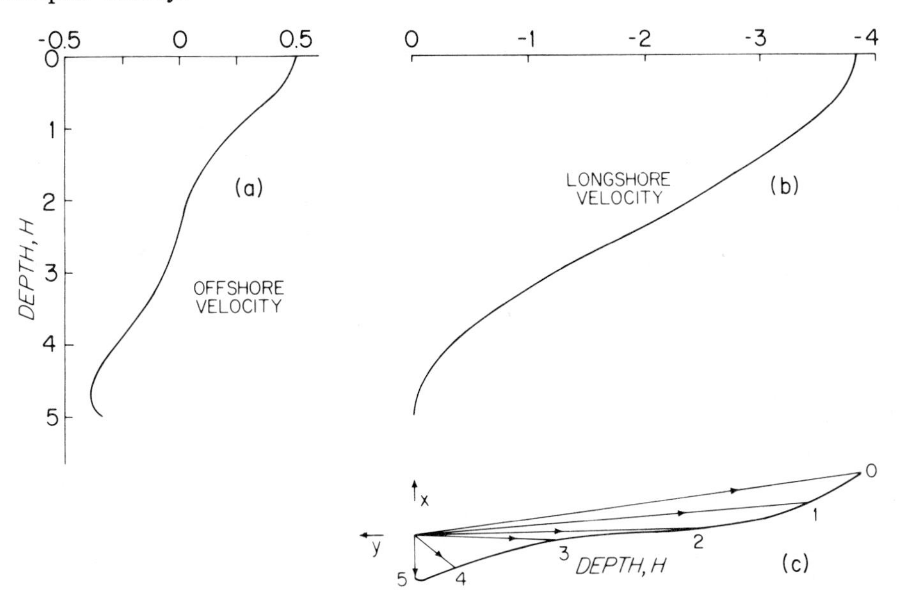

Fig. 7.7. Thermohaline circulation component velocities in moderately deep water subject to a horizontal (offshore) density gradient, showing velocity components versus depth (a), (b) and hodograph (c). The unit of depth is the Ekman depth D, of velocity $(\beta\sigma_0 gD/fL)$, where $\beta\sigma_0/L$ is the horizontal density gradient. From Csanady (1976a).

7.3.3. Determination of the Density Field

It is now possible to return to Equation (7.35) and determine the magnitude of the density gradient $\sigma_0/L = d\sigma_0/dx$. Suppose that the thermohaline circulation is present in isolation and that the depth H is much greater than the Ekman depth D. The transport $U(z)$ of Equation (7.42) is then the Ekman transport associated with the mid-water column shear, $U(z) = Kg\beta\sigma_0/f^2L$, over most of the water column, and the integral of $U^2(z)$ over the depth is the depth times the mid-water column value of U^2, times a constant differing from unity by a quantity of order D/H. Ignoring the latter correction, the effective horizontal diffusivity is

$$K_H = \frac{K^2(g\beta\sigma_0)^2}{K_S f^4 L^2} \, . \tag{7.44}$$

Substitution into Equation (7.35) yields a cubic equation for σ_0/L:

$$\left(\frac{\sigma_0}{L}\right)^3 = \frac{K_S Q S_E f^4}{H K^2 (g\beta)^2} \, . \tag{7.45}$$

Typical values for the continental shelf of the Mid-Atlantic Bight in winter are

K and K_S	10^{-2} m^2 s^{-1}
Q	10^{-2} m^2 s^{-1}
S_E	33‰
f	10^{-4} s^{-1}
H	100 m
$g\beta$	8 m s^{-2}.

These yield with Equation (7.45) $\sigma_0 H/L = 0.7 \times 10^{-5}$ or a salinity gradient of 1‰ per 15 km.

Consider now the circulation generated by a longshore wind stress of magnitude ρu_*^2, with no opposing pressure gradient or other forcing. In a deep water column $U(z)$ is the Ekman transport $u_*^2 f^{-1}$ over most of the depth, and

$$K_H \cong \frac{u_*^4}{f^2 K_S}$$

from Equation (7.34). With the above typical parameters for the Mid-Atlantic Bight, and $u_* = 0.01$ m s^{-1} one calculates $K_H = 100$ m^2 s^{-1}, and from Equation (7.35):

$$\frac{\sigma_0 H}{L} = 3.3 \times 10^{-6}$$

or a salinity gradient of 1‰ in 30 km, a scale twice as large as due to the thermohaline circulation on its own.

With longshore wind-induced and thermohaline circulation considered together, but otherwise with the same assumptions as above and $K_S = K$, one derives the following cubic equation for σ_0/L:

$$\frac{QS_E}{K} = \frac{\sigma_0 H}{L}\left(\frac{u_*^2}{fK} + \frac{g\beta}{f^2 H}\,\frac{\sigma_0 H}{L}\right)^2 . \tag{7.46}$$

Given the above typical values of the parameters, this equation yields

$$\frac{\sigma_0 H}{L} = 2.35 \times 10^{-6}$$

or a salinity gradient of 1‰ in 42 km.

Among their other features, the above results demonstrate the 'self-advection' of the density field by the thermohaline circulation in a relatively simple case. Equation (7.45) is cubic in the density gradient because the freshwater flux is that gradient times an effective diffusivity, the latter a quadratic quantity in the interior velocities. Because these velocities are themselves proportional to the density gradient in 'pure' thermohaline circulation, the cubic relationship follows. However, when other circulation components appear, the importance of the thermohaline circulation in determining the horizontal density gradients rapidly diminishes: adding just the longshore wind-driven flow, the effect of the thermohaline circulation becomes merely a minor increase in cross-shelf scale, as compared to the action of the wind alone. A reasonable conjecture is that when several other energetic flow components are present (e.g., tides and inertial oscillations, as well as wind-driven flow) the horizontal density gradient on a 'nearly homogeneous' shelf will be determined essentially independently of the thermohaline circulation. In other words, the density and velocity fields are in this case uncoupled, self-advection of density unimportant.

7.4. MEAN CIRCULATION ON A NEARLY HOMOGENEOUS SHELF

One is thus lead to consider the problem of mean circulation over a continental shelf in winter. The primary ('first order') observable flow is chaotic, consisting mainly of tides and of a random succession of winter storm episodes. However, average currents over the period of the order of a month, or the drift of tracers over a similar period, generally show a more orderly pattern. Such monthly or seasonal mean circulation is practically important in long-term dispersal problems and it is useful to consider its dynamics. Because the establishment of a winter density distribution is a slow process, the associated thermohaline circulation is a persistent contribution to different first-order flow episodes and may be expected to remain important in the averaged pattern.

The linearized equations of motion (7.23) may be supposed to apply to the mean circulation, if one neglects horizontal mean momentum advection by first-order flow episodes and similar phenomena. As discussed briefly early in Chapter 6, the neglect of residual effects of large flow fluctuations, not parameterized by the vertical momentum diffusion term in Equation (7.23), is not necessarily justifiable. However, an estimate of such effects on winter shelf circulation has shown them to be of moderate magnitude (Csanady, 1976a) unlikely to influence the circulation pattern in an important way. Thus Equations (7.23) are applied with surface and bottom stress boundary conditions

$$\left.\begin{aligned} \frac{\tau_x}{\rho} &= K\frac{\partial u}{\partial z}, \\ \frac{\tau_y}{\rho} &= K\frac{\partial v}{\partial z}, \end{aligned}\right\} \quad (z = 0)$$
$$\left.\begin{aligned} ru &= K\frac{\partial u}{\partial z}, \\ rv &= K\frac{\partial v}{\partial z}, \end{aligned}\right\} \quad (z = -H) \qquad (7.47)$$

where now τ_x, τ_y are components of the *mean* wind stress. K and r are eddy diffusivity and bottom friction coefficients, both determined primarily by the first-order fluctuating flow episodes. Thus K depends on the mean square wind stress and bottom stress, and r is a bottom drag coefficient times a typical fluctuation velocity magnitude near the bottom, both independent of the intensity of mean circulation.

7.4.1. Application of the Parallel Flow Model

If the mean circulation is supposed to be uniform in the longshore direction, and subject to the coastal constraint, $U = 0$ the model discussed in Section 7.3.2 applies. Under these conditions, the salt diffusion equation may be formulated as

$$\frac{\partial}{\partial x}\left(K_H \frac{\partial \sigma}{\partial x}\right) + K_z \frac{\partial^2 \sigma}{\partial z^2} = 0, \tag{7.48}$$

where K_H is now a mean value of the shear-diffusivity calculated in Equation (7.34), the cross-shelf transport $U(z)$ involved being mostly due to first-order fluctuations. A constant horizontal salt flux is imposed, so that Equation (7.35) yields the horizontal salinity gradient. Linearized equations of motion are taken to apply, so that the different contributions to circulation may be calculated without difficulty. The density driven flow component remains as calculated in Section 7.3.2, see the illustration, Figure 7.7.

Other contributions to the interior velocities may now be added to the results above. Tables 7.1 and 7.2 list the parameters assumed in such a calculation for a model purporting

TABLE 7.1
Quantitative data of Mid-Atlantic bight model

Parameter	Value
Coriolis parameter f, s^{-1}	0.9×10^{-4}
Longshore surface gradient	-1.4×10^{-7}
Length scale of density L, cm	10^7
Salinity variation, σ_0	10^{-3}
Bottom resistance coefficient r, cm s^{-1}	0.16
Offshore wind stress F_x, cm^2 s^{-2}	0.49
Longshore wind stress F_y, cm	0.60
Ekman depth D, m	16
Mid-shelf depth, m	48
Outer shelf depth, m	96

TABLE 7.2
Resultant velocities in mid-shelf model in centimeters per second
(x component, y component)

	Mid-shelf (48 m)	Outer shelf (96 m)
Surface	(6.0, −1.4)	(6.0, −17.8)
Mid-column	(−1.5, − 1.5)	(−1.5, − 15.2)
Bottom	(−1.0, − 0.3)	(1.2, − −4.4)

to represent mean circulation over the Mid-Atlantic Bight continental shelf off the Long Island coast, and the calculated velocities for the 'surface' (below surface wall layer), mid-depth, and 'bottom' (above bottom wall layer) along two isobaths, mid-shelf and outer shelf. The resultant velocities show strong offshore flow at the surface, weak onshore flow at mid-depth and cross-shore flow of opposite sign at the bottom, identifying a divergent bottom layer. The line of divergence where the bottom cross-shore velocity vanishes occurs with these quantitative parameters near the 70 m isobath. The principal physical reason for this divergence is the same as already discussed in Chapter 6, the increasing predominance in deepening water of the total pressure force associated with the longshore sea level slope.

A combined parallel flow model of the kind discussed realistically reflects the observed mean circulation over several Atlantic-type continental shelves, as discussed further in the next chapter.

7.5. SURFACE ELEVATION DISTRIBUTION

It remains to complete the discussion of thermohaline circulation by addressing the global problem, the determination of surface elevations over an extended region of a shallow sea. A depth integration of the equations of motion (7.23) and the equation of continuity yields:

$$-fV = -gH\frac{\partial\zeta}{\partial x} - g\beta\int_{-H}^{0}(z+H)\frac{\partial\sigma}{\partial x}\,dz + F_x,$$

$$fU = -gH\frac{\partial\zeta}{\partial y} - g\beta\int_{-H}^{0}(z+H)\frac{\partial\sigma}{\partial y}\,dz + F_y - rv_b, \tag{7.49}$$

$$\frac{\partial U}{\partial x} + \frac{\partial V}{\partial y} = 0,$$

where the x-component of bottom stress has been dropped for reasons discussed in Section 6.3. Although the bottom velocity v_b is not precisely equal to the bottom geostrophic velocity, as the discussion in Section 7.3.1 above has highlighted, the difference is minor compared to other sources of error in parameterizing bottom stress in a steady circulation problem. The geostrophic velocity at the bottom of a stratified water column is

$$v_b \cong v_{bg} = \frac{g}{f}\frac{\partial\zeta}{\partial x} + \beta\int_{-H}^{0}\frac{\partial\sigma}{\partial x}\,dz. \tag{7.50}$$

With this substitution, and after taking curl on (7.49), the equation of the shelf pressure field is obtained:

$$\frac{\partial^2 \zeta}{\partial x^2} + \frac{f}{r}\frac{dH}{dx}\frac{\partial \zeta}{\partial y} = -\frac{\beta f}{r}\frac{dH}{dx}\int_{-H}^{0}\frac{\partial \sigma}{\partial y}\,dz - \beta\frac{\partial}{\partial x}\int_{-H}^{0}\frac{\partial \sigma}{\partial x}\,dz. \tag{7.51}$$

More general forms of this equation have been derived by Hendershott and Rizzoli (1976), Hsueh and Peng (1978), and others. The discussion here follows Csanady (1979).

Equation (7.51) is the same parabolic equation that was discussed in considerable detail in Chapter 6, but now with two density-field related forcing terms on the right. If the new forcing terms are independent of the thermohaline flow field they themselves give rise to, the problem is relatively straightforward. The surface elevation distribution $\zeta(x, y)$ may then be regarded as a linear superposition of wind-driven and thermohaline components, the former calculated as in a homogeneous fluid, the latter in a very similar manner, with a given density related source term distribution.

The assumption that the momentum and density equations are uncoupled is, however, not realistic even for a nearly homogeneous (winter) shelf. A generalization of the approach of Section 7.3.1 with longshore variations of salinity considered leads to a simplified salt diffusion equation of the following form:

$$\frac{V}{H}\frac{\partial \sigma}{\partial y} = \frac{\partial}{\partial x}\left(K_H \frac{\partial \sigma}{\partial x}\right). \tag{7.52}$$

This form arises on the realistic suppositions that: (1) cross-shore salt transport is mostly shear-diffusion, effected by the first order flow episodes, which may be taken to be dominate mean cross-shore salt advection $U(\partial\sigma/\partial x)$; (2) contrariwise, longshore salt transport is mostly advection by the mean circulation. Given typical values of mean longshore transport V, much larger than cross-shore transport U, the advection term in the longshore direction dominates, unless an unrealistic value of shear diffusivity is assumed. Fischer (1980) has discussed some aspects of this problem, and calculated the salinity distribution on a winter shelf, caused by a few large point sources at the coast, representing major rivers entering the east coast shelf. Such calculations typically show that alongshore salinity gradients are about one order of magnitude smaller than cross-shore ones, a result which agrees with observation. The two terms in Equation (7.52) are thus of the same order, as are the two forcing terms on the right of the vorticity tendency equation (7.51). The thermohaline contribution to the longshore transport V under these circumstances is important, and the self-advection of the density field expressed by (7.52) cannot, in general, be ignored.

7.5.1. Two-Dimensional Density Field

Some insight is nevertheless gained by considering the hypothetical case of vanishing longshore density gradient

$$\frac{\partial \sigma}{\partial y} = 0. \tag{7.53}$$

The case may be thought to arise from uniform inflow of freshwater per unit length of coastline, much as supposed earlier in the parallel flow model. The cross-shore salinity gradient can then be calculated as discussed earlier, along with the interior velocity distribution. Here only the accompanying global surface elevation field is considered further, due to the freshwater inflow alone, with vanishing wind stress and freshening trapped in some coastal band. A first integral of Equation (7.51) is immediate:

$$\frac{\partial \zeta}{\partial x} = -\beta \int_{-H}^{0} \frac{\partial \sigma_a}{\partial x}\,dz = -\beta H \frac{\partial \sigma_a}{\partial x}. \tag{7.54}$$

The surface level slope is thus as calculated from the dynamic height referred to the bottom. The second integration over arbitrarily varying depth $H(x)$ requires some care (Csanady, 1979). The result may be written as

$$\zeta = -\beta H \sigma_a(x) - \beta \int_{H}^{\infty} \sigma_a(H)\,dH. \tag{7.55}$$

The first term is the dynamic height relative to the bottom, the second an integral with respect to depth, necessary for a consistent extension of dynamic height calculations into shallow water. At the coast, for example, $H = 0$ and only the second term remains:

$$\zeta_0 = -\beta \int_{0}^{\infty} \sigma_a(H)\,dH. \tag{7.55a}$$

The function $\sigma_a(H)$ is of course easily constructed from $\sigma_a(x)$ and $H(x)$. With the reference density chosen so that $\sigma_a \to 0$ as $x \to \infty$, σ_a is negative everywhere and it is easily seen that sea level is highest at the coast.

Physically, the integrals (7.54) and (7.55) of the pressure field equation (7.51) represent the simple case when bottom stress vanishes. Going back to the origin of this equation, with no longshore pressure gradients present there is no vortex stretching term, hence the curl of bottom stress must vanish. With the ocean supposed inert far from the coast, the bottom stress is zero there, hence it is zero everywhere. Zero bottom stress, given the parameterization scheme of (7.28) and (7.50), implies zero geostrophic velocity at the bottom, a result from which (7.54) follows immediately. That part of the thermohaline circulation forced by cross-shore density gradients is thus simply the longshore thermal wind, with vanishing velocity at the bottom (or near the bottom, see Section 7.3.1, and with other minor adjustments for internal friction near the free surface).

7.5.2. Qualitative Effects of Longshore Density Variations

When longshore density gradients are present, it is convenient to retain the solution of (7.55) as one component of the surface elevation field, which is associated with zero longshore geostrophic velocity at the bottom. Writing $\zeta = \zeta_1$ for the solution of (7.55) where $\sigma_a = \sigma_a(x, y)$, and $\zeta = \zeta_1 + \zeta_2$, one easily obtains the equation for the complementary or residual field $\zeta_2(x, y)$:

$$\frac{\partial^2 \zeta_2}{\partial x^2} + \frac{f}{r}\frac{\mathrm{d}H}{\mathrm{d}x}\frac{\partial \zeta_2}{\partial y} = \beta \frac{f}{r}\frac{\mathrm{d}H}{\mathrm{d}x}\int_H^\infty \frac{\partial \sigma_a}{\partial y}\,\mathrm{d}H, \tag{7.56}$$

where the integration is to be carried out along constant y.

In terms of the vorticity tendency balance the function of the residual field $\zeta_2(x, y)$ is as follows. The dynamic height calculation according to (7.55) gives rise to a pressure field, with which is associated the geostrophic cross-shore bottom velocity

$$u_{1g} = \frac{g\beta}{f}\int_H^\infty \frac{\partial \sigma_a}{\partial y}\,\mathrm{d}H. \tag{7.57}$$

The vortex stretching due to this velocity is represented by the right-hand side of (7.56). It must be balanced by a combination of cross-shore geostrophic flow u_{2g} (cancelling out u_{1g}) and the curl of bottom stress arising from the circulation cell of the ζ_2-field.

The circulation cell of the ζ_2-field is generated by a distributed forcing field, akin to wind stress curl. Where $\partial \sigma_a/\partial y$ is positive, the forcing is much like under cyclonic wind stress curl, the effects of which have been illustrated in Figures 6.10 and 6.11. Such a vorticity tendency is balanced either by onshore transport, or by cyclonic curl in the velocity, or, generally, a combination of the two. Cyclonic curl in the velocity may take the form of a concentrated longshore current to negative y, with high velocities nearshore.

Where a river enters the coastal waters of a continental shelf, the localized nearshore freshening gives rise to $\partial \sigma_a/\partial y > 0$ toward positive y, $\partial \sigma_a/\partial y < 0$ to negative y. A simple analytical solution could no doubt be devised to represent a typical situation of this sort, but has not so far been given. In a qualitative way one infers nevertheless that shoreward transport would occur over that portion of the coast where $\partial \sigma_a/\partial y > 0$, a coastal current toward negative y would be generated, which would then decay slowly in the negative y direction. In other words, the flow field would be qualitatively similar to that of a coastal 'mound' discussed in Section 6. A large mound generated by a river with high discharge would affect a considerable 'forward' portion of the coastline.

Further quantitative studies of such thermohaline circulation problems are clearly required in order to understand the presumably complex details. The longshore current generated by a coastal mound in all probability shapes the density field, by stretching out toward negative y the surfaces of constant density. It remains to be seen whether the effect is amenable to some simple parameterization scheme, such as could be devised for the cross-shore shear diffusion of salinity.

CHAPTER 8

Observed Quasi-steady Flow Patterns in Shallow Seas

8.0. INTRODUCTION

This final chapter will attempt to connect the simple steady-state models of circulation of Chapters 6 and 7 to observational evidence. Circulation features more persistent than weather related events are sometimes brought to light by averaging fixed-point current meter data for periods of the order of a month or longer, or by Lagrangian tracer studies involving horizontal displacements of drifters or of water masses over similar periods. A mean circulation pattern determined by such methods is not necessarily relatable to simple dynamical models of the kind discussed above, because cumulative effects of transient flow events may swamp those of steady, low-level forcing. In the long-term average equations of motion, in the heat and salt transport equations, and in stress or flux boundary conditions the cumulative effects are manifested by various mean products, akin to Reynolds stresses. Where transient flow events dominate, such mean products are difficult to quantify or to relate to the mean flow. One has to allow for the possibility, therefore, that transient flow events bring about some unexpected redistribution of heat, salt or momentum. A few concrete examples of this have been discussed elsewhere (Csanady, 1975, 1976a). Generally, redistribution effects are localized and do not modify the basic pattern of mean circulation. However, in confronting the mean circulation problem in a shallow sea previously unexplored, one would do well to keep an open mind in this regard.

Another important caveat is that, when averaged over long periods, particle (Lagrangian) mean velocities are likely to differ considerably from fixed point (Eulerian) means. The physically meaningful quantity in the circulation problem is the long term particle displacement, or its smoothed time rate of change, i.e., some running average Lagrangian velocity. Monthly mean velocities determined at fixed locations in a shallow sea may bear no relationship to the Lagrangian velocity, and represent no more than a few statistics among many, characterizing local current climatology. This is, however, not necessarily the case: a persistent circulation of reasonably large amplitude stands out in Eulerian as well as in Lagrangian observations, so that, again, each case must be judged on its merits.

Although the long-term mean circulation problem is the most obvious application of steady, frictionally controlled flow models, the same models have already been shown in Chapter 6 to be relevant also to storm-driven currents over Atlantic type shelves. A shallow water column responds to strong winds in a period of the order of the tidal cycle or less, so that currents averaged over a full tidal cycle may be expected to change at a slow enough rate for the acceleration terms to be negligible. The observations to be discussed in this chapter will therefore also include the 'detided' effects of strong winds on Atlantic type continental shelves.

Before embarking on the discussion of the observational evidence it is also appropriate to reflect, in what sense one may expect to find agreement with theory. The theoretical models developed earlier are all highly idealized and cannot be expected to apply directly and in all details to any given real-life situation in the coastal ocean. However, it may be possible to discern in the observational evidence certain key signatures of shelf circulation cells and of the associated coastally trapped pressure fields. These include, for example, preferential response to longshore rather than cross-shore wind stress in weather systems of large size, appearance of longshore pressure gradients of a magnitude about as expected from theory, significant cross-shore transport relatively close to shore, influence of a disturbance spreading in the 'forward' or negative y direction (in the northern hemisphere). If the observations conform to such general characteristics of the simple models, it is reasonable to conclude that an appropriate theoretical framework has been found for their interpretation.

8.1. MEAN CIRCULATION IN THE MID-ATLANTIC BIGHT

The circulation of the east coast shelf of North America north of Cape Hatteras, specifically that of the Mid-Atlantic Bight, is characterized by a persistent southwestward drift at velocities of order 10 cm s^{-1} (Bumpus, 1973; Beardsley *et al.*, 1976; Mayer *et al.*, 1979). On time averaging Eulerian or Lagrangian data over periods of one month or longer a clear and consistent pattern of longshore and cross-shore velocities emerges which is reasonably regarded as a steady state flow field, and compared with frictional models.

A successful steady flow model of the winter circulation of the Mid-Atlantic Bight may be constructed by taking into account forcing by wind stress, both longshore and cross-shore, freshwater influx and a longshore pressure gradient due to some unspecified 'distant' effect. It is reasonable to suppose that each of the component forces produces a pattern of its own and that the resultant pattern is a simple linear superposition of the component patterns. The details of such a model have been discussed above in Section 7.3.1. Over the middle and outer shelf it turns out to be sufficiently accurate to regard the longshore pressure gradient a 'distant' effect, i.e., to take it as constant with distance from shore. The freshwater influx is supposed to generate a two-dimensional density anomaly field, $\beta\sigma(x, y)$, represented over mid-shelf by $\partial\sigma/\partial x = \sigma_0 L^{-1}$ = constant, where L is an offshore scale of density variations.

The calculations in Section 7.3.1 yielded 'typical' magnitudes of the different velocity contributions, appropriate for conditions in the Mid-Atlantic Bight, at the surface, mid-column and bottom, at two different mid- and outer shelf locations of different depths. The magnitude of the longshore pressure gradient was inferred from a comparison with observed currents (Scott and Csanady, 1976; Csanady, 1976a).

The observed mean circulation of the Mid-Atlantic Bight, over the middle and outer portions of the continental shelf, conforms remarkably closely to this parallel flow-thermohaline circulation model. Figure 8.1 shows some mean velocities observed over·

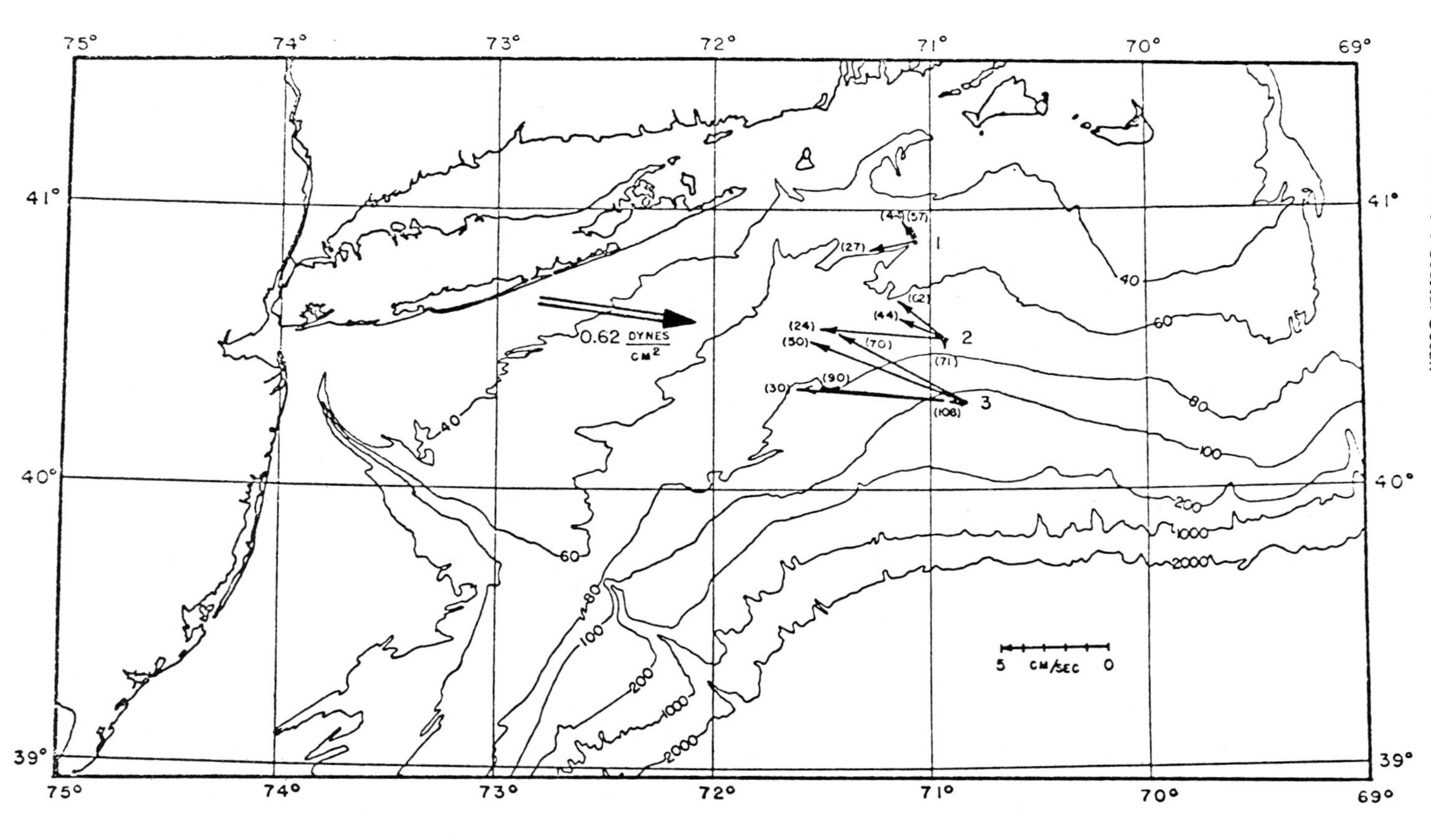

Fig. 8.1. Mean currents and wind stress during the 1974 MIT New England Shelf Dynamics Experiment; February 27 – April 3, 1974. Numbers in parenthesis are the instrument depths in meters. Isobaths are in meters. Note that mean flow opposes wind stress and that velocities increase significantly in deeper water. From Flagg (1977).

the New England shelf (averaging period a little over a month, March 1974, taken from Flagg (1977). A larger body of similar evidence was summarized by Beardsley *et al.* (1976). Other current meter studies have shown comparable results, demonstrating the persistence of this pattern in time, and its spatial extension over the entire Mid-Atlantic Bight. The principal mean flow features derived from current meter studies characterize motion in the middle of the water column, below the surface and above the bottom frictional layer. They are:

(1) longshore (long-isobath, more accurately) flow toward the southwest at an intensity of 3–10 cm s^{-1}, increasing noticeably with increasing distance from shore

(2) onshore (cross-isobath) flow over most of the water column, at an amplitude of 1–3 cm s^{-1}.

Where reliable current meter measurements close to the surface or close to the bottom are available, they confirm Bumpus' (1973) conclusions derived on the basis of surface and bottom drifter studies. These are:

(3) surface waters move to the southwest at mean speeds of 10–30 cm s^{-1}, and in an offshore direction, at 3–10 cm s^{-1}

(4) bottom waters diverge at about the 60 m isobath, moving in an onshore direction at 0–3 cm s^{-1} in shallower water, offshore at similar speeds in deeper water.

These observed facts may be understood in the framework of the model described above, as being a consequence of four mean circulation components. Specifically, the action of longshore sea level gradient is responsible for the increase of longshore velocity in increasing depth according to the model discussion earlier. The interaction of wind-stress and longshore sea level gradient explains the divergence of the bottom boundary layer at a specific depth. The high offshore velocities at the surface result from the two wind-stress related circulation components (due to longshore and cross-shore wind respectively) and from the thermohaline circulation.

The observational evidence suggests that any trapped cells affecting the mean flow beyond the 40 m isobath or so have a very long spatial scale. The key driving force, the longshore pressure gradient, is then indeed a distant or large-scale effect, possibly impressed upon the shelf by offshore oceanic gyres (Csanady, 1978a; Beardsley and Winant, 1979). At the edge of the shelf this longshore gradient is certainly as large as it is closer to shore, as evidenced by substantial mid-water column onshore flow, which is presumably geostrophic.

The magnitude of the longshore pressure gradient is not constant in time, however, but is subject to clear seasonal – and perhaps longer term – variation (Chase, 1979). Chase's analysis suggests that some of the seasonal varaitions may be related to freshwater inflow, but there is also an indication of deepwater effects, in that the position of the Gulf Stream is related to the longshore gradient variations.

The discussion of thermohaline circulation in Chapter 7 suggests that sea level gradients in the Mid-Atlantic Bight could be affected by variations of freshwater inflow much further north, notably in the Gulf of Maine and the Gulf of St. Lawrence. According to the models discussed, longshore variations of freshwater influx over such a long range (order 1000 km) affect more or less the entire outer shelf, which would make their effects difficult to distinguish from deepwater gyre effects. The magnitude of the longshore pressure gradient due to observed density variations may be estimated to be 10 cm in 1000 km (10^{-7}) during the spring runoff period only, and a much lower slope at other times of the year (Csanady, 1979). During the spring-early summer period the longshore sea level gradient due to freshwater sources is therefore of the same order of magnitude as required to account for the observed southwestward drift of shelf waters. This effect may in fact explain the seasonal variation of the longshore gradient. It should be added, however, that the estimation of a long-term mean density field is uncertain due to the paucity of the data, so that these conclusions must be regarded as tentative.

Over the inner shelf, given the complex geography of the coastline, there should be trapped cells according to theory. Some evidence in support of this proposition comes from nearshore studies in different locations, arriving at different magnitudes of the longshore pressure gradient (e.g., Bennett and Magnell, 1979). The details of these trapped cells have not been elucidated so far, however, not at any rate in connection with a long-term mean circulation pattern. An analysis of observations within 12 km of the south coast of eastern Long Island suggested that this is an outflow region of a trapped cell, the inflow coming on the north coast of the Island from Long Island sound, presumably as Ekman drift associated with the mean eastward wind stress (Pettigrew, 1981). More detailed and convincing evidence is available on inner shelf circulation cells caused by storms.

8.2. STORM CURRENTS OVER ATLANTIC TYPE SHELVES

From an economical point of view, the most important problem in applied oceanography is the prediction of storm surges which from time to time cause tremendous damage along coastlines adjacent to broad continental shelves, such as the North Sea, the U.S. Gulf Coast and the East Coast. Consequently, numerical models are well developed for the prediction of coastal sea levels associated with hurricanes and extra-tropical storms (Jalesnianski, 1965; Heaps, 1969; Welander, 1961). These models have been calibrated empirically and constitute today a useful practical tool. They do not, however, give a particularly realistic description of storm driven currents (Forristal *et al.*, 1977), not at least without considerable further development and calibration. In any event, the predictions of the models are almost as complex as the observational evidence, and it is desirable to understand the contribution of storms to the circulation problem in terms of simpler concepts.

As already discussed earlier, strong winds acting over shallow water rapidly establish frictional equilibrium flow so that steady state models should apply. The classical models of hurricane surge are of this kind (Freeman *et al.*, 1957; Bretschneider, 1966). Although wave-like 'resurgences' are sometimes important (Redfield and Miller, 1957), the bulk of the coastal sea level rise attributable to storms can be explained as steady state response. Associated with the storm surge intense longshore currents are present which give rise to large particle displacements. Boicourt and Hacker (1976) point out, for example, that most of the mean southwestward drift off Chesapeake Bay in the Mid-Atlantic Bight is generated by a few nor'easterlies, as illustrated in their paper vividly by progressive vector diagrams of observed currents.

Strong wind stress is exerted on the sea surface in extratropical storms and in hurricanes. Extratropical storms have a typical diameter (half wavelength) of about 100 km, and exert a stress of the order of 1 Pa. Hurricanes are typically 3 times smaller in diameter, but their maximum winds are much higher, exerting a stress of order 3 Pa. Mooers *et al.* (1976) discuss in some detail the average properties of extratropical storms over the Mid-Atlantic Bight, while Cardone *et al.* (1976) describe a model hurricane. Apart from weather system size and stress intensity, another important factor is the location of the storm track. In the Mid-Atlantic Bight, a southwestward driving extratropical storm has its center usually well offshore, a northeastward driving storm well on land, typically over the St. Lawrence valley. Beardsley and Butman (1974) illustrate two such prototype storms, see Figure 8.2 here. The intermediate possibility is that the storm center is right at the coast, and moves along it (Figure 8.3). In the latter case the wind stress is mostly directed cross-shore, while with the storm center well away from the coast the dominant influence comes from the longshore wind. A hurricane usually crosses rapidly from sea to land. Figure 8.4 from Redfield and Miller (1957) illustrates the wind field of a hurricane.

In a steady state model, coastal sea level rise is due to two effects: setup in response to onshore wind, and geostrophic adjustment to balance the Coriolis force of longshore currents. Freeman *et al.* (1957) refer to the resultant effect at the coast as the 'bathystrophic tide', because both effects are in some sense related to the bathymetry of the continental shelf, principally the width of the region of shallow water. In the detailed comparison of theory and observation it proves to be convenient to discuss separately the effects of longshore and cross-shore winds.

8.2.1. Statistical analyses of coastal sea level

A number of statistical analyses have recently been published of nontidal coastal sea-level changes associated with winds over Atlantic type shelves (Cragg *et al.*, 1982) of the West Florida shelf; Wang, 1979; Chase, 1979; Noble and Butman, 1979; and Sandstorm, 1980; of the North American east coast shelf north of Cape Hatteras). The main signal in such wind-induced sea level variations is due to extratropical storms passing over the area. The typical diameter of weather systems of this kind is of the order of 1000 km, their translational velocity of the order of 30 km hr^{-1}, northeastward over the northern east coast, southeastward over Florida. Because the sense of translation is thus opposite to the direction of propagation of topographic waves, no confusion arises between free and forced patterns of motion. In any case, wave-like propagation of flow events is barely detectable over Atlantic-type shelves, the sea level responding more or less directly to the local wind stress, so that a comparison with steady flow models should be appropriate.

As an extratropical storm passes over a long, more or less straight segment of the coast, each location is exposed to roughly the same history of forcing. In some configuration of coast and weather system the maximum response should be approached at most locations. Hence the statistically determined response amplitude should be comparable to the amplitudes ζ_a, v_a, etc. of the theoretical models more or less without regard to phase. However, at locations close to sharp changes in coastline orientation the geography

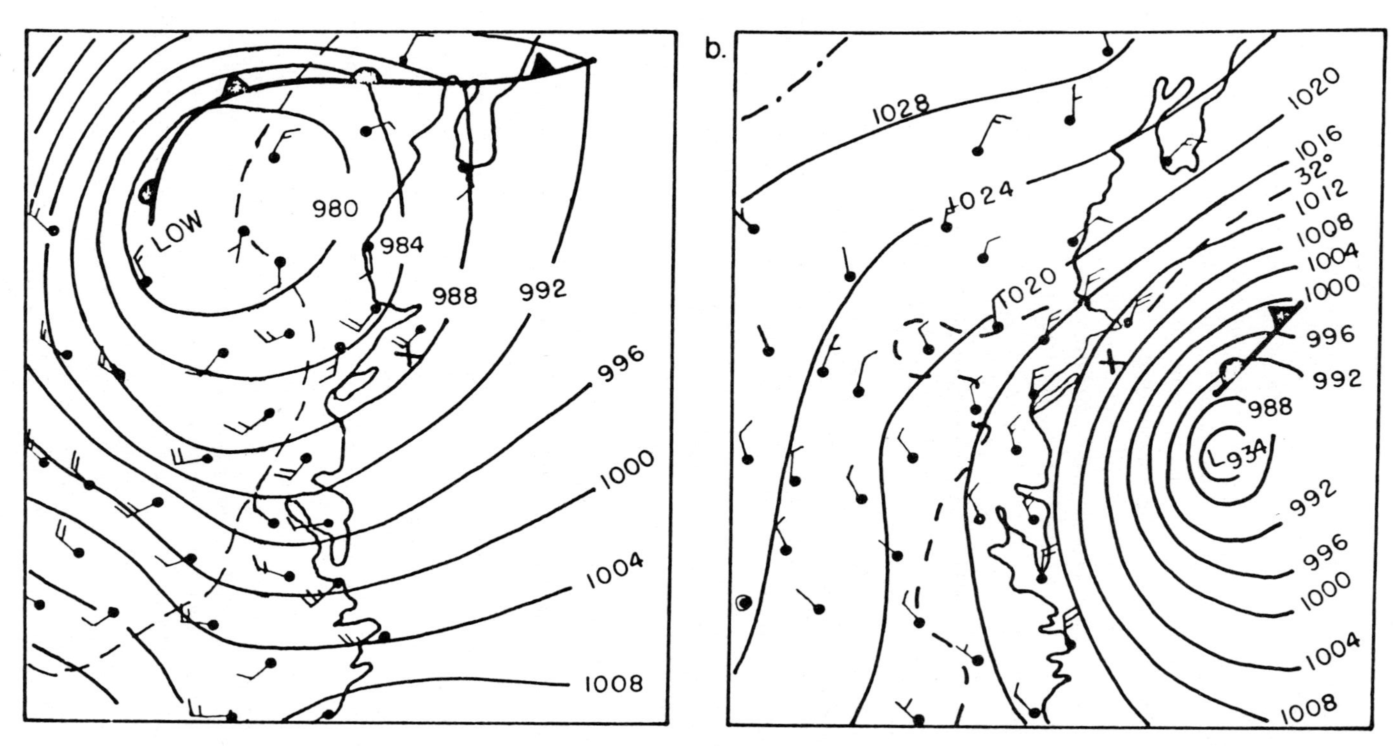

Fig. 8.2. Wind field of two prototype storms affecting the Mid-Atlantic Bight. When the storm center is over the St. Lawrence valley (left), longshore wind stress over the Bight is northeastward, when well offshore, southwestward. Cross-shore winds are, however, roughly the same, shoreward ahead of the storm center, offshore behind. From Beardsley and Butman (1974). Copyright American Geophysical Union.

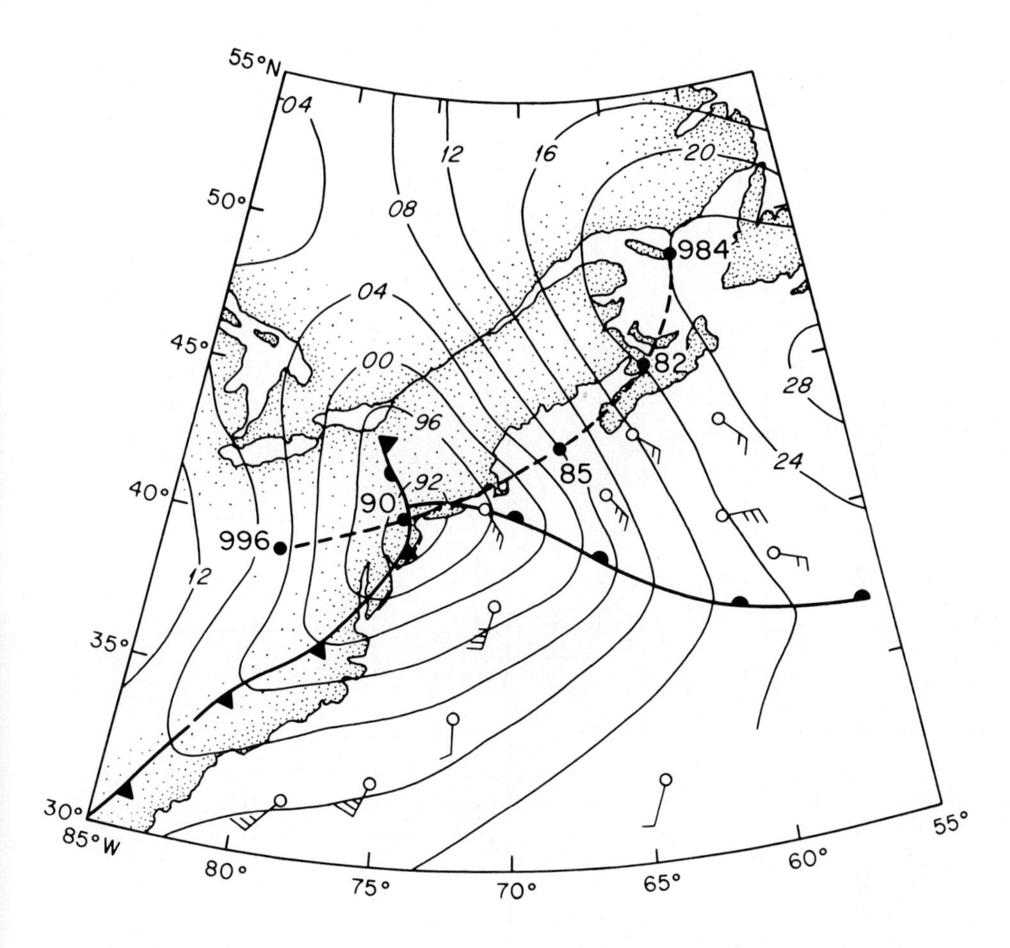

Fig. 8.3. Storm moving along northern coast of the Mid-Atlantic Bight, the center remaining roughly on the coast. Such storms cause mainly cross-shore wind stress. From Beardsley *et al.* (1977).

of the coast should become important. For example, near the tip of the Florida peninsula a wind blowing from the northwest is directed toward negative y (in a coordinate system defined analogously to the simple models) on the east coast, toward positive y on the west coast, with a rapid variation in between. Similar, and even more complex considerations apply in the northeast sector of the U.S. east coast, in the region of the Gulf of Maine or around Long Island. In such more complex situations one must think of at least two important longshore scales, one being the weather cycle scale, the other the distance to the closest sharp corner of the coast.

Observing the necessary caveats, it is nevertheless possible to distill from the statistical analyses important generalizations and to seek an interpretation of them in terms of the theory discussed earlier. For convenience, the amplitudes of sea level perturbation, longshore velocity and longshore pressure gradient, predicted by theory for the coast ($x = 0$)

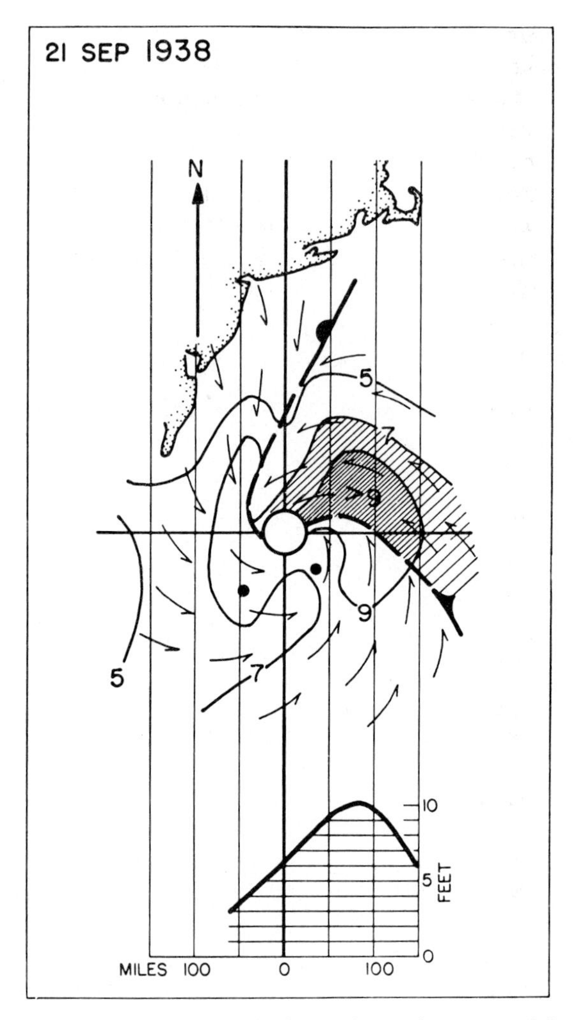

Fig. 8.4. Wind field of a hurricane crossing to land over the northern coast of the Mid-Atlantic Bight. Graph at bottom shows observed maximum storm surge at locations where vertical lines (parallel to the storm track) cross the coast. From Redfield and Miller (1957).

and for a sinusoidal alongshore distribution (cos ky) will be summarized here. The response to *longshore* wind should be characterized by

$$\zeta_a = \frac{u_*^2}{g}\frac{fL}{r} = \frac{u_*^2}{gs}\frac{2}{KL},$$

$$v_a = \frac{u_*^2}{r}, \tag{8.1}$$

$$\left.\frac{\partial\zeta}{\partial y}\right|_a = \frac{ku_*^2}{g}\frac{fL}{r} = \frac{2u_*^2}{gsL},$$

where

$$L = \left(\frac{2r}{fks}\right) \tag{8.2}$$

is the cross-shore scale of the coastal circulation cell generated by the wind stress field.

Typical values of the above parameters are: $u_*^2 = 1$ cm^2 s^{-2}, $f = 10^{-4}$, $r = 0.05$ cm s^{-1}, $s = 3 \times 10^{-3}$, $k = 2\pi/2000$ km. The above formulae give a trapping width scale L = 33 km and typical amplitudes $\zeta_a = 6.5$ cm, $v_a = 20$ cm s^{-1}, $\partial\zeta/\partial y|_a = 2 \times 10^{-7}$, or 2 cm in 100 km.

Physically, one may connect Equations (8.1) as follows. The longshore velocity-amplitude necessary to balance the wind stress by bottom stress is u_*^2/r; this gives rise to a cross-shore sea level gradient of amplitude $(f/g)\,(u_*^2/r)$ through geostrophic balance. The effective width of the current is L, hence the coastal sea level amplitude is (fL/r) (u_*^2/g). The longshore gradient in a sinusoidally varying field of wavenumber k is simply $k\zeta_a$.

A periodic cross-shore wind should generate a response at amplitudes

$$\zeta_a = \frac{u_*^2}{gs},$$

$$Va = \frac{u_*^2}{f}, \tag{8.3}$$

$$\left.\frac{\partial\zeta}{\partial y}\right|_a = \frac{ku_*^2}{gs}.$$

The sea level and pressure gradient amplitudes are similar to those for longshore wind, Equation (8.1), except for a factor of $kL/2$. A typical value of this quantity is 0.05, so that cross-shore winds should be much less effective in building coastal sea level than longshore winds. It is noted here, however, that the 'typical' k used in these estimates was small, appropriate to an extratropical storm affecting a long straight coast. With a much larger k, appropriate to a hurricane, the relative importance of cross-shore winds should increase.

8.2.2. Sea level and longshore wind

The proposition that de-tided sea level responds mainly to longshore wind is now well established by a considerable amount of early work, as well as by the more recent studies quoted above. Figure 8.5 illustrates an example of the association for a 20 day period over the Nova Scotia shelf from Sandstrom (1980). Even more impressive is Figure 8.6 due to Cragg *et al.* (1982), who have plotted sea level response against wind direction (for constant wind speed) for a location on the West Florida shelf. The peak occurs when the wind is parallel to the coast.

The amplitude of the sea level rise should, according to Equation (8.1), be proportional to wind stress, given otherwise the same parameters, mainly resistance coefficient r,

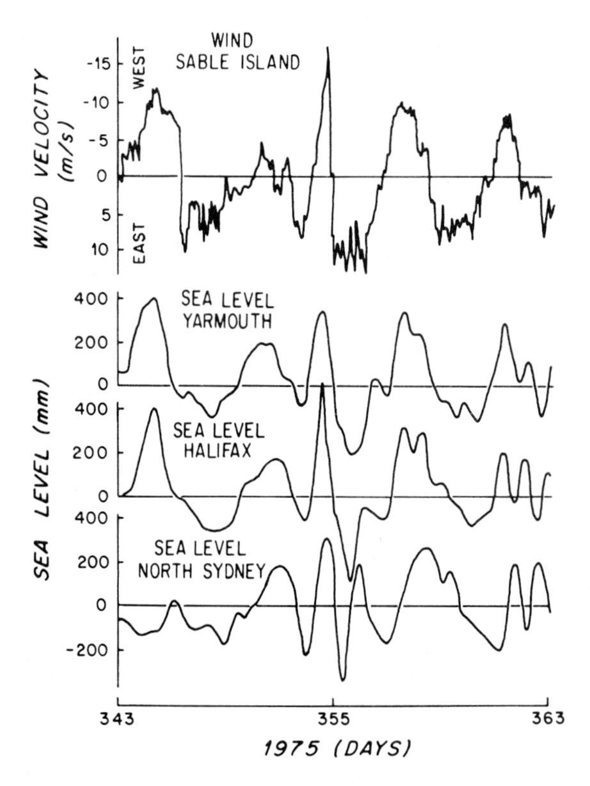

Fig. 8.5. Longshore wind and detided coastal sea level at several Nova Scotia tide gauges. From Sandstrom (1980).

wavenumber of forcing k, and bottom slope s. Noble and Butman (1979) demonstrate this in some detail, see Figure 8.7 here.

On forming the nondimensional coefficient:

$$a = \frac{g\zeta_a}{{u_*}^2} \tag{8.4}$$

the influence of wind stress is removed from coastal sea level amplitude, which should then depend on r, k, and s. Supposing that over the east coast and Florida shelves there are no systematic variations in either bottom friction coefficient r or weather system wavenumber k, the coefficient a should vary mainly with bottom slope s, and according to simple theory as $s^{1/2}$. This would apply to locations suitably distant from sharp corners of the coastline. Noble and Butman (1979) give a table of the coefficient a, times a dimensional factor. The same coefficient is also readily computed from the data of Cragg *et al.* (1982). These are plotted in Figure 8.8 against bottom slope s, the latter defined as 40 m depth divided by the distance of the 40 m isobath from the coast. This particular depth was chosen as a typical mid-shelf value, but other choices up to 100 m

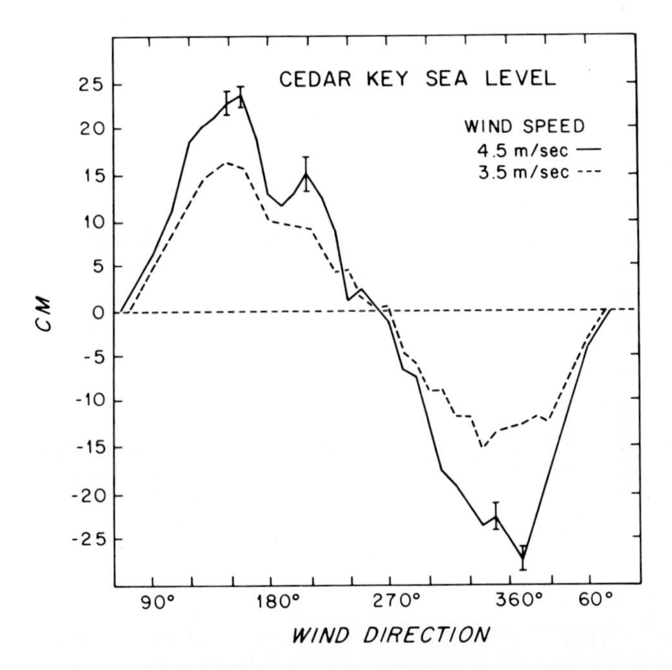

Fig. 8.6. Sea level elevation (tides taken out) and wind direction at a tide gauge on the West Florida shelf (Cragg *et al.*, 1982). The maximum occurs when the wind blows parallel to the coast.

give similar results, although somewhat different quantitative values for s. Some scatter is expected in such a graph for reasons that should be apparent from the above remarks, and yet a rough agreement with the theoretical trend is evident.

8.2.3. Longshore Pressure Gradients

Longshore gradients of sea level were seen earlier to be key features of coastally trapped pressure fields. Coastal sea level observations carried out at several tide gauges should in principle reveal such gradients. However, in practice the magnitude of the level-difference signal is small, typically a few centimeters, and its identification requires a particularly effective filtering out of sea level variations due to other causes. Along the U.S. east coast, Wang (1979) finds such sea level gradients to be significantly related to longshore wind stress only over the southern Long Island coast. Along the same piece of coastline, Chase (1979) and Pettigrew (1981) studied the longshore pressure gradient in greater detail. Figure 8.9, from Pettigrew, shows the two time series in question for a brief period. Even a visual inspection leaves little doubt about the relationship of the two. Regression analysis by Pettigrew yields the result that a 1 dyne cm^{-2} wind stress evokes a 3×10^{-7} longshore level gradient, opposing the wind stress. Cragg *et al.* (1982) find on the West Florida shelf a gradient of about 7.5×10^{-7} for 1 dyne cm^{-2} longshore wind. This is a larger signal and the association of longshore wind and opposing pressure gradient is correspondingly clearer than along the east coast.

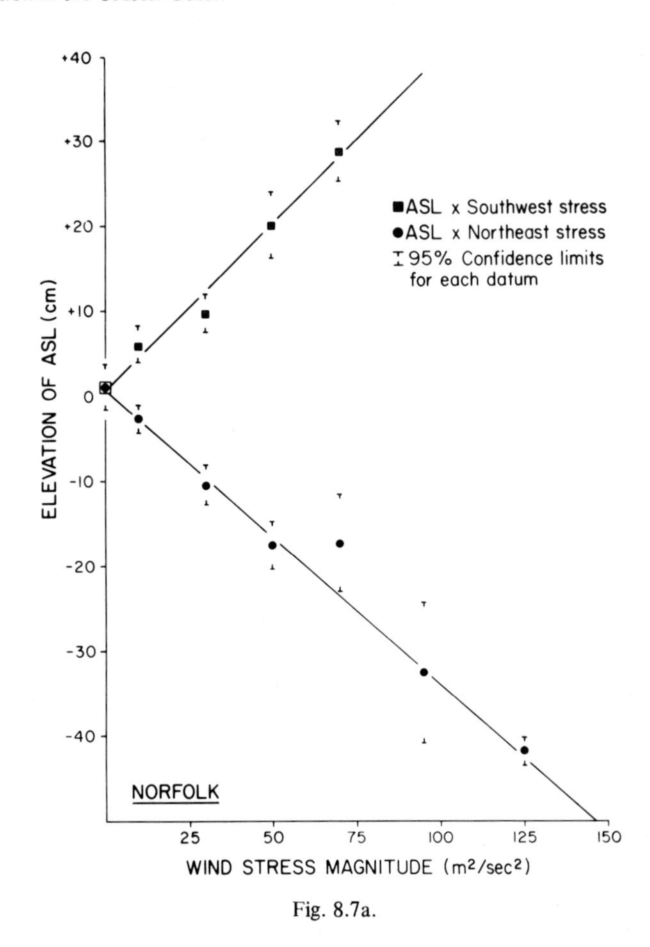

Fig. 8.7a.

Fig. 8.7a–b. Sea level elevation versus wind stress to positive and negative y (alongshore) at two east coast tide gauges. The asymmetry at Nantucket is presumably due to the complex geography of the 'backward' (positive y) segment of the coast at this location. From Noble and Butman (1979).

The quantitative relationships between wind stress and coastal sea level on the one hand and wind stress and longshore level gradient on the other hand can be used to calculate the longshore wavenumber k, supposing that the first and third of Equation (8.1) apply. More directly, if one writes for the amplitude of longshore level gradient:

$$\frac{\partial \zeta_a}{\partial y} = k\zeta_a \tag{8.5}$$

one has at once

$$\frac{g}{u_*^2} \frac{\partial \zeta_a}{\partial y} = k \frac{g\zeta_a}{u_*^2} = ak, \tag{8.6}$$

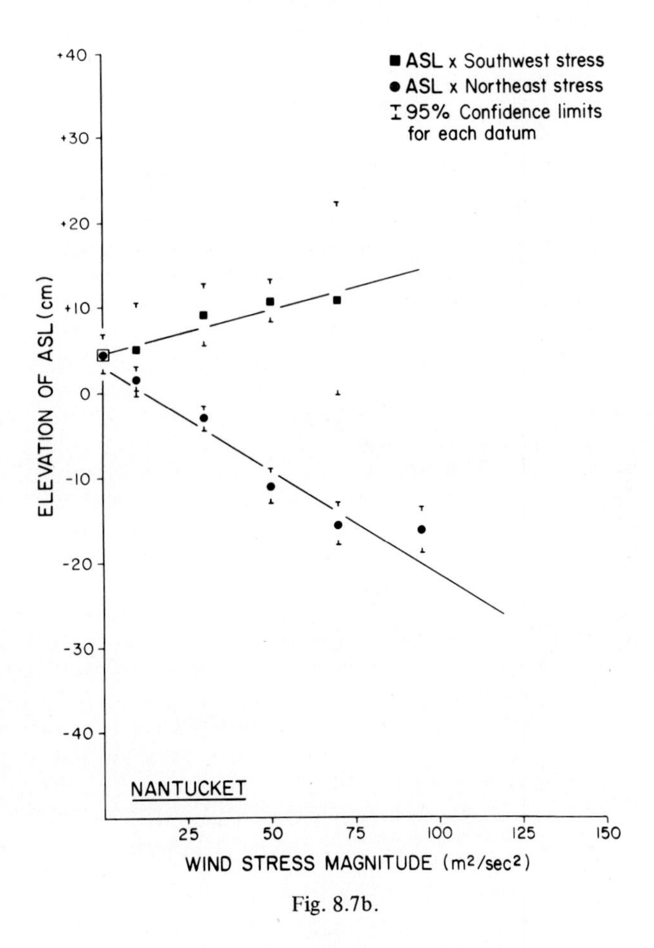

Fig. 8.7b.

where a is as defined above in Equation (8.4). The left-hand side of (8.6) is empirically found, as was a before, yielding the wavenumber k. With the data of Pettigrew (1981) one finds $k = 2 \times 10^{-8}$ cm^{-1}, with data of Cragg *et al.* (1982) 1.5×10^{-8} cm^{-1}, corresponding respectively to half wavelengths of 1500 and 2000 km. In order of magnitude, these dimensions agree with typical extratropical storm diameters, although they are on the high side.

One of the most revealing pieces of evidence on longshore pressure gradients is the asymmetric response of the Mid-Atlantic Bight to northeasterly versus southwesterly wind stress. This was documented clearly by Beardsley and Butman (1974) and was confirmed also by studies of Scott and Csanady (1976) and Bennett and Magnell (1979). Figure 8.2 above showed the isobars, i.e., approximately the wind stress field, of two storms producing respectively northeastward and southwestward longshore stress within the period of the coastal circulation study of Beardsley and Butman (1974). Figure 8.10 shows the effects of these two storms (March 18 and 22, respectively) on coastal sea levels at Nantucket and Sandy Hook (map, Figure 8.1). Sea level at Montauk was

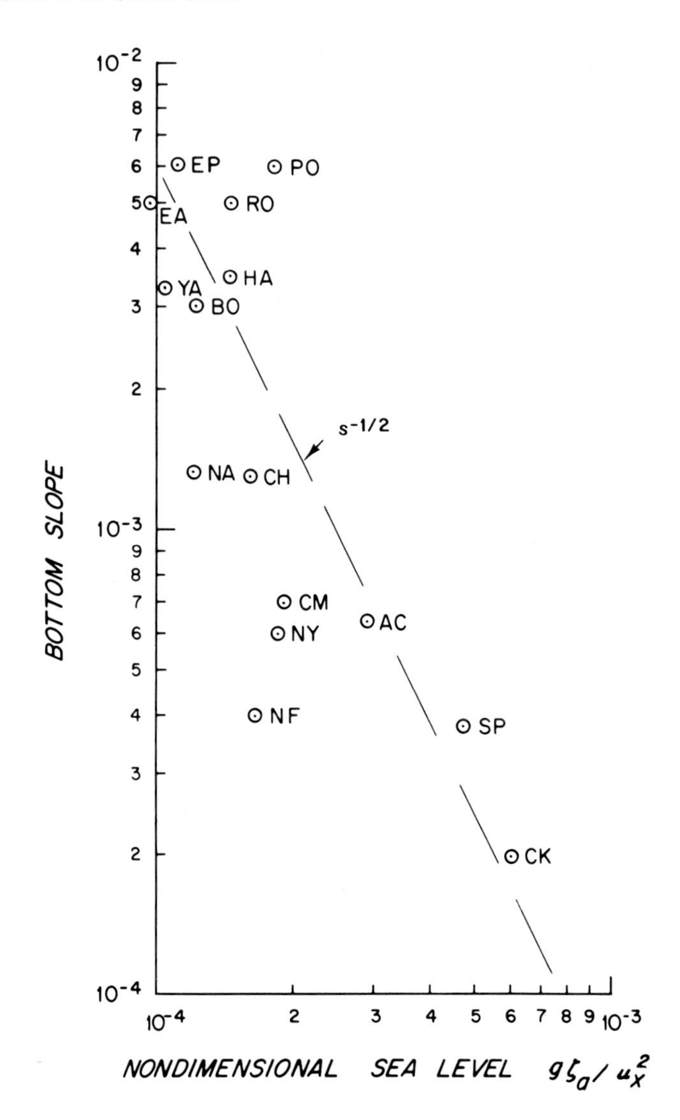

Fig. 8.8. Nondimensional sea level *a* (Equation (8.4)) versus bottom slope for a number of West Florida and east coast shelf tide gauges. Letters identify the gauges (HA = Halifax, CK = Cedar Key, etc.) which are listed in Noble and Butman (1979) and Cragg *et al.* (1982). From Csanady (1981b).

intermediate. There was clearly an elevation gradient opposing the wind during the first (northeastward driving) storm, of a magnitude of about 1.8×10^{-6}.

The second storm had more tightly packed isobars and presumably exerted a somewwhat stronger wind stress, although surface winds at Block Island were about of the same intensity as during the first storm. There was, however, now no opposing pressure

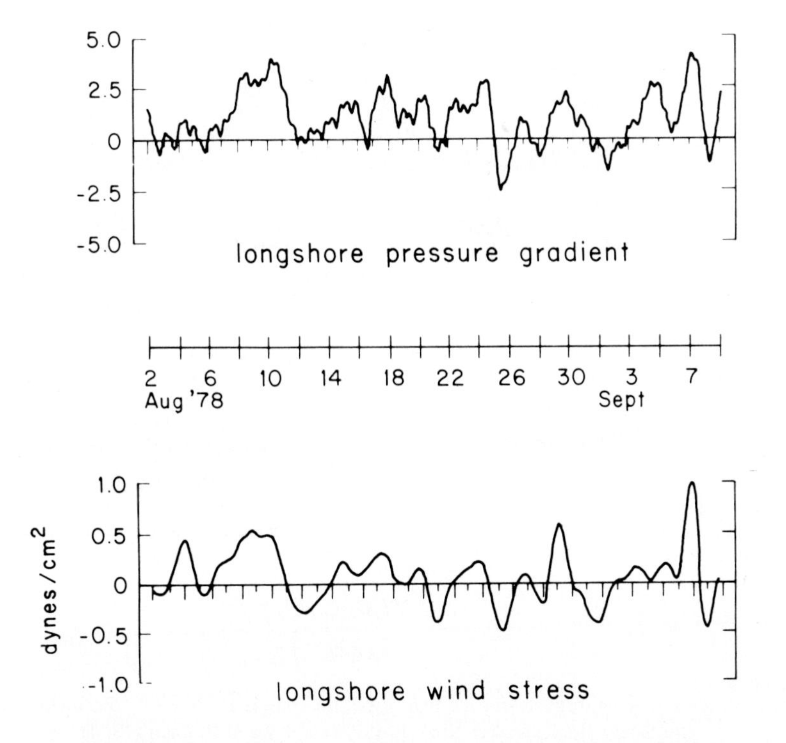

Fig. 8.9. Longshore pressure gradient between Montauk and Sandy Hook (detided), versus wind stress at Tiana Beach on southern coast of Long Island. From Pettigrew (1981).

gradient: the longshore gradient was weak (3×10^{-7}) and its sense was such that it helped the wind drive the water southwestward.

What was the cause of the spectacular asymmetry of the response to oppositely directed longshore winds? The principal observed difference was the sea level at Nantucket,* which showed a small response (10 cm) to the first storm, a large one (60 cm) to the second. According to the boundary layer model of shelf circulation, the sea level at a given location is the integrated result of forcing over the entire 'backward' portion of the coast. The proximte backward sector in the case of Nantucket Island is the coastline of the Gulf of Maine, beginning with Cape Cod Bay and Massachusetts Bay. From the spacing of the isobars it is reasonable to infer that the longshore winds in the Gulf of Maine were generally rather weaker during the first storm than the second. The aggregate effect is difficult to quantify, and yet a 6:1 increase in coastal sea level at Nantucket seems to be difficult to justify on this basis alone. Another possible hypothesis is that cross-shore wind stress also had a significant influence on sea levels. Because, in the Gulf of Maine, this was onshore on the occasion of both storms, it would have tended to

* Nantucket is the island crossed by the 70° W meridian, while Sandy Hook is the tongue of land at 40° 30′ N, 74° W. Montauk is at the eastern tip of Long Island.

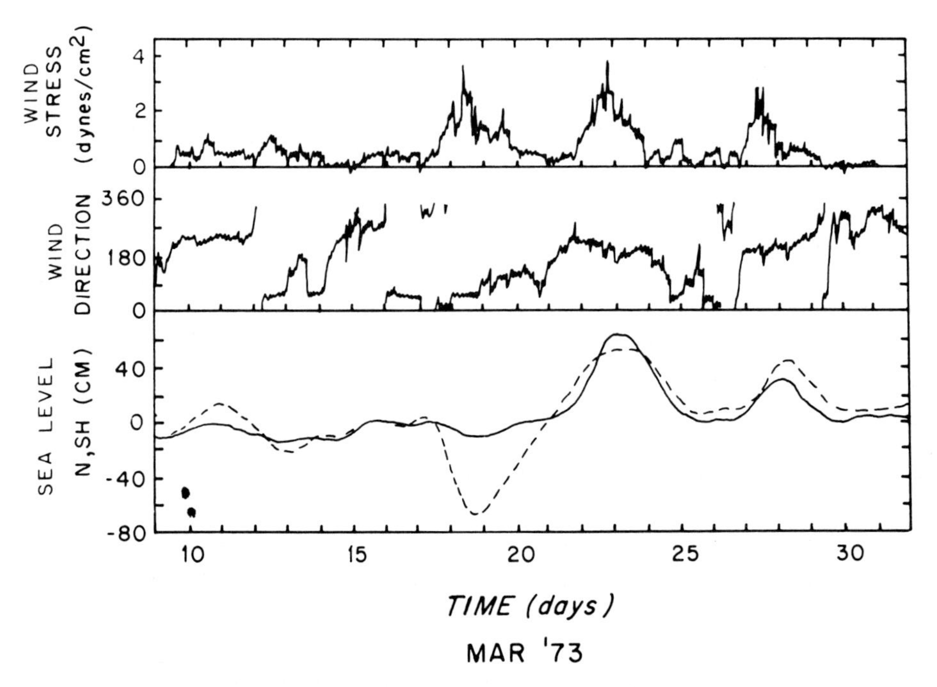

Fig. 8.10. Effects of prototype storms of Figure 8.2 on coastal sea levels at Nantucket (solid line) and Sandy Hook (broken line). From Beardsley and Butman (1974).

decrease the response during the first storm, increase it during the second. An analysis in greater detail is clearly called for, but one is also guided to a further examination of cross-shore wind effects.

8.2.4. Effects of Cross-Shore Winds

The boundary layer theory of shelf circulation predicts a relatively strong response to cross-shore winds when the diameter of a storm is suitably small. The quantitative criterion is kL = order one, which translates into k = order $(fs/2r)$. Given $r = 0.05$ cm s^{-1} and $f = 10^{-4}$ s^{-1} the criterion is equivalent to a longshore half-wavelength π/k of order 30 km for $s = 10^{-3}$, and correspondingly higher or lower for other bottom slopes. A half-wavelength of 100 km is small even for a hurricane. Therefore in the Gulf of Maine, where bottom slopes are 5×10^{-3}, cross-shore wind effects are likely to be important because cross-shore winds of storms of realistic size would be expected to contribute significantly to coastal sea levels. On the other hand, over the flat New Jersey shelf ($s = 0.5 \times 10^{-3}$) such effects would not easily be detected.

When the storm center happens to lie on the coastline, and to move alongshore upcoast or downcoast, longshore winds are negligible and any observable effects on coastal sea level should be attributable to cross-shore winds. Storms with their center moving along the northern coast of the Mid-Atlantic Bight are infrequent, but they do occur, and allow under favorable circumstances a study of cross-shore wind effects more or less in isolation. Such a storm was fortuitously observed during a series of coastal current studies off the south coast of Long Island (Csanady, 1980).

On this occasion a storm center passed eastward along the Long Island coast and was followed by a strong offshore wind stress impulse (0.6 Pa). A nontidal longshore surface level gradient developed 8–10 hr *prior* to the local appearance of the storm, driving southwestward, of a magnitude of about 10^{-6}. This was evident in tide gauge data from Montauk and Sandy Hook, as well as from the development of strong westward flow at all levels, including near the bottom.

These observations may be interpreted in terms of the periodic cross-shore wind stress model of Section 6.7 as follows. Consider the central circulation cell in Figure 6.11, which at large distances from the coast is bounded by $ky = \pm \pi/2$. At the shore, the same cell is phase-shifted to $ky = -0.4$ to $+2.7$, although the wind-stress is still positive (offshore) between $ky = \pm \pi/2$, peaking at $ky = 0$. Suppose this pattern slowly translating toward positive y, corresponding to the eastward movement of the storm along the Long Island coast. Nearshore flow to negative y arises some time prior to the arrival of offshore wind, due to the phase shift of the streamlines near the coast. The longshore flow is, of course, pressure gradient driven at the coast. The time of passage of the wind-stress cell may be estimated to be 27 hr (a speed of $ky = \pi$ radians in 27 hr along the y-axis). Accordingly the early westward current should arise about 9 hr prior to the offshore wind. This was pretty much what showed up in the observations, a rather puzzling fact at first sight.

Quantitatively, the sea level gradient calculated from Equation (8.3) with a stress of 0.6 Pa and an offshore bottom slope of 10^{-3}, for $k = \pi/300$ km is about 0.7×10^{-6}, or close to that observed.

Another instance of longshore gradient generation by cross-shore winds (also due to a storm of small dimensions moving along the New England coast, Figure 8.3) has been documented by Beardsley *et al.* (1977). In these observations bottom pressure transducers were deployed on the sea floor, giving more or less directly both cross-shore and longshore gradients. The results are generally similar to the ones just discussed, the indicated longshore elevation gradient having had an amplitude of about 2×10^{-6}.

Dramatic examples of longshore pressure gradients occur during hurricanes. Redfield and Miller (1957) showed a long time ago that sea level gradients associated with hurricane surges can be an order of magnitude larger than the values quoted above, presumably because they are evoked by a much higher wind stress. A recent article by Smith (1978) documents the development of longshore sea level gradient along the Texas coast, on the fringes of hurricane Anita. Figure 8.11, from Smith's article, shows the track of the hurricane, Figure 8.12 the wind stress history at Pt. Aransas, Figure 8.13 the longshore level difference between two neighboring tide gauges. Allowing for the apparent zero offset of the tide gauges of 25 cm, the maximum sea level difference, *opposing* the wind stress, late on September 1 was about 65 cm, over a 340 km distance, following the occurrence of a 20 cm difference *aiding* the wind stress. Longshore currents observed in 17 m deep water 21.5 km from the coast are shown in Figure 8.14. The longshore bottom stress associated with currents of this intensity had to be of the order of 2 Pa or much larger than the estimated longshore wind-stress (0.3 Pa, Figure 8.12) or the pressure

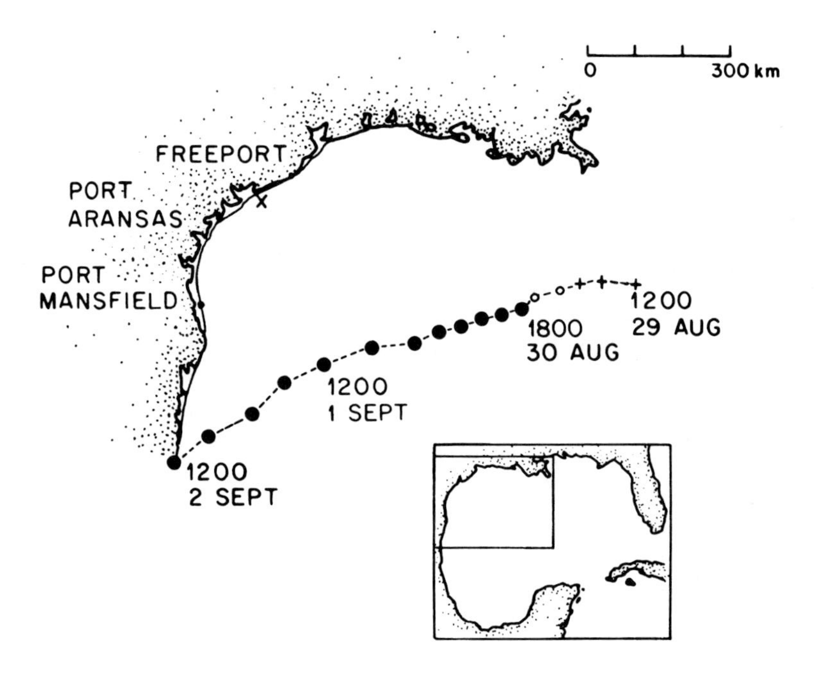

Fig. 8.11. Trajectory of Hurricane Anita, 1977, in the Gulf of Mexico. From Smith (1978).

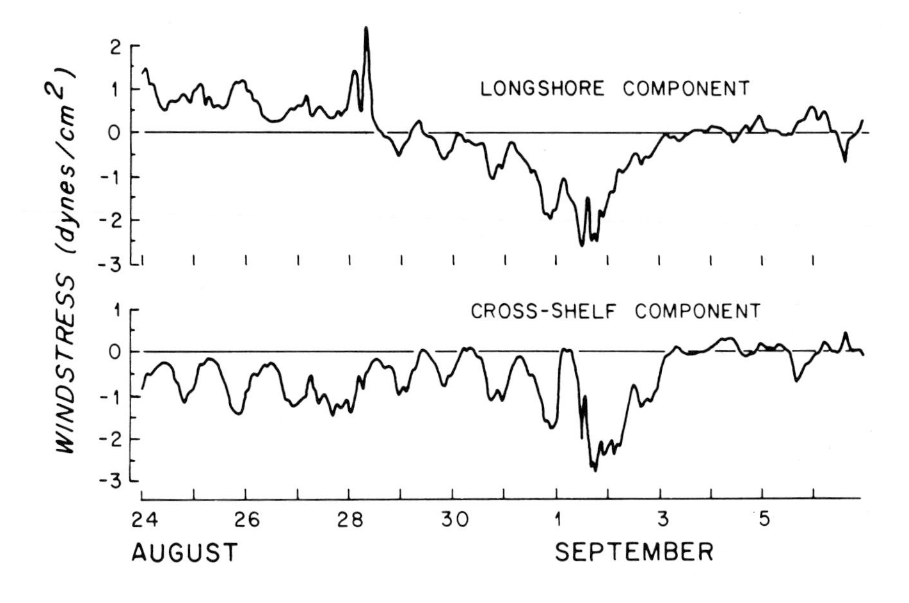

Fig. 8.12. Wind stress history at Port Aransas during passage of Hurricane Anita. From Smith (1978).

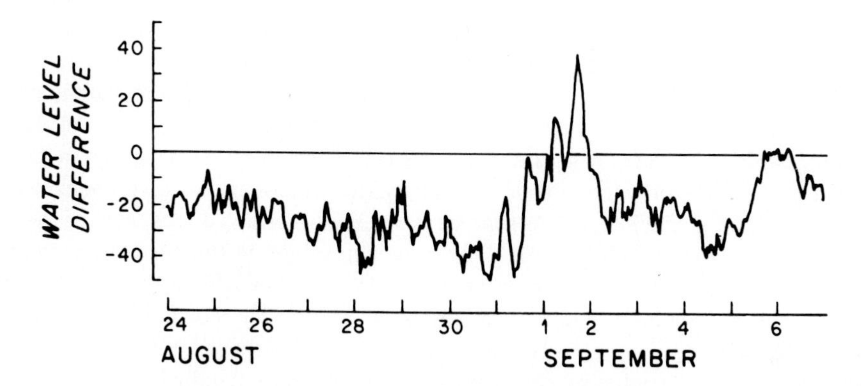

Fig. 8.13. Sea level difference between Freeport and Port Mansfield on the Texas coast during the passage of Hurricane Anita. From Smith (1978).

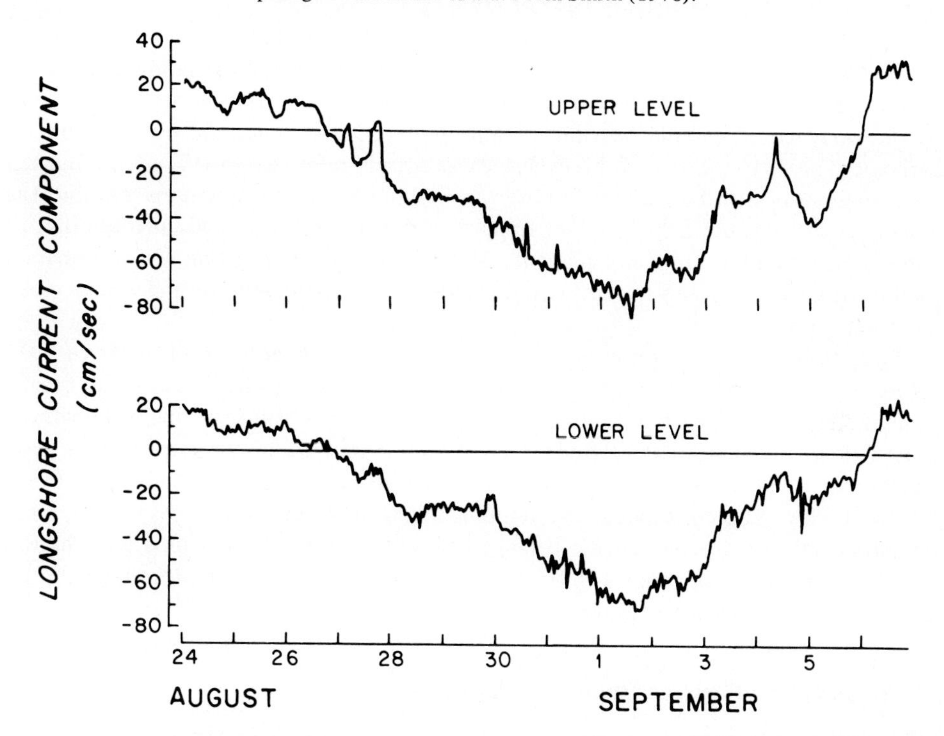

Fig. 8.14. Longshore currents observed 21.5 km from shore (cross in Figure 8.11), 2 and 10 m above bottom in 17 m deep water, during the passage of Hurricane Anita. From Smith (1978).

gradient force associated with the level difference aiding the wind stress (0.1 Pa). It is not quite clear how to resolve the questions regarding longshore momentum balance, but it is certain that during some phase of the hurricane's passage the longshore gradients were aiding the longshore wind in driving longshore currents. Similar conclusions may be derived from the study of Forristal *et al.* (1977) who also find bottom stresses at times considerably in excess of wind stress.

The frequent coincidence of longshore pressure gradient and longshore wind in driving currents under a hurricane may tentatively be ascribed to cross-shore wind effects. Returning again to Figures 6.10 and 6.11, the cell between $ky = 0$ and $ky = \pi$ may be regarded as a crude hurricane model, considering the cross-shore wind component only. A large longshore gradient is set up near $ky = \pi$, i.e., shoreward of the 'eye' of the hurricane. When the eye is well offshore, longshore winds drive in the same direction as the longshore gradient due to cross-shore winds. The longshore winds on their own generate mostly opposing longshore gradients, as discussed earlier. The combination of the two effects is likely to lead to a fairly complex history of coastal currents. Further, more detailed observational studies of similar phenomena are clearly required.

8.3. EVIDENCE OF MOORED INSTRUMENTS

Coastal sea level observations have been taken for many years and the various analyses of subtidal variations quoted above could have been carried out decades ago. More recent is the evidence of moored instrumentation, deployed over inner and outer regions of Atlantic type shelves. Some of this evidence was already referred to briefly above. Bottom pressure sensors placed at varying distances from shore can answer questions regarding the extent of the pressure field (whether trapped or not, for example) and the magnitude of cross-shore as well as longshore gradients. Moored current meter records give information on longshore velocity and transport, and somewhat less accurately, on the weaker cross-shore motions.

In a study of bottom pressure fluctuations in the Mid-Atlantic Bight in March 1974 Beardsley *et al.* (1977) have provided convincing evidence on the presence and coastal trapping of cross-shore pressure gradients. In the frequency band covering wind-induced fluctuations (2 to 13 day period) they found a root-mean-square cross-shore level gradient of about 2.2×10^{-6} between 0 and 30 km from the coast, and half that between 30 and 60 km. This corresponds to a trapping scale L of 43 km. The corresponding r.m.s. longshore velocity by geostrophic balance is about 25 cm s^{-1} close to shore. The rms longshore gradient was about 4×10^{-7}, or of the same order as deduced from coastal sea level observations discussed above.

8.3.1. Longshore Flow and Momentum Balance

The most readily observable result of longshore wind stress is longshore flow. This has been reported on a number of occasions, on the inner shelf for example by Scott and Csanady (1976) or Bennett and Magnell (1979), on the outer shelf by Beardsley and Butman (1974), Beardsley *et al.* (1976), or Boicourt and Hacker (1976). Figure 8.15, from Pettigrew (1981), shows the association of longshore wind stress and depth-average velocity at 3, 6, and 12 km from shore off Long Island. There is an underlying trend in the velocity record unrelated to the wind. However, there is also a clear response to wind, of an amplitude of about 20 cm s^{-1} for a 1 dyne cm^{-2} wind stress.

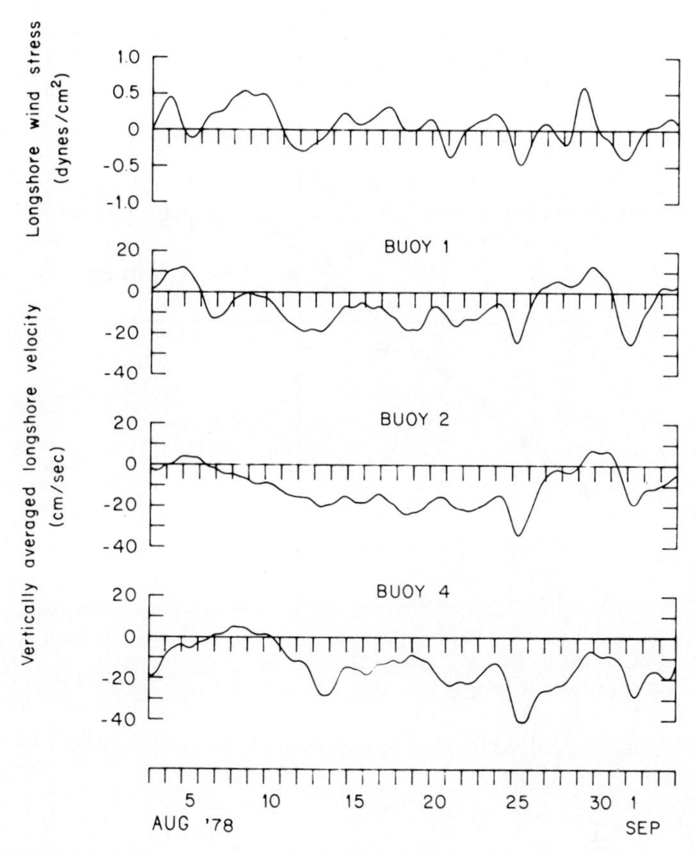

Fig. 8.15. Longshore wind stress and depth-average velocity in coastal boundary layer off south cost of Long Island. From Pettigrew (1981).

At such inner-shelf locations the longshore momentum balance allows some fairly definite deduction on the key parameters locally affecting the flow. A useful technique is to take periods of relatively constant wind, average over a number of tidal cycles, and compare longshore wind stress with near-bottom longshore velocity, the latter being a measure of bottom stress. A comparison of this kind, from Scott and Csanady (1976), is shown in Figure 8.16. The observations were taken 11 km from the coast, where the Coriolis force associated with cross-shore transport is reasonably neglected. The steady longshore momentum balance then reduces to

$$F_y = rv_b + gH \frac{\partial \zeta}{\partial y} \,. \tag{8.7}$$

The analysis of Pettigrew (1981) showed that fluctuations of the longshore gradient $\partial \zeta / \partial y$ were proportional to fluctuations of the wind stress in such a way that, at the coast, a positive change in F_y of 1 cm^2 s^{-2} gave rise to a change in the gradient $\partial \zeta / \partial y$ of

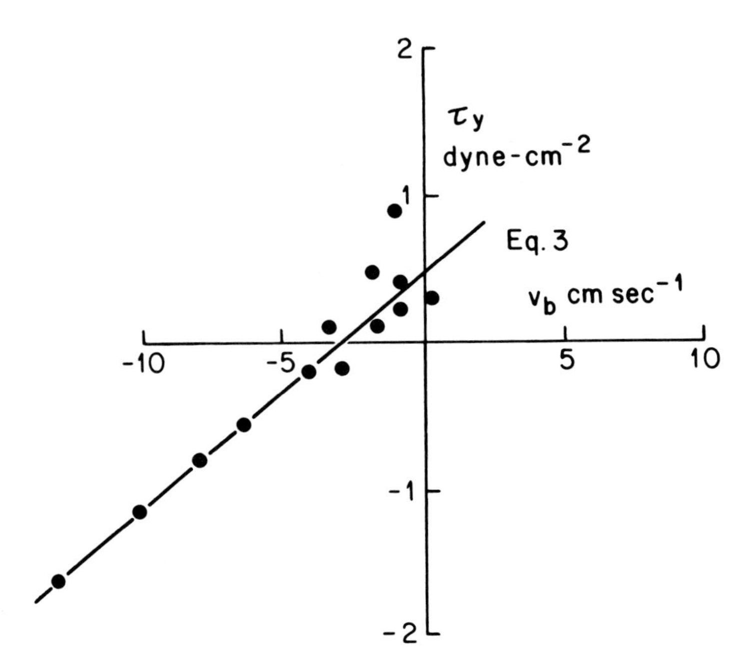

Fig. 8.16. Longshore wind stress τ_y, versus observed near-bottom velocity at a location on the southern coast of Long Island. Adapted from Scott and Csanady (1976).

3×10^{-7}. This linear relationship may be expressed in terms of an 'effective' depth H_e as

$$\frac{\partial \zeta}{\partial y} = \frac{F_y}{gH_e} + G, \qquad (x = 0) \tag{8.8}$$

where $H_e = 33$ m, G an unknown constant, which may be interpreted as sea level gradient due to some effect other than the fluctuating wind. At $x = 11$ km, assuming a trapping scale of about 45 km according to the data of Beardsley *et al.* (1977) the longshore gradient should reduce to some 70% of the shore value given by (8.8) Substituting such a reduced value of $\partial \zeta / \partial y$ into 8.7 one finds

$$F_y \left(1 - \frac{0.7H}{H_e} \right) = rv_b + gHG. \tag{8.9}$$

At the location where the observations shown in Figure 8.16 were taken the actual depth was 32.5 m or about equal to H_e. The equation of the empirical line in Figure 8.16 is

$$F_y = 0.46 + 0.158\, v_b \qquad [\mathrm{cm}^2\, \mathrm{s}^{-2}]. \tag{8.10}$$

Equating the coefficients in (8.9) and (8.10) one obtains

$$r = 0.05 \text{ cm s}^{-1},$$

$$G = 0.5 \times 10^{-7},$$

to one significant digit, in line with the crudeness of the estimates.

Taking this value of r, it is possible to attempt a reconciliation of the various parameters entering the theoretical formulae. The empirically determined value of a (Equation (8.4) and Figure 8.8) in the Mid-Atlantic is between 1 and 2 times 10^{-4}. The first of Equations (8.1) gives with this a and $r = 0.05$ a trapping width of 50 to 100 km, or rather higher than computed from the Beardsley *et al.* (1977) data, but of the same order. Given the crudeness of these estimates and the overidealizations of the simple model, better agreement could hardly be expected.

Substituting into the definition of L (Equation (8.2)), one finds that $r = 0.05$ and $a = 2 \times 10^4$ ($L = 100$ km) is consistent with the previously estimated longshore wavenumber of $k = 2 \times 10^{-8}$ cm and a slope of $s = 10^{-3}$, the latter being a more or less reasonable model of this portion of the shelf.

The long-term mean longshore level gradient of 0.5×10^{-7} estimated from the above data is of the correct order of magnitude and in fact agrees pretty closely with the best estimate mean given by Chase (1979) for the month of September, which was the period of observations used in preparing the graph of Figure 8.16.

8.3.2. Cross-Shore Transport

Some caution is required in relying on longshore momentum balance arguments based on Equation (8.7), because the neglect of the Coriolis force associated with cross-shore transport may not always be justified. In the model calculations of shelf circulation cells carried out in Chapter 6 one of the key signatures of the flow field is the appearance of cross-shore transport at significant amplitudes (comparable to the Ekman transport) even at a fraction of the trapping width from the coast. A cross-shore transport of the order of the Ekman transport of course makes an order one contribution to Equation (8.7). Depending on the phase of any local circulation cell, 10 km from the coast may not be close enough to exclude the possibility of such significant cross-shore transport.

In a series of studies of circulation of the flat continental shelf off the Texas Gulf coast, Smith (1978, 1979, 1980) has drawn attention to the fact that cross-shore transport at distances of the order of 10 km from the coast is dynamically significant in the above sense. To force a two-dimensional mass balance in the cases studied by Smith requires entirely unreasonable assumptions on the velocity distribution.

Much the same proposition was demonstrated by Pettigrew (1981) for the Long Island coast in greater detail. Figure 8.16 shows depth average cross-shore velocities at 3, 6, and 12 km in August, 1978. While the values at 3 km may be noise, the cross-shore transport at 12 km from the shore would be difficult to explain away, at least during some of the more pronounced episodes. Pettigrew attributes the relatively large cross-shore transports to the proximity of the tip of Long Island, which puts the observation site close to a location of longshore wind stress reversal. An inner boundary layer may be thought to grow westward from the tip of Long Island, with a trapping scale of perhaps 20 km near the observation site. Within this boundary layer strong offshore flow would be expected to accompany the prevailing eastward wind episodes.

Summarizing the evidence on wind driven currents over Atlantic type shelves it is clear that many complex and otherwise puzzling phenomena may be understood in the framework of the boundary layer model of shelf circulation. Some important gaps remain, especially relating to thermohaline effects, or more generally, to the response of a stratified shelf. Although important contributions to these outstanding problems have been made in recent years, the knowledge of spring and summer circulation over Atlantic type shelves does not yet appear ready for synthesis. The same is true of the long-term mean circulation in closed basins and Pacific type shelves. Nevertheless, to round out the analysis of these more difficult environments a brief discussion is included of their mean circulation in the remaining final sections of this monograph. This further discussion is taken largely from a recent review article (Csanady, 1981a).

8.4. MEAN CIRCULATION IN LAKE ONTARIO

In enclosed basins typical observationally determined mean velocities are about an order of magnitude weaker than wind-driven transient flow. One complication this gives rise to is a sampling problem: if a monthly average flow pattern is determined in a month which

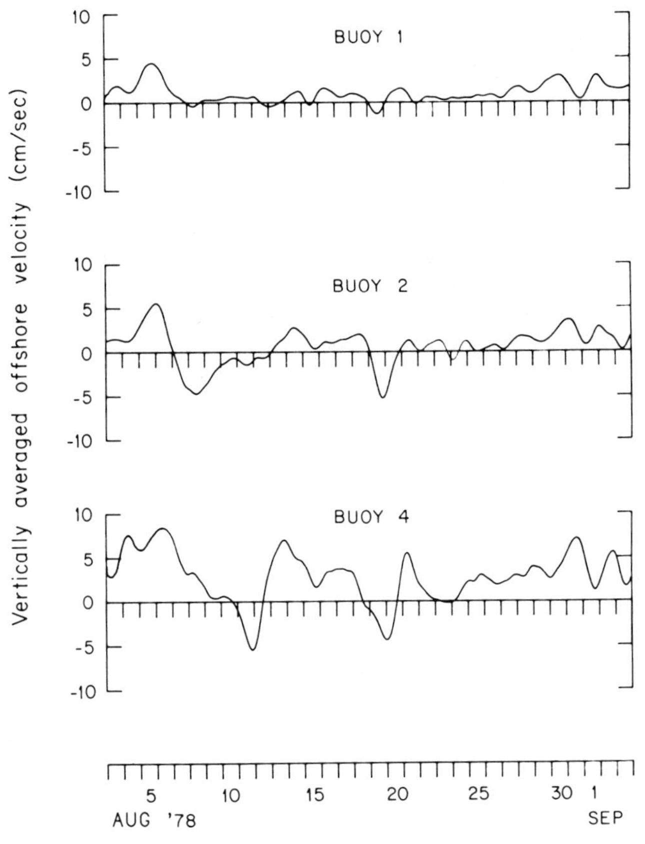

Fig. 8.17. Depth-average cross-shore velocities at a site on the southern coast of Long Island. From Pettigrew (1981).

happens to contain one or two episodes of unusually strong wind in the same directon, the 'mean' pattern will in fact be dominated by the response of the basin to those storms. Furthermore, the response of closed basins includes coastally trapped waves, in Lake Ontario of a period of 12–16 days, see Chapter 5. A monthly mean circulation pattern may therefore be expected to vary, depending on what precise phase of these waves is included in a month-long averaging period. Under summer conditions, storms bring about significant internal redistribution of mass in upwelling-downwelling episodes and subsequent wave or front propagation, the changes taking place again on a time scale much too close to a month not to distort montly averages. When the location of a current meter mooring is alternately occupied by warm and cold water during the month (for periods of the order of a week at a time) the Eulerian average current is a composite of cold and warm water average velocities. Since temperature can be regarded as a particle tracer in a first approximation, this at once shows that Eulerian and Lagrangian averages are likely to be quite different.

Particle average (Lagrangian) velocities in enclosed basins generally tend to show a cyclonic circulation pattern (Emery and Csanady, 1973). For Lake Ontario, this was found to be the case by Harrington (1895) and was confirmed by recent evidence on mirex deposits (Pickett and Dossett, 1979). Eulierian observations, however, do not always show the same pattern. Pickett and Richards (1975) have shown a weak cyclonic mean flow pattern for July 1972, Pickett (1977) a much more pronounced pattern of the same kind for November 1972. However, Pickett (1977) also illustrates the January 1973 mean flow pattern (Figure 8.18 here) which is quite different. Sloss and Saylor (1976) and Saylor and Miller (1979) show cyclonic mean flow patterns in Lake Huron, both for winter and summer, based on current meter studies.

The Lake Ontario mean circulation pattern of January 1973 (Figure 8.18) may be satisfactorily related to the boundary layer model of coastal circulation, driven by eastward wind stress. During this month, alone of the winter months studied during IFYGL,

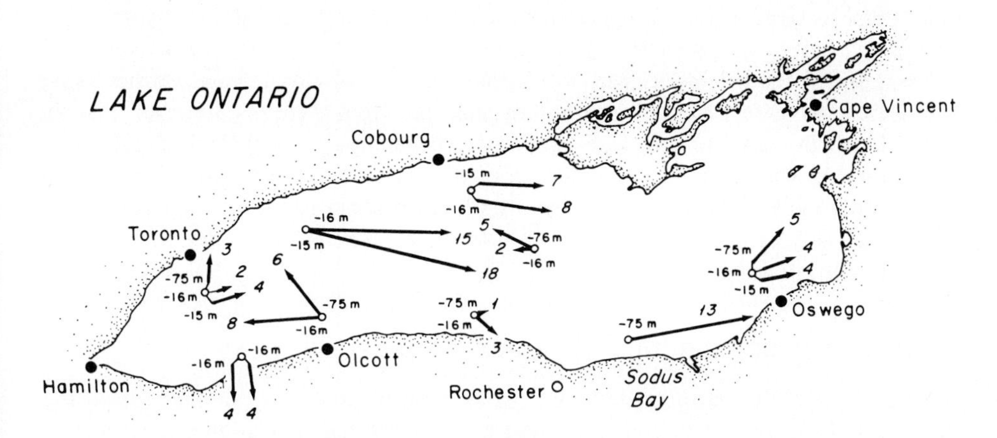

Fig. 8.18. January 1973 mean flow pattern in Lake Ontario, inferred from fixed point current meter observations (Pickett, 1977).

the average wind was strong and westerly. Along both north and south shores this wind apparently generated an eastward longshore current. The current meter data are consistent with the interpretation that the longshore currents increased in width in the cyclonic direction, and flowed around the ends of the basin. Specifically, the westward flow off Olcott may be interpreted as an extension of the north shore anticyclonic current around the western end of the lake, somewhat as a wrapped-around version of the theoretical pattern shown in Figure 6.10, or more directly, as the return flow leg in the circular basin model, Figure 6.13.

Under summer stratified conditions the mean flow pattern is more complex and cannot be satisfactorily described without a detailed exploration of the coastal boundary layer (Csanady and Scott, 1980). The monthly average flow for July 15–August 15, 1972 exhibited a coastal jet pattern similar to that suggested by an inertial model, Figure 2.3, along the south shore of the lake only. Eastward and westward jets in warm water were associated with a thermocline tilt both cross-shore, for geostrophic equilibrium, and longshore. The longshore thermocline tilt corresponded to a pressure gradient not quite sufficient to balance the wind stress. Some momentum was apparently transferred downward by interface friction, which generated a cold water current especially around the eastern end of the lake, again somewhat as suggested by the boundary layer model, Figure 6.13. The flow pattern along the north shore of the lake was quite different from the linear model of Figure 2.3, however, because the return flow of warm water apparently took place some distance offshore. This is also suggested by the July 1972 circulation pattern reported by Pickett and Richards, although that pattern applies to a period only partially overlapping. The July 15–August 15 mean flow was not cyclonic, containing a clear anticyclonic loop in the western basin of the lake, although this was rather smaller than the cyclonic loop occupying the eastern 2/3 of the basin.

The present status of understanding of the mean circulation in enclosed basins may perhaps be summed up as follows. There is a general tendency to cyclonic circulation, which shows up most clearly when the mean wind stress is weak. With a strong mean wind stress, a wind-driven two-gyre component overwhelms the cyclonic pattern, but the cyclonic half of the gyre remains stronger (Bennett, 1977). The boundary layer model and a stratified inertial model partially account for the observed characteristics of the wind-driven two-gyre flow. A conspicuous departure from theory is, however, that the coastal jet sometimes separates from the coast. The tendency to a cyclonic circulation in the absence of wind, and the strengthening of the cyclonic cell with stronger wind stress acting, is not explained satisfactorily by any of the mechanisms so far proposed. Attention should also be drawn to the fact that Eulerian mean currents may be physically meaningless in certain locations.

8.5. MEAN SUMMER CIRCULATION OVER THE OREGON SHELF

Many observations of currents, densities, etc. are available during the summer (upwelling) season off the Oregon shelf and from these a seasonal mean circulation pattern may be pieced together. Because of the relatively large cross-shore excursions of the density front, which generally intersects the surface but sinks sometimes below the surface during this season, there are difficulties in the determination of mean particle velocities from fixed point records. At the same time, surface heating and near-surface and nearshore mixing play an important role in determining the temperature and salinity structure of the coastal water mass, so that temperature and salinity anomalies can only be used with considerable discretion as short-term Lagrangian particle tracers. The main features of the mean summer circulation are nevertheless clear, and have been discussed by Smith *et al.* (1971), Mooers *et al.* (1976a), Smith (1974), Huyer *et al.* (1975), Bryden (1978), Halpern *et al.* (1978), and others.

The longshore components of mean velocities are southward at the surface, strong above the inclined density front, and northward below the surface, in depths of 100 m or more. Mooers *et al.* (1976a) give a schematic illustration of longshore velocity distribution in terms of a coastal jet above the upwelled pycnocline and a poleward undercurrent. The mean velocity of the coastal jet is given by Mooers *et al.* (1976a) as about 20 cm s^{-1}, but this is probably an underestimate, if jet velocity is defined as the average (Lagrangian) velocity of the warm layer above the (usually) upwelled density front. The northward flow below is referred to as the poleward undercurrent and this appears to be trapped over the upper slope, perhaps in depths less than 500 m. Mooers *et al.* also illustrate the typical appearance of the mean isopycnals, although these move about considerably, as already mentioned, and are not easily described in terms of a 'mean' field.

Cross-shore velocities present a more complex picture and have been the subject of considerable controversy. Some of this was no doubt caused by a confusion of Eulerian and Lagrangian means, an acute problem in an upwelling zone where some fixed point current meters sample widely different water masses in the course of upwelling-downwelling events. What is not in doubt is that over most of the water column the cross-shore velocity is directed shoreward most of the time and has an amplitude of about 2 cm s^{-1}. Across the 100 m isobath this implies onshore transport of about 2 m^2 s^{-1}, or about three times more than the offshore Ekman transport at the surface associated with the mean wind stress (Bryden, 1978). There is also some offshore Ekman transport in the bottom boundary layer associated with the poleward undercurrent (Kundu, 1977), but it is very unlikely that this is sufficient to maintain two-dimensional mass balance by transporting away most of the onshore flow arriving throughout the water column. Smith *et al.* (1971), Stevenson *et al.* (1974), and Mooers *et al.* (1976a) also convincingly demonstrate that some of the water drawn from deeper levels is heated at the surface and sinks along isopycnals of the pycnocline when the latter intersects the free surface. This implies offshore motion along some isopycnals, but it is not clear whether the quantity is sufficient to affect the cross-shore mass balance significantly.

A long-term mean onshore velocity of about 2 cm s^{-1}, constant with depth, implies a longshore sea level gradient of 2×10^{-7}, driving northward. Reid and Mantyla (1976) have demonstrated that a longshore gradient of this sign and magnitude may also be inferred from the density field of the North Pacific. In the yearly average this gradient is confined to latitudes south of 38° N, but in the summer it extends to 44° N and thus encompasses the Oregon shelf. The cross-shore sea surface slope associated with the density field extends to about 100 km from the shore, i.e., it coincides with the poleward undercurrent. The longshore sea level slope is associated with a corresponding pycnocline slope. The longshore momentum balance of the poleward undercurrent is thus dominated by a northward driving pressure gradient, balanced in the frictionless interior by onshore flow. Where the onshore flow runs into the continental slope, a northward current develops, with associated bottom friction. The vorticity tendency balance of this current is as discussed in connection with the boundary layer model of shelf circulation, as first pointed out by Pedlosky (1974). The discussion of Reid and Mantyla (1976) also suggests that the longshore isopycnal slope along the west coast of North America is part of a larger-scale response of the North Pacific to wind stress. Halpern *et al.* (1978) confirm this by showing that the undercurrent transports relatively warm and saline water northwards over a considerable range of latitude.

In the surface layers, offshore wind-driven Ekman drift is compensated by onshore flow in a layer of about 30 m depth, i.e., essentially above the main pycnocline. Onshore flow in the deeper layers turns seaward partly in the bottom boundary layer below the undercurrent, but this circulation is confined only to the trapping width of the boundary current ($\sim$50 km) and is in any case insufficient for two-dimensional mass balance. The rest of the inflow is presumably accommodated in a broadening of the boundary current. The need to view the Oregon shelf circulation problem in three dimensions has been pointed out already by O'Brien and his collaborators (O'Brien and Hurlburt, 1972; Thompson and O'Brien, 1973).

In contrast with the case of an enclosed basin, or an Atlantic type shelf discussed above, the mean circulation of the Pacific type shelf is seen to be significantly influenced by the deep ocean.

References

Allen, J. S. (1975), 'Coastal trapped waves in a stratified ocean', *J. Phys. Oceanogr.* **5**, 300–325.

Allen, J. S. and Kundu, P. K. (1978), 'On the momentum, vorticity and mass balance of the Oregon shelf', *J. Phys. Oceanogr.* **8**, 13–27.

Amorocho, J. and Devries, J. J. (1980), 'A new evaluation of the wind stress coefficient over water surfaces', *J. Geophys. Res.* **85**, 433–442.

Ayers, J. C., Chandler, D. C., Lauff, G. H., Power, C. F., Henson, E. B. (1958), *Currents and Water Masses of Lake Michigan*, Univ. Michigan, Great Lakes Res. Div. Publ. No. 3, 169 pp.

Babister, A. W. (1967), *Transcendental Functions*, MacMillan, New York, 414 pp.

Ball, F. K. (1965), 'Second-class motions of a shallow liquid', *J. Fluid Mech.* **23**, 545–561.

Batchelor, G. K. (1967), *An Introduction to Fluid Mechanics*, Cambridge Univ. Press, 615 pp.

Beardsley, R. C. and Butman, B. (1974), 'Circulation on the New England Continental Shelf: response to strong winter storms', *Geophys. Res. Letters* **1**, 181–184.

Beardsley, R. C. and Flagg, C. N. (1976), 'The water structure, mean currents and self-water/slope-water front of the New England continental shelf', *Mém. Soc. Roy. Sci. Liège* **10**, 209–225.

Beardsley, R. C. and Winant, C. D. (1979), 'On the mean circulation in the Mid-Atlantic Bight', *J. Phys. Oceanogr.* **9**, 612–619.

Beardsley, R. C., Boicourt, W. C., and Hansen, D. V. (1976), 'Physical oceanography of the Middle Atlantic Bight', in M. G. Gross (ed.), *Middle Atlantic Continental Shelf and the New York Bight*, American Society of Limnology and Oceanography, Special Symposia, 2, pp. 20–34.

Beardsley, R. C., Mofjeld, H., Wimbush, M., Flagg, C., and Vermersch, J. (1977), 'Ocean tides and weather-induced bottom pressure fluctuations in the Middle Atlantic Bight', *J. Geophys. Res.* **82**, 3175–3182.

Bennett, J. R. (1971), 'Thermally driven lake currents during the spring and fall transition periods', *Proc. 14th Conf. Great Lakes Res.*, Int. Assoc. Great Lakes Res., Ann Arbor, pp. 535–544.

Bennett, J. R. (1977), 'A three-dimensional model of Lake Ontario's summer circulation', *J. Phys. Oceanogr.* **7**, 591–601.

Bennett, J. R. (1978), 'A three-dimensional model of Lake Ontario's summer circulation II. A diagnostic study', *J. Phys. Oceanogr.* **8**, 1095–1103.

Bennett, J. R. and Magnell, B. A. (1979), 'A dynamical analysis of currents near the New Jersey coast', *J. Geophys. Res.* **84**, 1165–1175,

Birchfield, G. E. (1967), 'Horizontal transport in a rotating basin of parabolic depth profile', *J. Geophys. Res.* **72**, 6155–6163.

Birchfield, G. E. (1969), 'The response of a circular model Great Lake to a suddenly imposed wind stress', *J. Geophys. Res.* **74**, 5547–5554.

Birchfield, G. E. and Davidson, D. R. (1967), *A Case Study of Coastal Currents in Lake Michigan*, Proc. 10th Conf. Great Lakes Res., Intern. Assoc. Great Lakes Res., pp. 264–273.

Birchfield, G. E. and Hickie, B. P. (1977), 'The time-dependent response of a circular basin of variable depth to a wind stress', *J. Phys. Oceanogr.* **7**, 691–701.

Blanton, J. O. (1974), 'Some characteristics of nearshore currents along the north shore of Lake Ontario', *J. Phys. Oceanogr.* **4**, 415–424.

Blanton, J. O. (1975), 'Nearshore lake currents measured during upwelling and downwelling of the thermocline in Lake Ontario', *J. Phys. Oceanogr.* **5**, 111–124.

Boicourt, W. C. and Hacker, P. W. (1976), 'Circulation on the Atlantic continental shelf of the United States, Cape May to Cape Hatteras', *Mém. Soc. Roy. des Sci. de Liège* **5**(X), 187–200.

Boyce, F. M. (1974), 'Some aspects of Great Lakes physics of importance to biological and chemical processes', *J. Fish. Res. Board Can.* **31**, 689–730.

Boyce, F. M. (1977), 'Response of the coastal boundary layer on the north shore of Lake Ontario to a fall storm', *J. Phys. Oceanogr.* **7**, 719–732.

Bretschneider, C. L. (1966), 'Engineering Aspects of the Hurricance Surge', in A. T. Ippen (ed.), *Estuarine and Coastline Hydrodynamics*, McGraw Hill Book Co., pp. 231–256.

Bryden, H. L. (1978), 'Mean upwelling velocities on the Oregon continental shelf during summer 1973', *Est. Coast. Mar. Sci.* **7**, 311–327.

Bumpus, D. F. (1973), 'A description of the circulation on the continental shelf of the east coast of the United States', *Progress in Oceanography* **6**, 111–157.

Cahn, A. (1945), 'An investigation of the free oscillations of a simple current system', *J. Met.* **2**, 113–119.

Cardone, V. J. Pierson, W. J., and Ward, E. G. (1976), 'Hindcasting the directional spectra of hurricane-generated waves', *J. Pet. Tech.* **28**, 385–396.

Carrier, G. F. (1953), 'Boundary layer problems in applied mechanics', *Adv. Appl. Mech.* **3**, 1–19.

Carslaw, H. S. and Jaeger, J. C. (1959), *Conduction of Heat in Solids*, 2nd ed. Oxford University Press, 510 pp.

Chandrasekhar, S. (1961), *Hydrodynamic and Hydromagnetic Stability*, Oxford Univ. Press, 652 pp.

Charney, J. G. (1955), 'Generation of oceanic currents by wind', *J. Mar. Res.* **14**, 477–498.

Chase, R. R. P. (1979), 'The coastal longshore pressure gradient: temporal variations and driving mechanisms', *J. Geophys. Res.* **84**, 4898–4905.

Church, P. E. (1945), *The Annual Temperature Cycle of Lake Michigan. II. Spring Warming and Summer Stationary Periods, 1942.* Univ Chicago, Inst. Meteorol. Misc. Rep. 18. 100 pp.

Clarke, A. J. (1977), 'Observational and numerical evidence for wind-forced coastal trapped long waves', *J. Phys. Oceanogr.* **7**, 231–247.

Collins, C. A. and Patullo, J. G. (1970), 'Ocean currents above the continental shelf off Oregon as measured with a single array of current meters', *J. Mar. Res.* **28**, 51–68.

Collins, C. A., Mooers, C. N. K., Stevenson, M. R., Smith, R. L., and Patullo, J. G. (1968), 'Direct current measurements in the frontal zone of a coastal upwelling region', *J. Oceanogr. Soc. Japan* **24**, 295–306.

Courant, R. and Hilbert, D. (1953), *Methods of Mathematical Physics*, Vol. I, Interscience Publishers, 561 pp.

Courant, R. and Hilbert, D. (1962), *Methods of Mathematical Physics*, Vol. II, Interscience Publishers, 830 pp.

Cragg, J., Sturges, W., and Mitchum, G. (1982), 'Wind-induced surface slopes on the West Florida shelf', *J. Phys. Oceanogr.* **12**, in press.

Crépon, M. (1967), 'Hydrodynamique marine en regime impulsionnel', *Cah. Oceanogr.* **19**, 847–880.

Crépon, M. (1969), 'Hydrodynamique marine en regime impulsionnel', *Cah. Oceanogr.* **21**, 333–353.

Csanady, G. T. (1968), 'Wind-driven summer circulation in the Great Lakes', *J. Geophys. Res.* **73**, 2579–2589.

Csanady, G. T. (1970), 'Dispersal of effluents in the Great Lakes', *Water Res.* **4**, 79–114.

Csanady, G. T. (1971), 'Baroclinic boundary currents and long-edge waves in basins with sloping shores', *J. Phys. Oceanogr.* **1**, 92–104.

Csanady, G. T. (1972a), 'Frictional currents in the mixed layer at the fee surface', *J. Phys. Oceanogr.* **2**, 498–508.

Csanady, G. T. (1972b), 'Response of large stratified lakes to wind', *J. Phys. Oceanogr.* **2**, 3–13.

Csanady, G. T. (1973), 'Wind-induced barotropic motions in long lakes', *J. Phys. Oceanogr.* **3**, 429–438.

Csanady, G. T. (1974a), 'Barotropic currents over the continental shelf', *J. Geophys. Res.* **4**, 357–371.

Csanady, G. T. (1974b), 'Spring thermocline behavior in Lake Ontario during IFYGL', *J. Phys. Oceanogr.* **3**, 425–445.

Csanady, G. T. (1974c), 'Mass exchange episodes in the coastal boundary layer, associated with current reversals', *Rapp. P.-v. Rèun. Cons. Int. Explor. Mer.* **167**, 41–45.

Csanady, G. T. (1974d), *Reply J. Phys. Oceanogr.* **4**, 271–273.

Csanady, G. T. (1975), 'Lateral momentum flux in boundary currents', *J. Phys. Oceanogr.* **5**, 705–717.

Csanady, G. T. (1976a), 'Mean circulation in shallow seas', *J. Geophys. Res.* **81**, 5389–5399.

Csanady, G. T. (1976b), 'Topographic waves in Lake Ontario', *J. Phys. Oceanogr.* **6**, 93–103.

Csanady, G. T. (1977a), 'The coastal jet conceptual model in the dynamics of shallow seas', in E. D. Goldberg, I. N. McCave, J. J. O'Brien, and J. H. Steele (eds.), *The Sea*, Vol. 6, John Wiley, New York, pp. 117–144.

Csanady, G. T. (1977b), 'Intermittent "full" upwelling in Lake Ontario', *J. Geophys. Res.* **82**, 397–419.

Csanady, G. T. (1978a), 'The arrested topographic wave', *J. Phys. Oceanogr.* 8, 47–62.

Csanady, G. T. (1978b), 'Wind effects on surface to bottom fronts', *J. Geophys. Res.* **83**, 4633–4640.

Csanady, G. T. (1978c), 'Turbulent interface layers', *J. Geophys. Res.* **83**, 2329–2342.

Csanady, G. T. (1979), 'The pressure field along the western margin of the North Atlantic', *J. Geophys. Res.* **84**, 4905–4914.

Csanady, G. T. (1980), 'Longshore pressure gradients caused by offshore wind', *J. Geophys. Res.* **85**, 1076–1084.

Csanady, G. T. (1981a), 'Circulation in the coastal ocean', in B. Saltzman (ed.), *Advances in Geophysics* **23**, 101–183.

Csanady, G. T. (1981b), 'Shelf circulation cells', *Phil. Trans. Roy. Soc.* **A302**, 515–530.

Csanady, G. T. (1981c), 'Circulation in the coastal ocean', *EOS* **62**, 9–11, 41–43, 73–75.

Csanady, G. T. and Scott, J. T. (1974), 'Baroclinic coastal jets in Lake Ontario during IFYGL', *J. Phys. Oceanogr.* **4**, 524–541.

Csanady, G. T. and Scott, J. T. (1980), 'Mean summer circulation in Lake Ontario within the coastal zone', *J. Geophys. Res.* **85**, 2797–2812.

Cutchin, D. L. and Smith, R. L. (1973), 'Continental shelf waves: low frequency variations in sea level and currents over the Oregon continental shelf', *J. Phys. Oceanogr.* **3**, 73–82.

Eckart, C. (1960), *Hydrodynamics of Oceans and Atmospheres*, Pergamon Press, Macmillan, New York, 290 pp.

Ekman, V. W. (1905), 'On the influence of the earth's rotation on ocean-currents', *Arkiv för Matematik, Astronomi och Fysik* **2:11**, 52 pp.

Elliott, G. H. (1971), 'A mathematical study of the thermal bar', *Proc. 14th Conf. Great Lakes Res.*, Int. Assoc. Great Lakes Res., pp. 545–554.

Emery, K. O. and Csanady, G. T. (1973), 'Surface circulation of lakes and nearly land-locked seas', *Proc. Nat. Acad. Sci. USA* **70**, 93–97.

Fischer, H. B. (1980), 'Mixing processes on the Atlantic continental shelf, Cape Cod to Cape Hatteras', *Limnol. Oceanogr.* **25**, 114–125.

Flagg, C. N. (1977), 'The kinematics and dynamics of the New England continental shelf and shef/slope front', Ph.D. Thesis, Massachusetts Institute of Technology/Woods Hole Oceanographic Institution, WHOI Ref. 77–67, 207 pp.

Forristal, G. Z., Hamilton, R. C., and Cardone, V. J. (1977), 'Continental shelf currents in tropical storm Delia: observations and theory', *J. Phys. Oceanogr.* **7**, 532–546.

Freeman, J. C., Baer, L., and Jung, G. H. (1957), 'The bathystrophic storm tide', *J. Mar. Res.* **16**, 12–22.

Gill, A. E. and Schumann, E. H. (1974), 'The generation of long shelf waves by the wind', *J. Phys. Oceanogr.* **4**, 83–90.

Gonella, J. (1971), 'The drift current from observations on the Bouée Laboratoire', *Cah. Oceanogr.* **23**, 19–33.

Greenspan, H. P. (1968), *The Theory of Rotating Fluids*, Cambridge Univ. Press, 327 pp.

Halpern, D. (1974), 'Variations in the density field during coastal upwelling', *Tethys.* **6**, 363–374.

Halpern, D. (1976), 'Structure of a coastal upwelling event observed off Oregon during July 1973', *Deep-Sea Res.* **23**, 495–508.

Halpern, D., Smith, R. L., and Reed, R. K. (1978), 'On the California under-current over the continental slope off Oregon', *J. Geophys. Res.* **83**, 1366–1372.

Hamon, B. V. (1962), 'The spectrums of mean sea level at Sydney, Coff's Harbor and Lord Howe Island', *J. Geophys. Res.* **67**, 5147–5155.

Hansen, D. V. (1977), 'Circulation', *MESA New York Bight Atlas Monograph* **3**, New York Sea Grant Institute, Albany, N.Y., 23 pp.

Harrington, M. W. (1895), 'Surface currents of the Great Lakes, as deduced from the movements of bottle papers during the season of 1892, 1893, and 1894', *Bull. B.* (revised) *U.S. Weather Bur.*, Washington, D. C., 23 pp.

Heaps, N. S. (1969), 'A two-dimensional numerical sea model', *Phil. Trans. Roy. Soc. London* **265**, 93–137.

Hendershott, M. and Rizzoli, P. (1976), 'The winter circulation of the Adriatic Sea', *Deep-Sea Res.* **23**, 353–370.

Hickey, B. M. and Hamilton, P. (1980), 'A spin-up model as a diagnostic tool for interpretation of current and density measurements on the continental shelf of the Pacific Northwest', *J. Phys. Oceanogr.* **10**, 12–24.

Horne, E. P. W., Bowman, M. J., and Okubo, A. (1978), 'Cross frontal mixing and cabbeling', in M. J. Bowman and W. B. Esaias (eds.), *Oceanic Fronts in Coastal Processes*, Springer-Verlag, pp. 105–113.

Hsueh, Y. and Peng, C. Y. (1978), 'A diagnostic model of continental shelf circulation', *J. Geophys. Res.* **83**, 3033–3041.

Huang, J. C. K. (1971), 'The thermal current in Lake Michigan', *J. Phys. Oceanogr.* **1**, 105–122.

Huthnance, J. M. (1975), 'On trapped waves over a continental shelf', *J. Fluid Mech.* **69**, 689–704.

Huthnance, J. M. (1978), 'On coastal trapped waves: analysis and numerical calculation by inverse iteration', *J. Phys. Oceanogr.* **8**, 74–92.

Huyer, A. and Patullo, J. G. (1972), 'A comparison between wind and current observations over the continental shelf off Oregon, summer 1969', *J. Geophys. Res.* 77, 3215–3220.

Huyer, A., Smith, R. L., and Pillsbury, R. D. (1974), 'Observations in a coastal upwelling region during a period of variable winds (Oregon coast, July 1972)', *Tethys.* **6**, 391–404.

Huyer, A., Hickey, B. M., Smith, J. D., Smith, R. L., and Pillsbury, R. D. (1975), 'Alongshore coherence at low frequencies in currents observed over the continental shelf off Oregon and Washington', *J. Geophys. Res.* **80**, 3495–3505.

Huyer, A., Smith, R. L., Sobey, E. J. C. (1978), 'Seasonal differences in low-frequency current fluctuations over the Oregon continental shelf', *J. Geophys. Res.* **83**, 5077–5089.

Huyer, A., Sobey, E. J. C. and Smith, R. L. (1979), 'The spring transition in currents over the Oregon continental shelf', *J. Geophys. Res.* **84**, 6995–7011.

Irbe, J. G. and Mills, R. J. (1976), 'Aerial surveys of Lake Ontario water temperature and description of regional weather conditions during IFYGL – January, 1972 to March, 1973', CLI 1–76, Atmos. Env. Service, Env. Canada, 151 pp.

Jeffreys, H. and Jeffreys, B. S. (1956), *Methods of Mathematical Physics*, Cambridge Univ. Press, 714 pp.

Jelesnianski, C. P. (1965), 'Numerical computations of storm surges without bottom stress', *Monthly Weather Rev.* **93**, 343–358.

Kao, T. W. Pao, H. P. and Park, C. (1978), 'Surface intrusions, fronts and internal waves: a numerical study', *J. Geophys. Res.* **83**, 4641–4650.

Keulegan, G. H. (1966), 'The mechanism of an arrested saline wedge', A. T. Ippen (ed.), *Estuary and Coastline Hydrodynamics*, McGraw-Hill, New York.

Krauss, W. (1966), *Interne Wellen*, Geb. Borntraeger, Berlin, 248 pp.

Kundu, P. K. (1977), 'On the importance of friction in two typical continental shelf waters: off Oregon and Spanish Sahara', in J. C. G. Nihoul (ed.), *Bottom Turbulence*, Elsevier Scientific Publishing Co., pp. 187–208.

Kundu, P. K. and Allen, J. S. (1976), 'Some three-dimensional characteristics of low-frequency current fluctuations near the Oregon coast', *J. Phys. Oceanogr.* **6**, 181–199.

Kundu, P. K. Allen, J. S., and Smith, R. L. (1975), 'Modal decomposition of the velocity field near the Oregon coast', *J. Phys. Oceanogr.* **5**, 683–704.

Lamb, H. (1932), *Hydrodynamics*, 6th ed. Dover, New York, 738 pp.

Lighthill, M. J. (1969), 'Dynamic response of the Indian Ocean to onset of the southwest monsoon', *Phil. Trans. Roy. Soc.* **A265**, 45–92.

Longuet-Higgins, M. S. and Stewart, R. W. (1964), 'Radiation stresses in water waves; a physical discussion with applications', *Deep-Sea Res.* **11**, 529–562.

Malone, F. D. (1968), 'An analysis of current measurements in Lake Michigan', *J. Geophys. Res.* **73**, 7065–7081.

Marmorino, G. O. (1978), 'Intertial currents in Lake Ontario, Winter 1972–73 (IFYGL)', *J. Phys. Oceanogr.* **8**, 1104–1120.

Marmorino, G. O. (1979), 'Low-frequency current fluctuations in Lake Ontario, Winter 1972–73 (IFYGL)', *J. Geophys. Res.* **84**, 1206–1214.

Mayer, D. A., Hansen, D. V., and Ortman, D. A. (1979), 'Long term current and temperature observations on the Middle Atlantic shelf', *J. Geophys. Res.* **84**, 1776–1792.

Monin, A. S. and Yaglom, A. M. (1971), *Statistical Fluid Mechanics*, MIT Press, Cambridge, Mass., 769 pp.

Mooers, C. N. K. and Smith R. L. (1968), 'Continental shelf waves off Oregon', *J. Geophys. Res.* **73**, 549–557.

Mooers, C. N. K., Collins, C. A., and Smith R. L. (1976a), 'The dynamic structure of the frontal zone in the coastal upwelling region off Oregon', *J. Phys. Oceanogr.* **6**, 3–21.

Mooers, C. N. K., Fernandez-Partagas, J., and Price, J. F. (1976b), 'Meteorological forcing fields of the New York Bight', Technical Report, Rosenstiel School of Marine and Atmospheric Science, University of Miami, TR76–8, Miami, Florida, 151 pp.

Morse, P. M. and Feshbach, (1953), *Methods of Theoretical Physics*, McGraw-Hill, New York, 2 Vols., 1978 pp.

Mortimer, C. H. (1963), 'Frontiers in physical limnology with particular reference to long waves in rotating basins', Publ. No. 10, Great Lakes Res. Div., Univ. Michigan, pp. 9–42.

Mortimer, C. H. (1968), 'Internal waves and associated currents observed in Lake Michigan during the summer of 1963', Special Rep. No. 1, Ctr. for Great Lakes Studies, Univ. Wisc., Milwaukee.

Murthy, C. R. (1970), 'An experimental study of horizontal diffusion in Lake Ontario', *Proc. 13th Conf. Great Lakes Res.*, Int. Assoc. Great Lakes Res., pp. 477–489.

Murty, T. S. and D. B. Rao (1970), 'Wind-generated circulations in Lakes Erie, Huron, Michigan, and Superior', *Proc. 13th Conf. Great Lakes Res.*, Intern. Assoc. Great Lakes Res., pp. 927–941.

Myers, E. P. (1981), in E. P. Myers (ed.), *Ocean Disposal of Municipal Wastewater: Impact on Estuarine and Coastal Waters*, MIT Press, approx. 500 pp.

Mysak, L. A. (1980), 'Topographically trapped waves', *Ann. Rev. Fluid Mech.* **12**, 45–76.

Noble, M. and Butman, B. (1979), 'Low frequency wind-induced sea level oscillations along the east coast of North America', *J. Geophys. Res.* **84**, 3227–3236.

O'Brien, J. J. and Hurlburt, H. E. (1972), 'A numerical model of coastal upwelling', *J. Phys. Oceanogr.* **2**, 14–25.

Pedlosky, J. (1974), 'Longshore currents, upwelling, and bottom topography', *J. Phys. Oceanogr.* **4**, 214–226.

Pedlosky, J. (1979), *Geophysical Fluid Dynamics*, Springer-Verlag, New York, 624 pp.

Pettigrew, N. R. (1980), 'The dynamics and kinematics of the coastal boundary layer off Long Island', Ph. D. Thesis, Massachusetts Institute of Technology/Woods Hle Oceanography Institution, 262 pp.

Pickett, R. L. (1977), 'The observed winter circulation of Lake Ontario', *J. Phys. Oceanogr.* **7**, 152–156.

Pickett, R. L. and Dossett, D. A. (1979), 'Mirex and the circulation of Lake Ontario', *J. Phys. Oceanogr.* **9**, 441–445.

Pickett, R. L. and Richards, F. P. (1975), 'Lake Ontario mean temperatures and mean currents in July 1972', *J. Phys. Oceanogr.* **5**, 775–781.

Platzman, G. W. (1972), 'Two-dimensional free oscillations in natural basins', *J. Phys. Oceanogr.* **2**, 117–138.

Platzman, G. W. and Rao, D. B. (1964), 'The free oscillations of Lake Erie', in Yoshida (ed.), *Studies on Oceanography*, Univ. Washington Press, 359–382.

Prandtl. L. (1965), *Führer durch die Strömungslehre*. Fr. Vieweg & Sohn, Braunschweig, 523 pp.

Pritchard-Carpenter (1965), 'Drift and Dispersion Characteristics of Lake Ontario's Nearshore Waters, Rochester, N.Y. to Sodus Bay, N.Y.', Unpublished Report, Pritchard-Carpenter, 208 MacAlpine Rd., Elliott City, Maryland.

Proudman J. (1953), *Dynamical Oceanography*, Methuen, London, and John Wiley & Sons, New York, 409 pp.

Rao, D. B. and Murthy, T. S. (1970), 'Calculation of the steady-state wind-driven circulation in Lake Ontario', *Arch. Meteor. Geophys. Bioklim.* **A19**, 195–210.

Redfield, A. C. and Miller, A. R. (1957), 'Water levels accompanying Atlantic coast hurricanes', *Meteorol. Monogr.* **2**(10), 1–23.

Reid, J. L. and Mantyla, A. W. (1976), 'The effects of the geostrophic flow upon coastal sea elevations in the northern Pacific Ocean', *J. Geophys. Res.* **81**, 3100–3110.

Robinson, A. R. (1964), 'Continental shelf waves and the response of sea level to weather systems', *J. Geophys. Res.* **69**, 367–368.

Rockwell, D. C. (1966), 'Theoretical free oscillations of the Great Lakes', *Proc. 9th Conf. Great Lakes Res.*, Publ. No. 15, Great Lakes Res. Div., Univ. Michigan, pp. 352–368.

Rodgers, G. K. (1965), 'The thermal bar in Lake Ontario, Spring 1965 and Winter 1965–66', *Proc. 9th Conf. Great Lakes Res.*, Great Lakes Res. Div. Publ. No. 15, Univ. Michigan, pp. 369–374.

Roll, H. U. (1965), *Physics of the Marine Atmosphere*, Academic Press, New York, 426 pp.

Rossby, C. -G. (1938), 'On the mutual adjustment of pressure and velocity distributions in certain simple current systems, II', *J. Mar. Res.* **1**, 239–263.

Saltzman, B. (ed.) (1962), *Selected papers on the Theory of Thermal Convection*, Dover Publications, 461 pp.

Sandstrom, H. (1980), 'On the wind-induced sea level changes on the Scotian Shelf', *J. Geophys. Res.* **85**, 461–468.

Sato, G. K. and Mortimer, C. H. (1975), 'Lake currents and temperatures near the western shore of Lake Michigan', Univ. Wisconsin, Milwaukee, Center for Great Lakes Studies. Spec. Rep. No. 22.

Saylor, J. H. and Miller, G. S. (1979), 'Lake Huron winter circulation', *J. Geophys. Res.* **84**, 3237–3252.

Saylor, J. H., Huang, J. C. K., and Reid, R. O. (1980), 'Vortex modes in southern Lake Michigan', *J. Phys. Oceanogr.* **10**, 1814–1823.

Scorer, R. S. (1978), *Environmental Aerodynamics*, Ellis Harwood Ltd., John Wiley & Sons, 488 pp.

Scott, J. T. and Csanady, G. T. (1976), 'Nearshore currents off Long Island', *J. Geophys. Res.* **81**, 5401–5409.

Simons, T. J. (1973), 'Comparison of observed and computed currents in Lake Ontario during Hurricane Agnes, June 1972', *Proc. 16th Conf. Great Lakes Res.*, Intern. Assoc. Great Lakes Res., Ann Arbor, pp. 831–844.

Simons, T. J. (1974), 'Verification of numerical models of Lake Ontario: Part I Circulation in spring and early summer', *J. Phys. Oceanogr.* **4**, 507–523.

Simons, T. J. (1975), 'Verification of numerical models of Lake Ontario: II Stratified circulations and temperature changes', *J. Phys. Oceanogr.* **5**, 98–110.

Simons, T. J. (1976), 'Verification of numerical models of Lake Ontario: III Long term heat transports', *J. Phys. Oceanogr.* **6**, 372–378.

Sloss, P. W. and Saylor, J. H. (1976), 'Large-scale current measurements in Lake Huron', *J. Geophys. Res.* **81**, 3069–3078.

Smith, N. P. (1978), 'Longshore currents on the fringe of Hurricane Anita', *J. Geophys. Res.* **83**, 6047–6051.

Smith, N. P. (1979), 'An investigation of vertical structure in shelf circulation', *J. Phys. Oceanogr.* **9**, 624–630.

Smith, N. P. (1980), 'Temporal and spatial variability in longshore motion along the Texas Gulf Coast', *J. Geophys. Res.* **85**, 1531–1536.

Smith, R. L. (1974), 'A description of current, wind and sea-level variations during coastal upwelling off the Oregon coast, July–August 1972', *J. Geophys. Res.* **79**, 435–443.

Smith, R. L., Patullo, J. G., and Lane, R. K. (1966), 'Investigation of the early stage of upwelling along the Oregon coast', *J. Geophys. Res.* **71**, 1135–1140.

Smith, R. L., Mooers, C. N. K., and Enfield, D. B. (1971), 'Mesoscale studies of the physical oceanography in two coastal upwelling regions: Oregon and Peru', in Costlow (ed.), *Fertility of the Sea*, Vol. 2, Gorden and Breach, New York, pp. 515–535.

Stevenson, M. R., Garvine, R. W., and Wyatt, B. (1974), 'Lagrangian measurements in a coastal upwelling zone off Oregon', *J. Phys. Oceanogr.* **4**, 321–330.

Stommel, H. (1965), *The Gulf Stream: A Physical and Dynamical Description*, 2nd ed. Univ. of California Press, Berkeley, 248 pp.

Stommel, H. and Leetmaa, A. (1972), 'Circulation on the continental shelf', *Proc. Nat. Acad. Sci., USA* **69**, 3380–3384.

Sverdrup, H. U., Johnson, M. W., and Fleming, R. H. (1942), *The Oceans: Their Physics, Chemistry, and General Biology*, Prentice-Hall, Englewood Cliffs, N.J., 1087 pp.

Taylor, G. I. (1920), 'Tidal oscillations in gulfs and rectangular basins', *Proc. London Math. Soc.* **20**, 148–181.

Taylor, G. I. (1954), 'The dispersion of matter in turbulent flow through a pipe', *Proc. Roy. Soc. London* **A223**, 446–467.

Thompson, J. D. and O'Brien, J. J. (1973), 'Time-dependent coastal upwelling', *J. Phys. Oceanogr.* **3**, 33–46.

Tikhomirov, A. I. (1964), 'The thermal bar of Lake Ladoga (in Russian)', *Izv. Akad. Nauk SSSR, Ser. Geogr.* **95**, 134–142. (*Sov. Hydrol. Selec. Pap.*, Engl. Trans., No. 2, pp. 182–191).

Verber, J. L. (1966), 'Inertial currents in the Great Lakes', *Proc. 9th Conf. Great Lakes Res.*, Great Lakes Res. Div. Publ. No. 15, Univ. Michigan, pp. 375–379.

Walsh, J. J., Whitledge, T. E., Barvenik, F. W., Wirick, C. D., Howe, S. O., Esaias, W. E., and Scott, J. T. (1978), 'Wind events and flood chain dynamics within the New York Bight', *Limnol. Oceanogr.* **23**, 659–683.

Wang, D.-P. (1975), 'Coastal trapped waves in a baroclinic ocean', *J. Phys. Oceanogr.* **5**, 326–333.

Wang, D.-P. (1979), 'Low frequency sea level variability in the Middle Atlantic Bight', *J. Mar. Res.* **37**, 683–697.

Wang, D. P. and Mooers, C. N. K. (1976), 'Coastally-trapped waves in a continuously stratified ocean', *J. Phys. Oceanogr.* **6**, 853–863.

Weiler, H. S. (1968), 'Current measurements in Lake Ontario in 1967', *Proc. 11th Conf. on Great Lakes Res.*, Univ. Michigan, Ann Arbor, pp. 500–511.

Welander, P. (1957), 'Wind action on a shallow sea: some generalizations of Ekman's theory', *Tellus* **9**, 45–52.

Welander, P. (1961), 'Numerical prediction of storm surges', *Adv. Geophys.* **8**, 316–379, Academic Press, New York and London.

Yamagata, T. (1980), 'On cyclonic propagation of a warm front in a bay of a northern hemisphere', *Tellus* **32**, 73–76.

Yih, C. S. (1965), *Dynamics of Nonhomogeneous Fluids*, McMillan.

Yoshida, K. (1967), 'Circulation in the eastern tropical oceans with special reference to upwelling and undercurrents', *Japan J. Geophys.* **4**, 1–75.

Index of Names

Index of Subjects